Green Energy and Technology

For further volumes:
http://www.springer.com/series/8059

Djamila Rekioua · Ernest Matagne

Optimization of Photovoltaic Power Systems

Modelization, Simulation and Control

Springer

Djamila Rekioua
LT.I.I Laboratory
University of Bejaia
Route de Terga Ouzemour
06000 Bejaia
Algeria
e-mail: dja_rekioua@yahoo.fr

Ernest Matagne
Université Catholique de Louvain
Place de l'Université 1
1348 Louvain-la-Neuve
Belgium

Additional material to this book can be downloaded from http://extra.springer.com

ISSN 1865-3529 e-ISSN 1865-3537
ISBN 978-1-4471-2348-4 e-ISBN 978-1-4471-2403-0
DOI 10.1007/978-1-4471-2403-0
Springer London Dordrecht Heidelberg New York

British Library Cataloguing in Publication Data
A catalogue record for this book is available from the British Library

Library of Congress Control Number: 2011942409

Printed on acid-free paper

Springer is part of Springer Science+Business Media (www.springer.com)

Introduction

Solar energy which is free and abundant in most parts of the world has proven to be an economical source of energy in many applications. The energy that the Earth receives from the Sun is so enormous and so lasting that the total energy consumed annually by the entire world is supplied in as short a time as half an hour. The sun is a clean and renewable energy source, which produces neither green-house effect gas nor toxic waste through its utilization.

Photovoltaic (PV) is a technology in which radiant energy from the sun is converted to direct current (DC) electricity. The most important advantages of photovoltaic systems are:

- The photovoltaic processes are completely solid state and self contained.
- There are no moving parts and no materials consumed or emitted.
- They are non-polluting emissions.
- They require no connection to an existing power source or fuel supply.
- They may be combined with other power sources to increase system reliability.
- They can withstand severe weather conditions, including cloudy weather.
- They consume no fossil fuels - their fuel is abundant and free.
- They can be installed and upgraded as modular building blocks; more photovoltaic modules may be added as power demand increases.

The watt peak power price is considerably decreased since the seventies. This leads to a large-scale application of photovoltaic systems in several promising areas. Compared with conventional fossil energy sources, small scale stand-alone photovoltaic (PV) systems are the best option for many remote applications around the world. Small-scale Stand-alone photovoltaic (PV) systems now provide power for hundreds of thousands of installations throughout the world. They have the potential to be used in millions more, particularly in developing countries where two billion people still do not have access to electricity.

Aims of the Book

Many books currently on the market are based around discussion of the solar cell as semiconductor devices rather than as a system to be modeled and applied to real-world problems.

The main objective of this book is to enable all students including graduation and post graduation, especially in the field of electrical engineering, to quickly understand the concepts of photovoltaic systems, provide models, control and optimization of some stand alone photovoltaic applications, such as rural electrification, pumping and desalination. Mathematical models are given for each system and a corresponding example under MATLAB™/SIMULINK™ package is given at the end of each section. The book is accompanied by Springers Extras available online containing each application scheme for an eventual implementation under DSPACE package. Some electrical machine control approaches, such as vector control and direct torque control are introduced in different drive systems used. Furthermore, in order to optimize the photovoltaic array operation, intelligent techniques are developed. By writing this book, we complete the existing knowledge in the field of photovoltaic and the reader will learn how to make the modeling and the optimization of the most used stand alone photovoltaic applications by applying different control strategies.

How the Book is Organized?

The book is organized through seven chapters. The first chapter is intended as an introduction to the subject. It defines the photovoltaic process, introduces the main meteorological elements, the solar irradiance and presents an overview of PV systems (stand alone systems and grid connected systems). This chapter also includes pre-sizing and maintenance of PV systems.

Chapter 2 focuses on an explicit modeling of solar irradiance and cells. Different models describing the operation and the behavior of the photovoltaic generator are presented. Some programs are given under MATLAB™/SIMULINK™.

Chapter 3 is devoted to power electronics modeling. The different structures of converters used in PV systems are presented.

In Chap. 4, a detailed review on the most used algorithms to track the maximum power point is presented. Some simple MATLAB™/SIMULINK™ examples are given.

In Chap. 5, a description and modeling of the storage device is showed. The study describes a usual battery bank and provides an explicit modeling and experimental scheme of the lead-acid battery.

Chapter 6 fulfils these tasks for a photovoltaic pumping system based on both DC and AC machines. Each component is modeled individually before connecting subsystems for simulation. Several control algorithms such as scalar, vector and direct torque control are well described. In addition, classic optimization algorithms are applied and an analysis of economic feasibility of PV pumping system in comparison with systems using diesel generators is presented. This chapter includes also environmental aspects of PV power pumping system.

The Chap. 7 is devoted to hybrid photovoltaic systems. The chapter describes the different configurations and the different combinations of hybrid PV systems. Different synoptic schemes and simulation applications are also presented.

Contents

Chapter 1
Photovoltaic Applications Overview

Symbols

A_{pv}	Solar cell surface (m^2)
b	Coefficient equal to $\sqrt{3}$ in 3-phase and equal to 2 in single phase lines
c	Velocity of light (m/s)
$C_{\text{batt,u}}$	Capacity of a battery unit (Ah)
$C_{\text{batt,min}}$	Minimum capacity of the battery bank (Ah)
$\cos(\varphi)$	Power factor (φ is the phase shift between AC current and voltage)
DOD	Depth of discharge
E_i	Energy dissipated by Joule losses in the conductor i
E_L	Total energy produced by the photovoltaic generator which supplies the load
$E_{L,m}$	Monthly energy required by the load
$\overline{E_L}$	Annual mean of the monthly load power consumption
E_{pv}	The electrical energy produced by a photovoltaic generator
$E_{\text{pv,m}}$	Monthly energy produced by the system per unit area (kW h/m^2)
$\overline{E_{pv}}$	Annual mean of the monthly PV contribution (kWh/m^2)
f_b	Fraction of the energy which passes through the batteries
FF	Fill factor
ff	Fraction of load supplied by the photovoltaic energy
FT	Temperature factor
G	Solar global irradiance (W/m^2)
h	Plank's constant
I_0	Saturation current of the diode
I_d	Diode-current (A)
I_i	(quadratic) Mean current of conductor i (A)

D. Rekioua and E. Matagne, *Optimization of Photovoltaic Power Systems*,
Green Energy and Technology, DOI: 10.1007/978-1-4471-2403-0_1,

I_{max}	Maximum current (A)
I_{max}	Maximum current when panels are in parallel (A)
I_{mpp}	Current at maximum power point (A)
I_{ph}	Light-generated current (A)
I_{pv}	Output-terminal current (A)
I_{Rsh}	Shunt-leakage current (A)
$I_{sc\text{-}Tjref}$	Short-circuit current at the reference temperature (A)
L_I	Length of cable (m)
l_c	Cable length (m)
N_{batt}	Number of batteries to be used
N_j	Days of autonomy (backup days)
$N_{maximal pv\_serial}$	Maximal number of photovoltaic modules in series
$N_{minimal pv\_serial}$	Minimum number of photovoltaic modules in series
N_{pv}	Number of photovoltaic generators
$N_{pv\text{-}serial}$	Maximum number of photovoltaic modules in series
R_i	Resistance of conductor I (Ω)
R_{serial}	Serial resistance (Ω)
R_{sh}	Shunt resistance (Ω)
S	Cable section (m^2)
S_1	Conductor section (mm^2)
T_{dur}	Considered time duration (in hours is the energy is expressed in Wh)
T_j	Temperature cells (°K)
T_{jref}	Reference temperature of the PV cell (°K)
U_{batt}	Battery voltage (V)
V_{max}	Maximum admissible input voltage (V)
V_{mpp}	Voltage at maximum power point (V)
V_n	Rated voltage (V)
$V_{oc\text{-}Tjref}$	Open-circuit voltage at the reference temperature (V)
α_{sc}	Relative temperature coefficient of short-circuit current (/°K) as found from the data sheet
β_{oc}	Relative voltage temperature coefficient (/°K) as found from the data sheet
γ_{MPP}	Relative MPP power temperature coefficient (/°K) as found from the data sheet
ΔV	Voltage drop (V)
ε	Admissible voltage drop
η_1	Efficiency of the PV panel
η_2	Efficiency due to the junction temperature
η_3	Efficiency due to the power losses by Joule effect in the cables
η_4	Efficiency due to losses in the inverter
η_5	Efficiency is related to the maximum power point tracking

η_{batt}	the battery energy efficiency
λ	Wavelength
λ_c	Reactance of conductor (Ω/m)
ρ_c	Resistivity of the cable (Ω. m)
ρ_1	Resistivity of the conductive material (copper or aluminum)

1.1 Photovoltaic Definitions

Photovoltaic is the direct conversion of light into electricity. It uses materials which absorb photons of lights and release electrons charges. It can be used for making electric generators. The basic element of these generators is named a PV cell.

1.1.1 Irradiance and Solar Radiation

Irradiance is an instantaneous quantity describing the flux of solar radiation incident on a surface (kW/m^2). The density of power radiation from the sun at the outer atmosphere is 1.373 kW/m^2 [1], but only a peak density of 1 kW/m^2 is the final incident sunlight on earth's surface. Irradiation measures solar radiation energy received on a given surface area in a given time. It is the time integral of irradiance. For example, daily irradiation can be given into kWh/m^2 per day. Insolation is another name for irradiation. Referring to a standard irradiance of 1000 W/m^2, insolation is usually given in hours. Figure 1.1. gives the relation between irradiance and insolation.

Solar radiation consists of photons carrying energy E_{ph} which is given by the following equation:

$$E_{ph} = h\frac{c}{\lambda} \tag{1.1}$$

where λ is the wavelength, h Plank's constant and c is the velocity of light.

Global radiation comprises three components:

- direct solar radiation: The sun radiation received directly from the sun.
- diffuse radiation scattered by the atmosphere and clouds.
- reflected radiation from the ground.

The measurements of solar irradiance are taken using either a pyranometer for global radiation or a pyrheliometer for direct radiation. The integral of solar irradiance over a time period is solar irradiation.

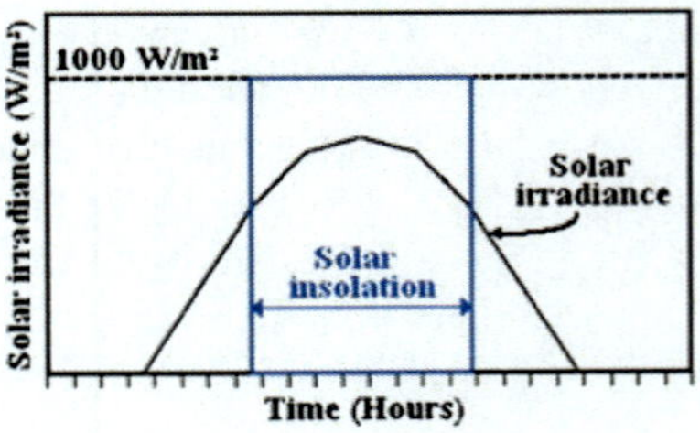

Fig. 1.1 Solar irradiance and insolation

1.1.2 Photovoltaic Cells Technologies

The basic element of a photovoltaic system (PV) is solar cells which convert the sunlight energy directly to direct current. A typical solar cell consists of a PN junction formed in a semi-conductor material similar to a diode. Semi-conductor material most widely used in solar cells is silicon. Each material gives different efficiency and has different cost. There are several types of solar material cells:

- monocrystalline silicon (c-Si)

It is the widely available cell material. Fig. 1.2

Its efficiency is limited due to several factors. The highest efficiency of silicon solar cell is around 23%, by some other semi-conductor materials up to 30%, which is dependent on wavelength and semiconductor material.

We give in Fig. 1.3 the efficiency development of crystalline silicon from 1977 to 2010 [2].

- polycrystalline cells

It is also called polysilicon. In this case, the molten silicon is cast into ingots. Then it forms multiple crystals. These cells have slightly lower conversion efficiency compared to the single crystal cells. Monocrystalline and polycrystalline silicon modules are highly reliable for outdoor power applications. The market share of crystalline silicon is represented in Fig. 1.4 [3].

- Thin films

Thin-film solar cell (TFSC), also called a thin-film photovoltaic cell (TFPV), is a solar cell made by thin film materials with a few μm or less in thickness. Thin-film solar cells usually used are [4]:

1. Amorphous silicon (a-Si) and other thin-film silicon (TF-Si). The efficiency of amorphous solar cells is typically between 10 and 13%. Their lifetime is shorter than the lifetime of crystalline cells.
2. Cadmium Telluride (CdTe) which is a crystalline compound formed from cadmium and tellurium and its efficiency is around 15%.
3. Copper indium gallium selenide (CIS or CIGS) is composed of copper, indium, gallium and selenium. Its efficiency is around 16.71%.

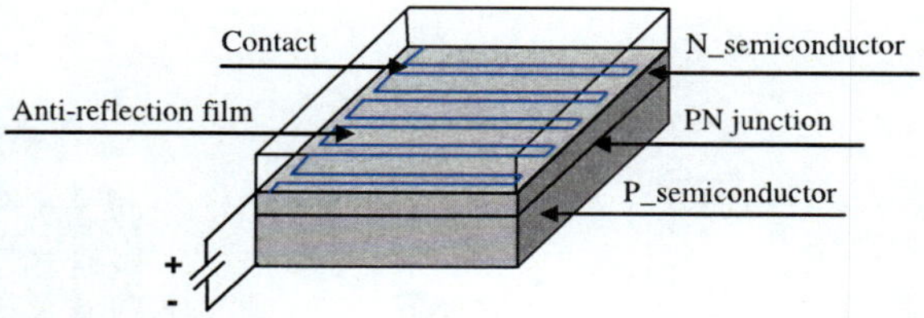

Fig. 1.2 Monocrystalline silicon cell [2]

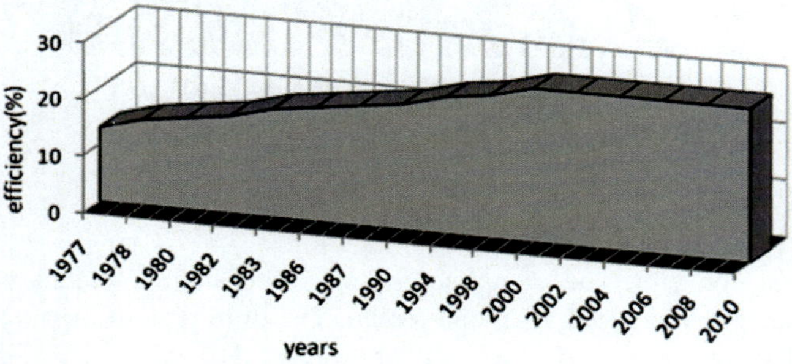

Fig. 1.3 Efficiency development of crystalline silicon

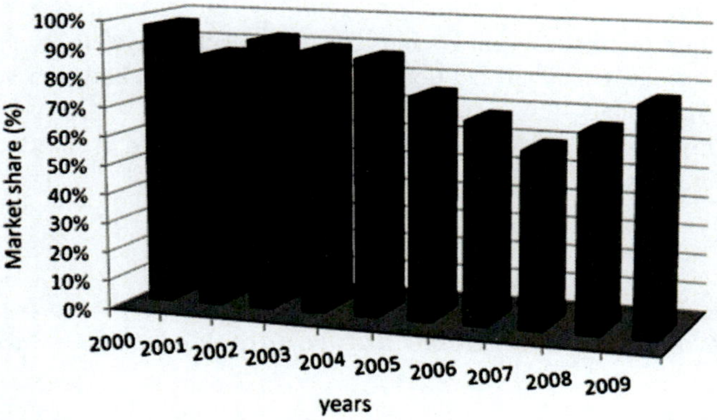

Fig. 1.4 Market share of crystalline silicon

4. Dye-sensitized solar cell (DSC) is formed by a photo-sensitized anode and an electrolyte. Its efficiency is around 11.1%.

Thin-film cells cost less than crystalline cells. The market share of thin films is represented in Fig. 1.5 [3].

- Other new technologies:

1. Organic solar cells (OSC) are made of thin layers of organic materials. Three different types of organic solar cells are known: the organic semiconducting

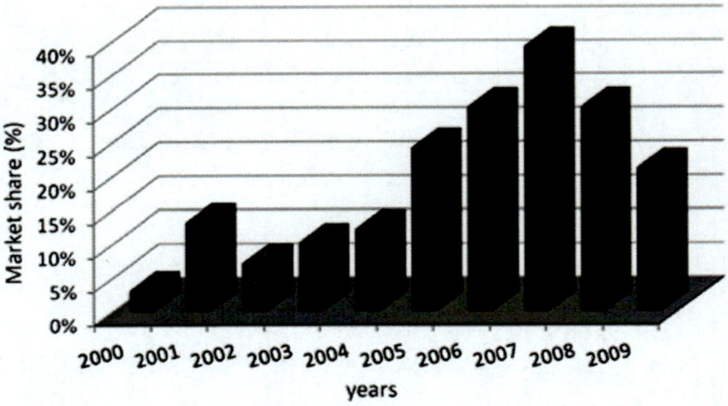

Fig. 1.5 Market share of thin-film

material can either be comprised of so-called small molecules (SM solar cells) or polymers (polymer solar cells). The third type of organic solar cells is called dye-sensitized solar cell (or Grätzel cell) [5].

2. Tandem or stacked cells: in this case, different semi-conductor materials, which are suited for different spectral ranges, will be arranged one on top of the other.
3. Concentrator cells use mirror and lens devices. This system uses only direct radiation and needs an additional mechanism for tracking the sun. Its efficiency is around 42.4% of direct radiation.
4. MIS Inversion Layer cells: the inner electrical field is produced by the junction of a thin oxide layer to a semiconductor.

1.1.3 Photovoltaic Cells and Photovoltaic Modules

1.1.3.1 Important Definitions

Cells and Panels

For obtaining high power, numerous cells are connected in series and parallel circuits. The photovoltaic module is comprised of several individual photovoltaic cells connected and encapsulated in factory. It is the commercial unit. A panel consists of one or several modules grouped together on a common support structure (Fig. 1.6).

Orientation and tilt of these panels are important design parameters, as well as shading from surrounding obstructions. By adding cells or identical modules in series, the current is the same but the voltage increases proportionally to the number of cells (modules) in series. By adding identical modules in parallel, the

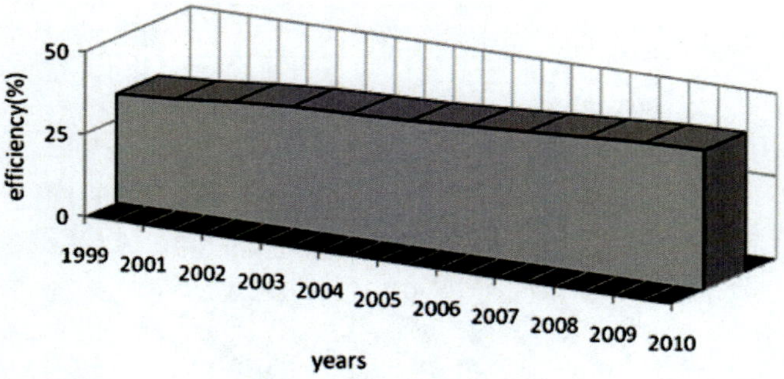

Fig. 1.6 Efficiency development of concentrator cells [4]

voltage is equal to the voltage of each module and the intensity increases with the number of modules in parallel (Fig. 1.7, Fig. 1.8).

Current Versus Voltage Characteristic

All other quantities being constant, the current I_{PV} supplied by a photovoltaic cell depends on the voltage V_{PV} at its terminals. The graph of that characteristic has typically the form shown in Fig. 1.9. The current decreases as the voltage is increasing, and the curve concavity is directed to the bottom.

Open Circuit Voltage and Short Circuit Current

Open circuit voltage and short circuit current are two parameters widely used for describing the cell electrical performance (Fig. 1.9). The short circuit current I_{sc} is measured by shorting the output terminals. It is the current at zero voltage ($V_{pv} = 0$). The open circuit voltage is the voltage at zero current ($I_{pv} = 0$).

The values of I_{sc} and V_{oc} obtained in standard conditions are named $I_{sc\text{-}ref}$ and $V_{oc\text{-}ref}$. Those values are given in the datasheet of the cell or module.

Maximum Power Point

The power supplied by a photovoltaic generator is

$$P_{pv} = V_{pv} I_{pv} \tag{1.2}$$

This power is positive for the part of the I_{PV}–V_{PV} curve included between the open-circuit point and the short-circuit point, thus for values of V_{PV} satisfying the condition

$$0 < V_{pv} < V_{oc} \tag{1.3}$$

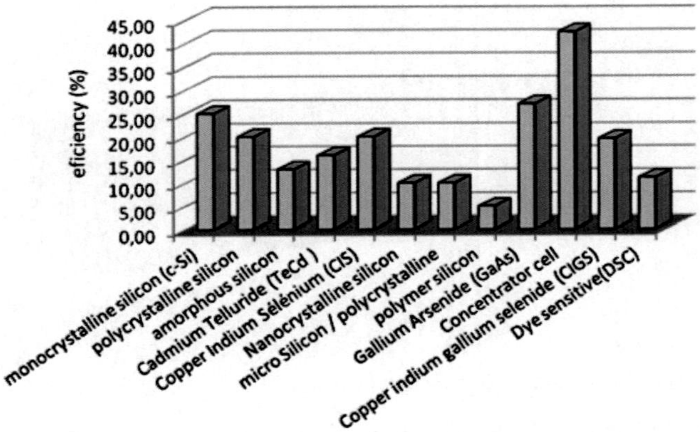

Fig. 1.7 Efficiency of different material cells in laboratory [2]

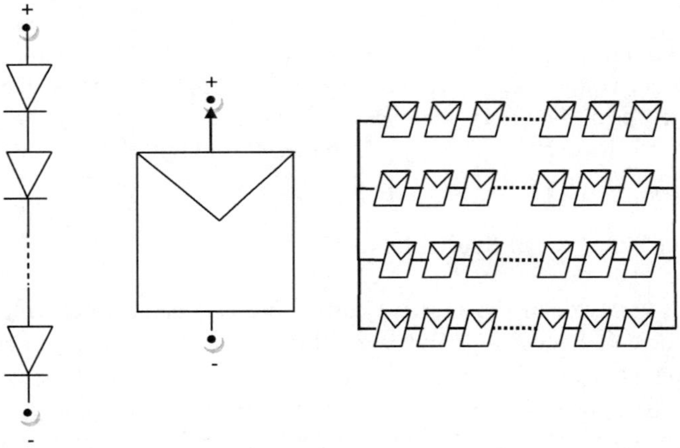

Fig. 1.8 Cells, photovoltaic module and panel [44]

Fig. 1.9 Typical form of I_{PV}-V_{PV} characteristic

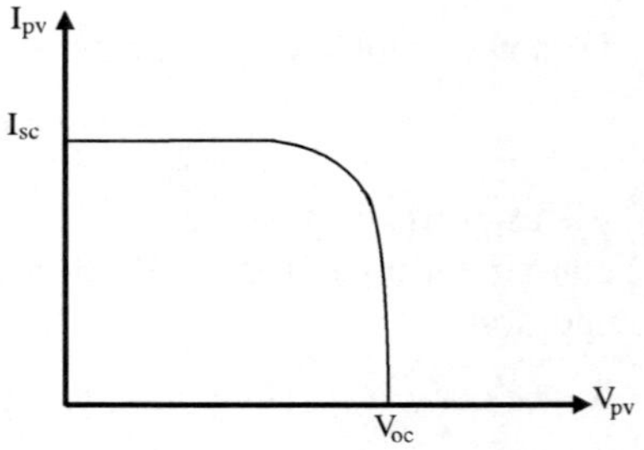

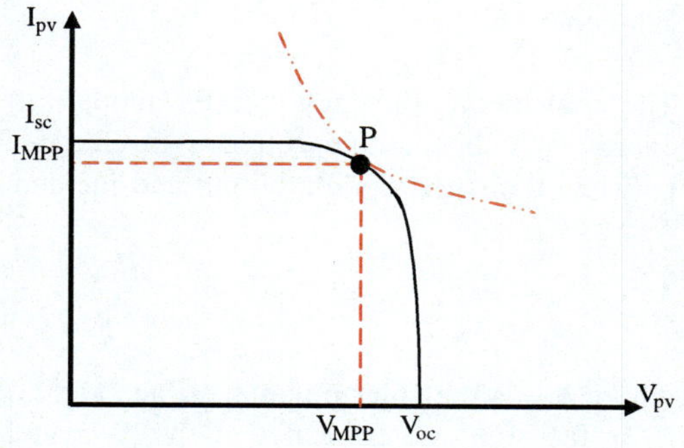

Fig. 1.10 Current versus voltage *I*pv–*V*pv characteristic for a solar cell

Outside the interval defined by Eq. 1.3, the power P_{pv} is negative: the PV device receives the power from the external electric circuit. This case is not considered here.

The power P_{PV} is null when $V_{PV} = 0$ (short-circuit point) by Eq. 1.2. Similarly, the power P_{pv} is null when $V_{pv} = V_{oc}$ (open-circuit point) since, then $I_{pv} = 0$ and thus, by Eq. 1.2, P_{pv} is also null. Then, in the interval defined by Eq. 1.3, P_{PV} reaches a maximum value. This arrives at a point named Maximum Power Point (MPP). The corresponding values of V_{pv} and I_{pv} are named respectively V_{MPP} and I_{MPP} (Fig. 1.10). At that point $P(V_{MPP}, I_{MPP})$, the power P_{pv} supplied by the photovoltaic generator is maximum and denoted P_{MPP}. We have:

$$P_{MPP} = V_{MPP} I_{MPP} \tag{1.4}$$

In standard conditions, the quantities P_{MPP}, V_{MPP} and I_{MPP} take respectively the values $P_{MPP\text{-}ref}$ $I_{MPP\text{-}ref}$ and $V_{MPP\text{-}ref}$.

The MPP is reached when

$$0 = \frac{\partial P_{PV}}{\partial V_{PV}} \tag{1.5}$$

i.e., owing to Eq. 1.2,

$$0 = \frac{\partial (V_{pv} I_{pv})}{\partial V_{pv}} = I_{pv} + V_{pv} \frac{\partial I_{pv}}{\partial V_{pv}} \tag{1.6}$$

or equivalently,

$$-\frac{\partial V_{pv}}{\partial I_{pv}} = \frac{V_{pv}}{I_{pv}} \tag{1.7}$$

The left member of Eq. 1.7 is the incremental internal resistance of the PV generator (the minus sign is due to the choice for that device of the generator reference directions). The right member is the apparent resistance of the load. Thus, one can consider Eq. 1.7 as the equation defining the resistance adaptation of the load the internal resistance of the PV generator.

Efficiency

The conversion efficiency of a PV module is the proportion of received sunlight energy that the module converts to electrical energy. It is defined as the ratio between the solar module output and incident light power.

$$\eta_1 = \frac{P_{\text{out}}}{P_{\text{in}}} = \frac{V_{\text{pv}}.I_{\text{pv}}}{A_{\text{pv}}.G} \tag{1.8}$$

where A_{pv} is the solar module surface and G the irradiance.

In fact, the true efficiency of the PV panel is given by:

$$\eta_{pv} = \eta_1.\eta_2.\eta_3.\eta_4.\eta_5 \tag{1.9}$$

where η_1 is the efficiency of the PV panel above calculated (Eq. 1.8),

η_2 is due to the junction temperature increase since a part of received solar flux is not converted in electric power but dissipated as heat inside the module. The temperature increase is higher in case of poor ventilation of the photovoltaic modules $(0.8\langle\eta_2\langle0.9)$.

η_3 is due to the power losses by Joule effect in the cables. In order to reduce those losses, the cable section is sized versus to a voltage drop in the cables $(\eta_3 \approx 0.98)$.

η_4 is due to losses in the inverter $(\eta_4 \approx 0.95)$

η_5 is related to the maximum power point tracking. If the losses of the converter which carries that tracking are included in η_4, then η_5 takes into account only the imperfections of the maximum power point traking ($\eta_5 \approx 0.98$). η_5 is lower if it takes also into account the losses of the tracking converter ($\eta_5 \approx 0.95$). Finally, if there is none maximum power point tracking, η_5 takes into account the consequence of that lack ($\eta_5 \approx 0.8$).

Fill Factor

It describes how square the $I_{\text{pv}}-V_{\text{pv}}$ curve is. The fill factor is defined as follows

$$FF = \frac{P_{\text{MPP}}}{V_{\text{oc}} \cdot I_{\text{sc}}} = \frac{V_{\text{MPP}}.I_{\text{MPP}}}{V_{\text{oc}} \cdot I_{\text{sc}}} \tag{1.10}$$

1.1.3.2 Characteristic Curves of Solar Cells

The electrical characteristic of the PV cell is generally represented by the current versus voltage ($I_{\text{pv}}-V_{\text{pv}}$) curve and power versus voltage ($P_{\text{pv}}-I_{\text{pv}}$) for different conditions.

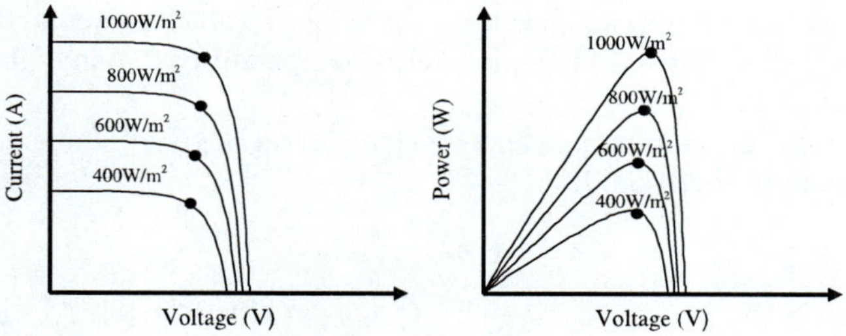

Fig. 1.11 Irradiance effect on electrical characteristic

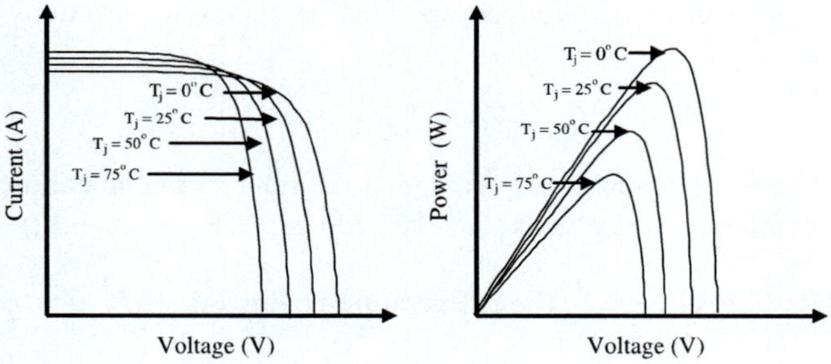

Fig. 1.12 Temperature effect on electrical characteristic [31]

Irradiance Effect

Figure 1.11 shows the current–voltage characteristics $I_{pv}-V_{pv}$ and power–voltage $P_{pv}-V_{pv}$ of the PV cell for different levels of radiation. We note that the current I_{sc} increases quasi linearly with irradiance and that the voltage V_{oc} increases slightly. Then, the maximum electric power P_{MPP} increases faster than the irradiance, i.e. the efficiency is better for high irradiance.

The reference conditions are generally chosen with an irradiance of 1,000 W/ m^2. In practice, the irradiance on PV without light concentration is lower, and thus the efficiency is lower than its rated value.

Temperature Effect

When the internal temperature T_j increases, the short circuit current I_{sc} increases slightly due to better absorption of light (as an effect of the gap energy decrease with temperature) but the open-circuit voltage strongly decreases with temperature. The maximum electric power also strongly decreases with temperature (Fig. 1.12).

The standard conditions are generally chosen for a value of internal temperature T_j equal to 25°C. Under sunshine, the internal temperature is often higher and thus the efficiency lower.

The short-circuit current I_{sc} can be calculated at a given temperature T_j, for small temperature variation, by:

$$\Delta T = T_j - T_{jref}. \tag{1.11}$$

$$I_{sc} = I_{sc-T_{jref}}[1 + \alpha_{sc}.\Delta T] \tag{1.12a}$$

where α_{sc} is the relative temperature coefficient of short-circuit current (/°K) as found from the data sheet, T_{jref} is the reference temperature of the PV cell (°K), $I_{sc\text{-}Tjref}$ is the short-circuit current at the reference temperature.

Similarly, the open-circuit voltage, for small temperature variations can be also expressed as:

$$V_{oc} = V_{oc-T_{jref}}[1 + \beta_{oc}.\Delta T] \tag{1.12b}$$

where $V_{oc-Tjref}$ is the open-circuit voltage at the reference temperature and β_{oc} is the relative temperature coefficient of that voltage (/°K) as found from the data sheet.

Often, datasheet also gives the temperature coefficient of P_{MPP}:

$$P_{MPP} = P_{MPP-Tjref}\left[1 + \gamma_{MPP}.\Delta T\right] \tag{1.12c}$$

where $P_{MPP-Tjref}$ is the maximum power at the reference temperature, γ_{MPP} is the relative maximum power temperature coefficient (/°K) as found from the data sheet.

Note on Spectral Effect

White light can be considered as a sum of radiations with different wave length (colors). The efficiency of PV generators is not the same for each wave length. For that reason, the standard conditions used for cells and modules rating include a constraint on the light spectrum. The standard spectrum commonly used is that one named AM 1.5. As a consequence, the PV cells and modules are sometimes optimized for that standard spectrum. In real condition, the light spectrum can be different, and that has also an effect on the PV efficiency.

1.2 Introduction to PV Systems

A PV system converts sunlight into electricity. A PV system contains different components including cells, electrical connections, mechanical mounting and a way to convert the electrical output. The electricity generated can be kept in a standalone system, stored in batteries or can feed a greater electricity power grid. It

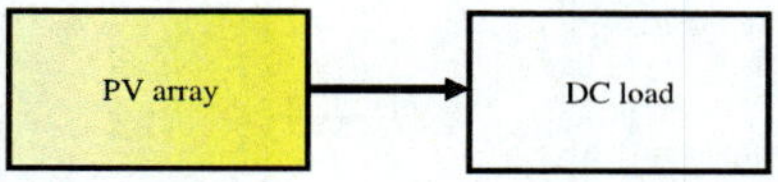

Fig. 1.13 Direct-coupled PV system

is interesting to include electrical conditioning equipment. This one ensures the PV system to operate under optimum conditions. In this case, we use special equipment to follow the maximum power of the array. This equipment is known as maximum power point tracking (see Chap. 4).

1.2.1 Stand Alone PV Systems

1.2.1.1 Direct-Coupled PV System

Stand-alone PV systems are designed to operate independent of the electric utility grid, and are generally designed and sized to supply certain DC and/or AC electrical loads. The simplest type of stand-alone PV system is a direct-coupled system, where the DC output of a PV module is directly connected to a DC load (Fig. 1.13).

In direct-coupled systems, the load only operates during sunlight hours. The common applications for this system are such as ventilation fans, water pumps and small circulation pumps for solar thermal water heating systems.

1.2.1.2 Stand-Alone PV System with Battery Storage Powering DC and AC Loads

In standalone PV applications, electrical power is required from the system during night or hours of darkness. Thus the storage must be added to the system. Generally, batteries are used for energy storage. Several types of batteries can be used such as lead-acid, nickel–cadmium, lithium zinc bromide, zinc chloride, sodium sulfur, nickel-hydrogen, redox and vanadium batteries. Different factors are considered in the selection of batteries for PV application (see Chap. 5). The inverter uses an internal frequency generator to obtain the correct output frequency (see Chap. 3).

A charge controller must keep the battery at the highest possible state while protecting it from overloaded by the photovoltaic generator and from over-discharge by loads. There are several types of charge controller [6]

- Shunt controller: the function is to regulate the charging of battery. This controller is basically connected in parallel with array and battery [2].
- Series controller: this controller is commonly used in small PV system and connected in series between PV array and battery.
- Tracking controller: This controller tracks the maximum power point of PV array (see Chap. 4).

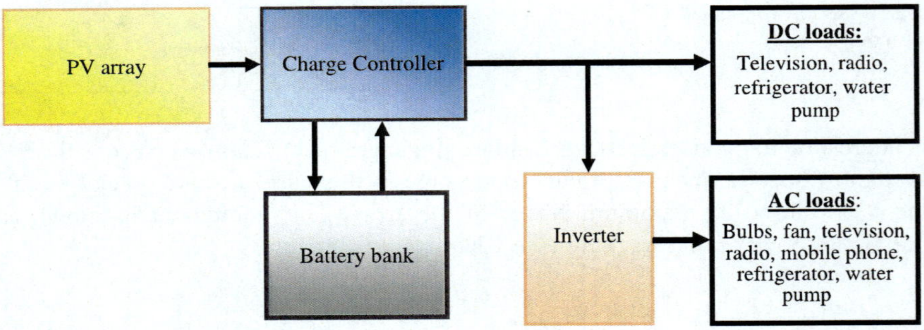

Fig. 1.14 Stand-alone PV system with battery storage powering DC and AC loads

Figure 1.14 shows a diagram of a typical stand-alone PV system powering DC and AC loads [7].

1.2.1.3 Applications

PV is widely used in many applications

Lighting

PV system is an ideal source for feeding lighting needs. In this case, the battery storage is essential because lighting demands are more important at night. The different applications are [8]:

- lighting homes, and solar power products for home use
- solar advertising/billboard
- piers and camping/lantern light
- flashlight
- solar lawn lightings
- solar street, highway information signs, and caboose lighting for trains
- parking
- marinas (especially during the busy summer season),
- mountain cabins.

Remote Site Electrification

Photovoltaic systems are good solutions to provide electricity in some areas far of the network. In these cases, other renewable energy sources can be used with the PV system to ensure availability of electric energy. Some examples of remote site electrification are for [8]:

- rural homes,
- water supply in rural areas,
- parks,
- mountain cabins,
- remote farms
- island electrification,
- mobile clinics for remote rural areas,
- solar highway,
- facilities at public beaches,
- campgrounds,
- military installations.

Communications

Some examples of communications are:

- radio telephone equipment,
- radio,
- television,
- telecommunication systems,
- military usage for telecommunications,
- relay towers or repeater stations,
- portable computer systems,
- highway callboxes,
- fire lookout tower.

Remote Monitoring

Some examples of remote monitoring

- power source monitoring,
- meteorological measurement systems,
- highway/traffic conditions,
- structural conditions,
- seismic recording,
- irrigation control,
- scientific research in remote locations.

Water Pumping and Control

These systems may be either:

- direct systems supplying water only when the sunlight is sufficient,
- pumping water to an elevated storage tower during sunny hours to provide available water at any time.

PV powered water pumping is used to provide water for

- campgrounds,
- irrigation,
- remote village water supplies,
- livestock watering

Charging Vehicle Batteries

PV systems may be used to

- directly charge vehicle batteries,
- or to provide a "trickle charge" for maintaining a high battery state of charge on little-used vehicles.

Some examples are:

- fire-fighting and snow removal
- equipment and agricultural machines such as tractors or harvesters
- direct charging is useful for boats and recreational vehicles
- solar stations may be dedicated to charging electric vehicles.

Refrigeration

PV systems are excellent for remote or mobile storage of medicines and vaccines.

Consumer Products

There is a variety of consumer products. PV is used to power

- small DC appliances for recreational vehicles,
- watches,
- lanterns,
- calculators,
- radios,
- televisions,
- flashlights,
- outdoor lights,

- security systems,
- gate openers.

1.2.1.4 Advantages of Stand-Alone PV System

The most important advantages of PV system are [8]:

- The reliable supply of the load with electricity during operating time,
- a long lifetime,
- expenses for maintenance must be low.

1.2.2 Grid-Connected PV Systems

Utility-interactive PV power systems mounted on residential and commercial buildings are likely to become important source of electric generation. Grid-connected PV systems offer the opportunity to generate significant quantities of high-grade energy near the consumption point, avoiding transmission and distribution losses. These systems operate in parallel with existing electricity grids, allowing exchange of electricity to and from the grid. Grid-connected PV system can be subdivided into two systems:

- Decentralized grid-connected PV systems
- Central grid-connected PV systems.

1.2.2.1 Decentralized Grid-Connected PV Systems

In these systems, energy storage is not necessary because solar radiation provides power in the houses and if there is surplus energy it can be injected into the grid (Fig. 1.15). In this case, the inverter must integrate harmoniously with the energy (voltage and frequency) provided by the grid.

During night or at instants when the PV power is inadequate, the grid can be used as a storage system and will feed the houses.

1.2.2.2 Central Grid-Connected PV Systems

It is a central photovoltaic power station and it is installed to systems up to the MW range. With this system, we can obtain medium or high voltage grid (Fig. 1.16).

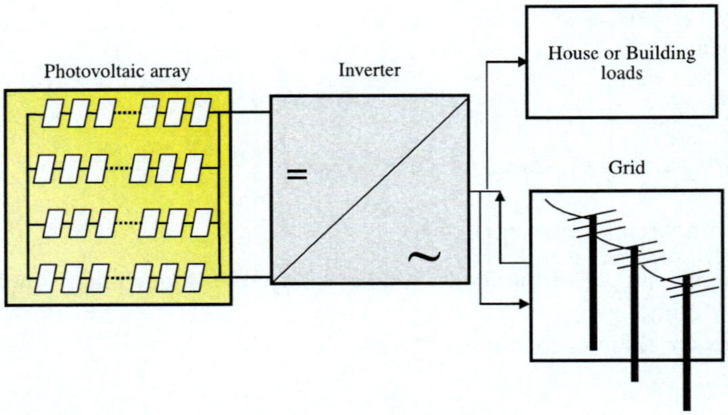

Fig. 1.15 Decentralized grid-connected PV systems

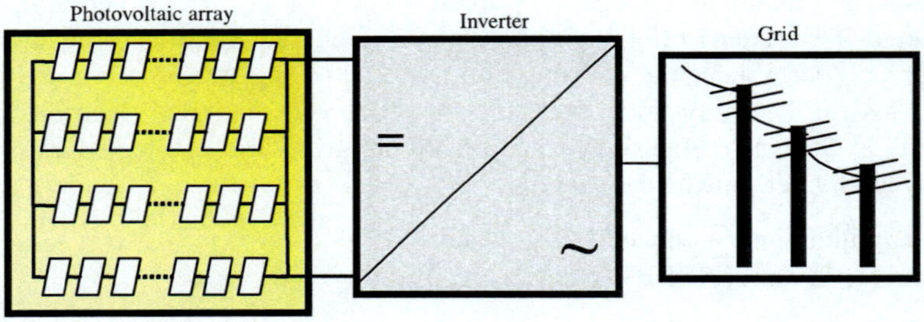

Fig. 1.16 Central grid-connected PV systems

1.3 System Pre-Sizing

In order to ensure acceptable operation at minimum cost, it will be necessary to determine the correct size. It is noting that the design should be done on meteorological data, solar irradiance and on the exact load profile of consumers over long periods.

1.3.1 Determination of Load Profile

Nature of loads and the loads profile are the two important parameters that we have firstly to need.

1.3.2 *Analysis of Solar Radiation*

We have to need information on latitude and longitude, on weather data and on the different constraints on installation system.

1.3.3 *Calculation of Photovoltaic Energy*

The energy produced by a photovoltaic generator is estimated using data from the global irradiation on an inclined plane, ambient temperature and the data sheet of the photovoltaic module given by the manufacturer. The electrical energy produced by a photovoltaic generator is given by:

$$E_{pv} = \eta_{pv} A_{pv} G[1. - f_b + f_b \eta_{batt}] \tag{1.13}$$

where f_b is the fraction of the energy which passes through the batteries and η_{batt} the battery energy efficiency ($\eta_{batt} \approx 0.8$)

1.3.4 *Size of PV*

1.3.4.1 First Method

The monthly energy produced by the system per unit area is denoted: $E_{pv,m}$ (kWh/m^2) and $E_{L,m}$ is the monthly energy required by the load (where m = 1,2,..., 12 represents the month of the year.). The minimum surface of the generator needed to ensure full (100%) coverage load (E_L) is expressed by [9]:

$$A_{pv} = \max_m \frac{E_{L,m}}{E_{pv,m}} \tag{1.14}$$

Surface larger than Eq. 1.14 can be needed for taking into account the limited size of the batteries or for including a security factor.

For systems including a grid connection or alternative energy sources, the sizing can be achieved on annual basis.

The total energy produced by the photovoltaic generator which supplies the load can be expressed by:

$$E_L = E_{pv}.A_{pv} \tag{1.15}$$

The calculation of the photovoltaic generator size (A_{pv}) is established from the annual mean of the monthly contribution $\left(\overline{E_{pv}}\right)$. The load is represented by the average annual monthly $\overline{E_L}$.

$$A_{pv} = ff \cdot \frac{\overline{E_L}}{\overline{E_{pv}}} \quad (1.16)$$

where *ff* is the fraction of load supplied by the photovoltaic energy.

The number of photovoltaic generator is calculated using the surface of the system unit $A_{pv,u}$ taking the entire value:

$$N_{pv} = \text{ENT}\left[\frac{A_{pv}}{A_{pv,u}}\right] + 1 \quad (1.17)$$

1.3.4.2 Second Method (see Sect. 1.3.6)

1.3.5 Size of Battery Bank

Always, before tackling the calculations, we start by identifying:

- the electricity usage per day
- number of days of autonomy
- depth of discharge limit
- ambient temperature at battery bank.

1.3.5.1 Electrical Usage

Firstly, we have to know the amount of energy we will be consuming per day $E_{L,max}$ (Wh/day).

1.3.5.2 Number of Autonomy Days

In the second step, we have to identify days of autonomy N_j (backup days). We multiply $E_{L,max}$ by this factor N_j.

$$\text{RES} = E_{l,\max} . N_j \quad (1.18)$$

1.3.5.3 Depth of Discharge Limit

We have to identify depth of discharge (DOD) and convert it to a decimal value. Divide Eq. 1.18 by this value (DOD).

$$\text{ANT} = \frac{\text{RES}}{\text{DOD}} = \frac{E_{l,\max} . N_j}{\text{DOD}} \quad (1.19)$$

Table 1.1 Factor FT calculation [10]

Temperature (°C)	Temperature (°F)	Factor (FT)
+26	80+	1.00
+21	70	1.04
+15	60	1.11
+10	50	1.19
+4	40	1.30
−1	30	1.40
−6	20	1.59

1.3.5.4 Ambient Temperature at Battery Bank

We have to derate battery bank for ambient temperature effect. We have to select the multiplier corresponding to the lowest average temperature that batteries will be exposed to. This multiplier depends on the battery type (Table 1.1 gives an example of such data).

We multiply Eq. 1.19 by this factor (*FT*). And then we obtain the minimum capacity of battery bank (Wh).

$$C_{\mathrm{batt,min}}(\mathrm{Wh}) = \frac{\mathrm{ANT} \cdot \mathrm{FT}}{\mathrm{N_m} \cdot \eta_{\mathrm{batt}}} \tag{1.20}$$

Finally we divide the minimum capacity of battery bank by battery voltage V_{batt} and we obtain the minimum capacity (Ah) of the battery bank.

$$C_{\mathrm{batt,\,min}}(A.h) = \frac{C_{\mathrm{batt,min}}(W.h)}{\mathrm{U_{batt}}} \tag{1.21}$$

The battery capacity of storage can be written as:

$$C_{\mathrm{batt,min}}(A.h) = \frac{\mathrm{E_{L,max}.N_j.FT}}{\mathrm{U_{batt}.DOD.N_m}.\eta_{\mathrm{batt}}} \tag{1.22}$$

where $\mathrm{U_{batt}}$ is the battery voltage, DOD is the depth of discharge, η_{batt} is the battery efficiency, $\mathrm{N_M}$ is the number of days in the month which has the maximum energy consumed.

The number of batteries to be used is determined from the capacity of a battery unit $C_{\mathrm{batt,u}}$ is given by:

$$N_{\mathrm{batt}} = \mathrm{ENT}\left[\frac{C_{\mathrm{batt,min}}}{C_{\mathrm{batt},u}}\right] \tag{1.23}$$

1.3.6 Inverter Size

The selection and number of inverters is based on three criteria: the voltage compatibility, the current compatibility and the power compatibility. From these three criteria, the design of inverters will impose how to wire the photovoltaic modules together

1.3.6.1 Voltage Compatibility

Maximum Admissible Input Voltage V_{max}

An inverter is characterized by a maximum admissible input voltage V_{max}. If the voltage delivered by the PV is greater than V_{max}, the inverter will be damaged. Exceeding the value V_{max} for the input voltage is also the only cause damaging the inverter. Moreover, as the PV voltages in series are added, the value of V_{max} will therefore determine the maximum number of modules in series. This will obviously depend on the voltage delivered by the photovoltaic modules. We will consider that the voltage delivered by a PV is its open circuit voltage V_{oc}. Thus, the maximum number of photovoltaic modules in series is calculated by the following simple equation:

$$N_{pv\_serial} = \mathrm{ENT}\left[\frac{V_{max}}{V_{oc} * 1.15}\right] \tag{1.24}$$

The coefficient 1.15 is a safety factor.

Maximum Power Point Tracking Voltage Range

We can also calculate the minimum and maximum number of photovoltaic modules in series according to the Maximum Power Point Tracking (MPPT) voltage. Indeed, the inverter must at all times track their maximum power modules. The MPPT system works only for a range of input voltage inverter defined by the manufacturer and specified on the inverter datasheet. When the input voltage of the inverter DC side is less than the MPPT minimum voltage, the inverter continues to operate but provides the power corresponding to the minimum voltage MPPT. We must therefore ensure that the voltage delivered by the PV system is in the range of the inverter voltage MPPT which it is connected. If this is not the case, there will be no damage to the inverter, but only a loss of power.

The minimum and maximum number of photovoltaic modules in series is calculated by the following equation [9]:

$$N_{minimal_{pv\_serial}} = \mathrm{ENT}\left[\frac{V_{mpp,min}}{V_{mpp} * 0.85}\right] \tag{1.25}$$

$$N_{maximal_{pv\_serial}} = \mathrm{ENT}\left[\frac{V_{mpp,max}}{V_{mpp} * 1.15}\right] \tag{1.26}$$

The coefficient 1.15 is a coefficient of increase to calculate the MPP voltage at −20°C.

The coefficient 0.85 is a reduction factor to calculate the MPP voltage at 70°C.

1.3.6.2 Compatibility with Current

As currents are added when panels are in parallel, the value of the current I_{max} will determine the maximum number of parallel panels. This will obviously depend on the current delivered by a PV system. In the design sizing, it is assumed that the current delivered by a PV system is equal to the short-circuit current (I_{sc}) given on the datasheet. The maximum number of panels in parallel is calculated by the following equation:

$$N_{pv\_parallel} = \mathrm{ENT}\left[\frac{I_{max}}{I^{*}_{sc} 1.25}\right] \tag{1.27}$$

The coefficient 1.25 is a safety factor.

1.3.6.3 Compatibility in Power

The value of the maximum power input of the inverter will limit the number of panels connected. Indeed, we must ensure that the power of a PV system does not exceed the maximum allowable power. As the power delivered by the PV system varies with radiation and temperature, we can consider for the sizing that the calculated power is less than the maximum allowable power by the inverter. Ideally, the power delivered by the PV system must be substantially equal to the maximum allowable power inverter.

1.3.7 Sizing of DC Wiring

The array cabling ensures that energy produced by PV array is transferred efficiently to the load. In theory, connections are made up of perfect current conductors with a zero resistance. In practice, a conductor is not perfect. It works like a resistance (Fig. 1.17).

The resistance of an electric conductor is very low but not zero. We have the following expression:

$$R = \frac{\rho \cdot l_c}{S} \tag{1.28}$$

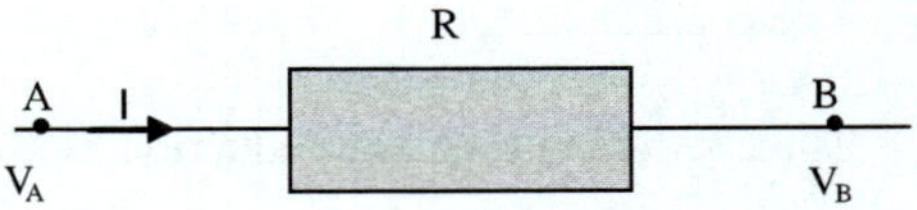

Fig. 1.17 Modeling of a cable [11]

Table 1.2 Material resistivity

ρ_c (Ω.m)	Material
2.7×10^{-8}	Aluminum cable
1.7×10^{-8}	Copper cable
1.6×10^{-8}	Silver cable

with l_c the conductor length (m), S the cross-section area (m^2), ρ_c (Ω. m) the resistivity of the cable. It depends on the material [11]:

The conductor resistance, defined above, will cause a potential drop between conductor input and the conductor output. We have:

$$U = V_A - V_B = R.I$$

Thus, if the conductor is perfect, we have:

$$R = 0$$
$$U = 0$$

Then:

$$V_A = V_B$$

But as $R > 0$ for a non-perfect conductor, we have$V_A \rangle V_B$ this corresponds to a potential drop. Table 1.2

The voltage drop in a DC conductor is related to power losses. We have:

$$E_{J,i} = R_i.I_i^2.T_{\text{dur}} \tag{1.29}$$

where E_i is the energy dissipated by Joule losses in the conductor i, R_i and I_i are the resistance and the (quadratic) mean current of that conductor and T_{dur} the considered time duration (in hours is the energy is expressed in Wh). Of course, the total Joule losses of the DC cabling are, replacing each R_i by its value from Eq. 1.28:

$$E_J = \sum_i \rho_c \frac{L_i}{S_i} I_i^2 T_{\text{dur}} = \rho_c T_{\text{dur}} \sum_i \frac{L_i}{S_i} I_i^2 \tag{1.30}$$

It is easy to proof that, in order to low the losses for a given conductor volume, we have to keep for all conductors the same ratio

$$\lambda_c = \frac{S_i}{I_i} \tag{1.31}$$

Thus, the Joule losses in the DC cabling are

$$E_J = \frac{\rho T}{\lambda_c} \sum_i L_i I_i \quad (1.32)$$

In practice, we limit the DC cabling losses to a fraction ε of the energy produced $N_{pv}\,E_{pv}$ ($\varepsilon \approx$ ...1% ... 3%). Thus, we find from Eq. 1.32:

$$\lambda_c = \frac{\rho_c T_{dur}}{E_J} \sum_i L_i I_i = \frac{\rho_c T_{dur}}{\varepsilon N_{pv} E_{pv}} \sum_i L_i I_i \quad (1.33)$$

Finally, we obtain from Eq. 1.31 the minimum section of each conductor

$$S_i = \lambda_c I_i \quad (1.34)$$

Of course, the security rules in force in the concerned country need also to be respected.

1.3.8 *Sizing of AC Cables*

The voltage drop in an AC electrical circuit is calculated as follows:

$$\Delta V = b\left(\rho_{c1}.\frac{L_1}{S_1}.\cos\phi + \lambda_c.L_1.\sin\phi\right).I_{max} \quad (1.35)$$

where ΔV is the voltage drop. In the three-phase case, currents are expressed as line currents and voltages are expressed as line-to-line voltages. b a coefficient equal to $\sqrt{3}$ in 3-phase and equal to 2 in single phase, $\rho_{c1}$is the resistivity of the conductive material (copper or aluminum), L_1 is the length of line (m), S_1 is the conductor section (mm^2), cos(φ) is the power factor (φ is the phase shift between current and voltage AC), I_{max} is the maximum current and λ_c is the reactance of conductor (Ω/m).

The reactance of the conductors, denoted λ_c, depends on the arrangement of conductors between them.

In the case of photovoltaic systems, the power factor cos(φ) is currently often equal to unity. This means that sin(φ) = 0. Therefore, the second term of the Eq. 1.35 is zero, whatever the value of the reactance. Thus, it is not necessary to know the reactance of the conductors to calculate the voltage drop on the AC side. It can be calculated as follows:

$$\Delta V = b\left(\rho.\frac{L_1}{S_1}.\cos\phi\right).I_{max} \quad (1.36)$$

We have

$$\varepsilon = \frac{\Delta V}{V_n} \tag{1.37}$$

where V_n is the rated voltage

Thus

$$S_1 = b\left(\rho_1 . \frac{L_1}{\varepsilon} . \cos\phi\right) . \frac{I_{max}}{V_n} \tag{1.38}$$

with $\varepsilon \approx 0.01$. Of course, the security rules in force in the concerned country need also to be respected.

1.3.9 Sizing of DC Fuses

In a photovoltaic system, fuses have to protect the photovoltaic modules against the risk of overload. The information needed to define a good protection against over current by fuses is:

- $N_{pv\text{-}Serial}$ Serial number of modules: in a photovoltaic system, panels are connected in series to obtain the desired DC voltage.
- $N_{pv\text{-}parallel}$, the number of PV in parallel: Up to three panels in parallel ($N_{pv\text{-}parallel} \leq 3$), protection against overcurrent is not necessary. From four panels in parallel ($N_{pv\text{-}parallel} \geq 4$), the over current, can heat the cables and damage photovoltaic panel. It must be eliminated with a fuse placed at each panel.
- I_{sc}, the current short-circuit (under standard test conditions STC).
- The fuse rating current should be between 1.5 and 2 times the current I_{sc}.
- V_{oc}, the open circuit voltage (under standard test conditions STC).

The operating voltage of a fuse should be 1.15 times the open circuit voltage ($1.15 \times V_{co} \times N_{pv\text{-}Serial}$).

Generally, fuses and switching equipment should be rated for DC operation.

1.4 Feasibility of Photovoltaic Systems

We can make an application of the typical stand-alone PV system powering DC and AC loads represented in Fig. 1.14. We represent it in the following figure (Fig. 1.18.)

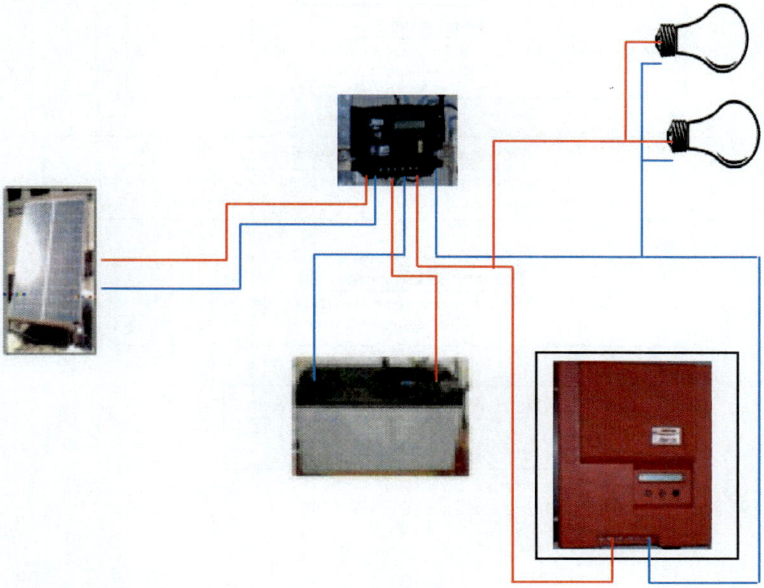

Fig. 1.18 Schematic diagram

1.4.1 Estimating the Size of a Photovoltaic System

A load includes anything that uses electricity from the power source (televisions, radios or batteries). Then, you must determine the daily amount of sunlight in your region. And finally we will determine PV array size (see Sect. 1.3.4) and battery bank size (see Sect. 1.3.5). The following flowchart will explain how to estimate the size of a PV array and battery bank. Fig. 1.19

1.4.2 Estimating of PV System Costs

When we will buy modules, verify that the modules meet electrical safety standards, and long-term warranties. Generally, in PV systems we use flooded lead acid batteries (see Chap. 5). We have to use an inverter which is needed to convert to AC power (see Chap. 3). Besides PV modules and batteries, complete PV systems also use wire, switches, fuses and connectors. Generally, we use a factor of 20% to cover balance of system costs [12] (Fig. 1.20).

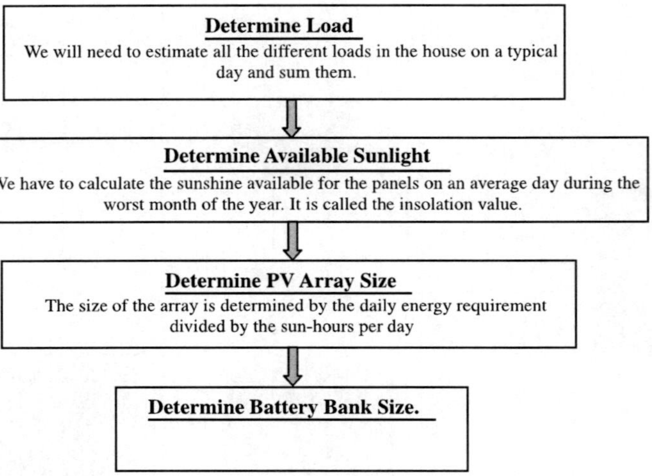

Fig. 1.19 Flowchart of estimating the size of a PV [11]

Fig. 1.20 Flowchart of estimating of PV system costs [11]

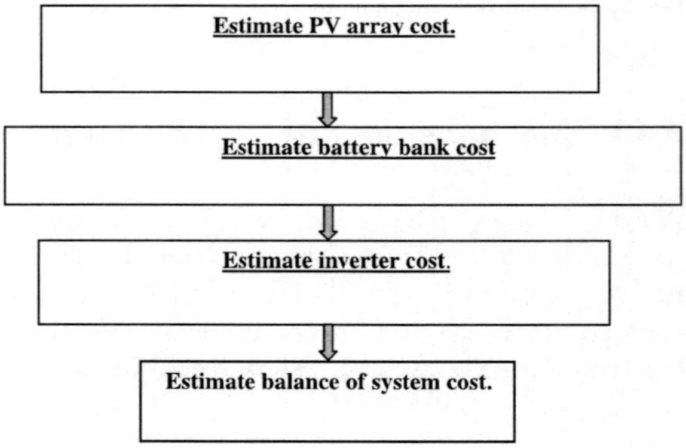

1.5 Maintenance of Photovoltaic Systems

Generally, a PV system requires little maintenance, but is important sometimes to clean panels. It is also necessary to control electrical connections to eliminate the problem of corrosion. And finally, the battery bank needs regular maintenance.

1.5.1 Panels Cleaning

We have to wash PV array, when there is a noticeable buildup of soiling deposits. But in desert area, there is dust on the modules. In this case, it is necessary to clean more frequently. Generally we use with ambient-temperature de-mineralized cleaning solution, to prevent any glass-shock or hard-water spots. We have also to clean dust and dirt from the electrical combiner box and from the DC-to-AC inverter(s)

1.5.2 Verification of Supports

We have to verify periodically the system with all its supports. Also, it is important to verify if the system performances are close to the previous ones.

1.5.3 Regular Maintenance of Batteries

We have to control batteries for any imperfection, especially corrosion or leakage and if necessary adding electrolyte and equalizing charging.

1.5.4 Inverters Control

We have only to verify if the inverter is properly matched to the panels.

Chapter 2
Modeling of Solar Irradiance and Cells

Symbols

A	Diode ideality factor
$C(N)$	Distance correction
CT	Civil time
$d\Omega$	Elementary solid angle
E	Normal irradiance of a beam radiation
E_t	Time equation
f_{circ}	The circumsolar fraction
G	Global irradiance on a plane (W/m^2)
G_{ref}	Reference irradiance (1000 W/m^2)
g	Asymmetry factor of the phase function $P(\Theta)$
g_h	Asymmetry factor of the hemispherical phase function
GMT	Greenwich Mean Time
$H\_$	Global irradiance on a horizontal surface
HA	Sun hour angle
h	Apparent Sun elevation
h_{astr}	Astronomic (real) sun elevation
H_d	Direct irradiance
$H_{d'}$	Directional irradiance
$H_{d'-}$	Directional irradiance on a horizontal surface
$H_{d'n}$	Directional Sun irradiance on a plane perpendicular to the Sun beam
H_{dn}	Direct Sun irradiance on a plane perpendicular to Sun beam
$H^*{}_{dn}$	Value of H_{dn} by sunshine time
$H^e_{dn\lambda}$	Normal direct irradiance outside the atmosphere for wave length λ
H^e_{dn}	Normal direct irradiance outside the atmosphere
H^e_0	Solar constant
H_s	Diffuse (scattered) irradiance

D. Rekioua and E. Matagne, *Optimization of Photovoltaic Power Systems*, 31
Green Energy and Technology, DOI: 10.1007/978-1-4471-2403-0_2,

H_{sg}	Hemispherical irradiance coming from ground
H_{sh}	Hemispherical irradiance
$H_{sh\ sky}$	Hemispherical irradiance coming from the sky
$H_{sh\ sky-}$	Hemispherical irradiance on a horizontal surface coming from the sky
I_0	Reverse saturation current of a diode (A)
$I_{app}(\Omega)$	Global apparent radiance of the sky coming from the direction Ω
I_d	Current shunted through the intrinsic diode
$I_{s\ app\ hem}$	Hemispherical diffuse apparent radiance of the sky
I_{ph}	Photocurrent
I_{Rsh}	Current of the shunt resistance
I_{sc}	Short-circuit current
$I_{sc\text{-}tjref}$	Short-circuit current at rated temperature
i	Angle between a direction Ω and the normal to a plane
$J(\mu, \varphi)$	Radiance source due to multiple scattering
$J_0(\mu,\varphi)$	Radiance source due to first scattering
k	Extinction coefficient in atmospheric model
K	Boltzman constant ($K = 1.381 \times 10^{-23}$ J/K) in cell model
k'	Directional extinction coefficient
k_λ dm	Elementary relative extinction coefficient at wave length λ
k_0 (m)	Value of k corresponding to the standard atmosphere
(lat)	North latitude of the place
(long)	East longitude of the place
m	Relative (without physical dimension) Air Mass
m'	Absolute air mass
$m'_{z\ ref}$	Absolute air mass of a standard atmosphere in the vertical direction (between TOA and the sea level)
MST	Local mean solar time
N	Number of the day of the year
$P(\Theta)$	So-called phase function
P_1, P_2 and P_3	Constant parameters
p	Atmospheric pressure in hPa (mbars)
p_0	Pressure used in the definition of the standard atmosphere ($p_0 = 1013.25$ hPa)
$p_{sea\ level}$	Pressure at the sea level
q	Quantum of charge (1.602×10^{-19} C)
R_{loc}	Local ground albedo (the fraction of the light received by the ground which is reflected)
R_{reg}	Regional soil albedo
R_s	Series resistance
R_{sh}	Shunt resistance
RST	Real solar time at the place
STC	Standard conditions
T	Temperature

T_j	Junction temperature
$T_{j\ ref}$	Reference temperature
T_{Linke}	Linke turbidity factor
TD	Time difference (which is defined for each country, with in some countries a seasonal change)
TOA	Top of Atmosphere
UT	Universal Time
V_{pv}	Voltage across the PV cell
z	Altitude
α_{sc}	Temperature coefficient of short-circuit current found from the datasheet (absolute or relative)
β_{oc}	Temperature coefficient of open-circuit voltage found from the datasheet (absolute or relative)
γ_1 and γ_2	Coefficients
δ	Declination
Θ	Angle between the incident and scattered light
θ_0	Apparent Sun zenithal angle
$\theta_{0\ astr}$	Astronomic (real) Sun zenithal angle φ_0
θ_p	Inclination of a plane
λ	Wave length
μ, μ_0, μ_p	Cosines of the corresponding zenithal angles
φ	Azimuth
φ_0	Sun azimuth
φ_p	Orientation of a plane
ω	Ratio of the scattering coefficient to the sum of the scattering and absorption coefficients
ω'	The ratio of the hemispherical scattering coefficient to the sum of the hemispherical scattering and absorption coefficients
$\varpi_0, \varpi_1, \varpi_2$	Coefficients of the decomposition of I_h in Legendre polynomials

In order to obtain a realistic view of the behavior of a photovoltaic system, it is necessary to achieve computer simulations. For that purpose, the most important data is the light irradiance of the photovoltaic array at a small time scale (some minutes) during a significant duration (one year or more). Unfortunately, complete experimental data (irradiance for all module inclination and orientation) are never available. For example, the only available measurement result is often hourly or daily global irradiance on a horizontal plane. Sky modeling is thus necessary in order to deduce from the available partial data, a realistic estimation of the module irradiance and some spectral characteristics of that irradiance. Of course, that first result is useful only in conjunction with a model of the photovoltaic modules, in order to deduce from it the electrical power generation for varying irradiance, spectrum, and temperature.

When one is concerned by optimization of a system, it is important to use models well suited in order to achieve the performance evaluation of each tested configuration in a short time, and so to have the possibility of comparing a large number of possibilities. The first part of this chapter is devoted to irradiation estimation using simplified sky or atmosphere models. The second part is devoted to module modeling.

2.1 Irradiance Modeling

2.1.1 Principles and First Simplifying Assumption

2.1.1.1 Sun Light Travel

In order to reach photovoltaic modules, sunlight must go through the atmosphere, where it is subject to absorption and scattering. A part of sunlight reaches the module without undergoing these phenomena: it is named the direct radiation. During its travel through atmosphere, a part of the light is scattered by air molecules, aerosols (dust), water drops or ice crystals, and also by the ground surface. That light has still a chance to arrive on the photovoltaic module after one or several scatterings. That part of module irradiation is named the diffuse fraction. By clear sky, the main part of irradiance is the direct one. By overcast sky, global irradiation is lower and the diffuse to global ratio is higher. So, the light which reaches a photovoltaic module can come from Sun by a variety of ways, as schematically shown on Fig. 2.1.

In this chapter, we do not consider the infrared radiation coming from the atmosphere, since this one has wavelengths too larger for inducing photovoltaic generation. However, we consider "light" as also the infrared radiation coming from Sun, because the part of that radiation which is not stopped by atmosphere has wavelengths able to cause photovoltaic effect.

2.1.1.2 Angles Definition

In order to correctly describe the different directions, we define a spherical coordinate system called horizontal coordinates, as shown in Fig. 2.2, where ZN is the local vertical.

Then, any direction Ω can be specified by the two angles θ and φ. In particular, the normal to the receiving plane can be specified by θ_p and φ_p, which are respectively the inclination of the plane and its orientation.

A solid angle is defined as the surface intersected by a cone on a unit radius sphere centered on its top. In spherical coordinate, the elementary solid angle is

$$\mathrm{d}\Omega = \sin\,\theta\,\mathrm{d}\theta\,\mathrm{d}\varphi \tag{2.1}$$

and is expressed in steradians (sr) if θ and φ are expressed in radians (rad).

Defining global irradiance G on a plane as the power received from light by unit area, we have then

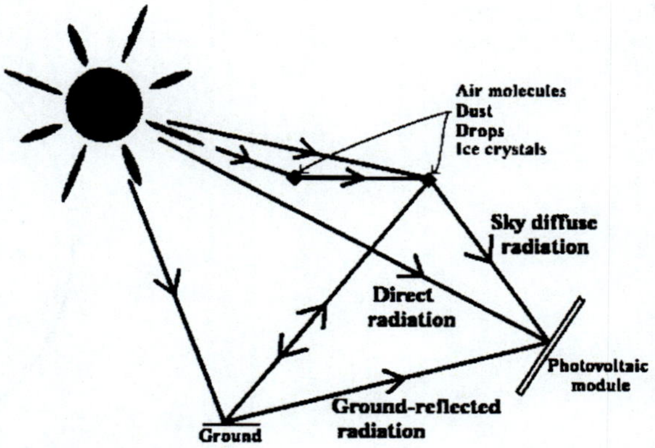

Fig. 2.1 Ways of the solar radiation

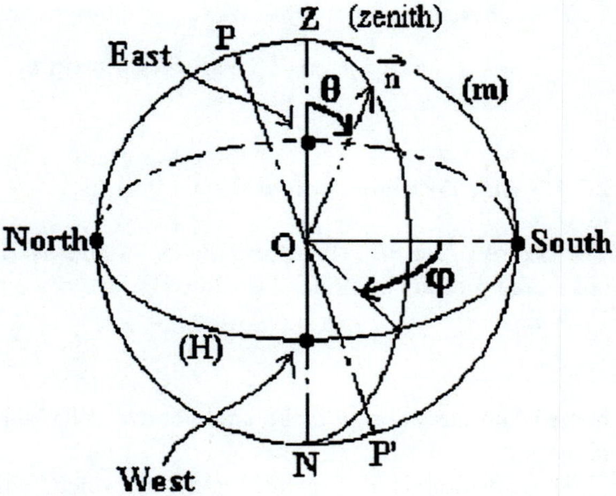

Fig. 2.2 Definition of horizontal coordinates. PP′ is the direction of the Earth rotation axis

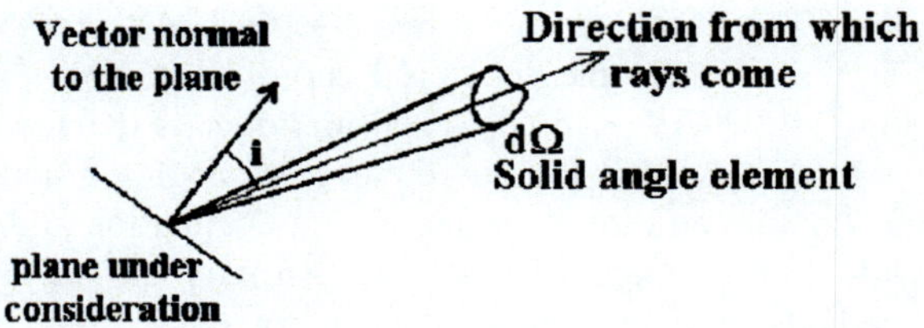

Fig. 2.3 Angle *i* definition

$$G = \iint_{i<90^\circ} I_{app}(\Omega)\,d\Omega = \iint_{i<90^\circ} I_{app}(\cos\theta, \phi)\cos i \sin\theta\, d\theta\, d\varphi \tag{2.2}$$

where $I_{app}(\Omega)$ is the global apparent luminance of the sky coming from the direction Ω and i the angle between that direction and the normal to the plane (Fig. 2.3).

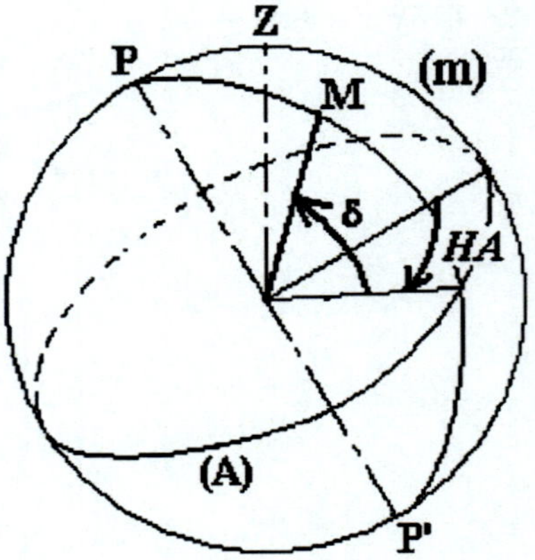

Fig. 2.4 Equatorial coordinates definition

The spherical geometry allows to compute the angle i by

$$\cos i = \cos\theta\cos\theta_p + \sin\theta\sin\theta_p\cos(\varphi - \varphi_p) \tag{2.3}$$

2.1.1.3 Sun Position Computation

The Sun position is defined by angles θ_0 and φ_0, respectively the Sun zenithal angle and the Sun azimuth. Instead of θ_0, we often use angle

$$h = \frac{\pi}{2} - \theta_0 \tag{2.4}$$

named Sun elevation. h is the angle between the sun direction and the horizontal plane.

Sun position is in practice never measured, since it is easy to compute it knowing date and time, longitude and latitude of the place, and some astronomic data. The computation is made easier in a coordinates system different from Fig. 2.2, namely the equatorial coordinates. These ones are defined as shown at Fig. 2.4, where (m) is the local meridian as defined on Fig. 2.2.

In that system, the coordinates are angles δ and HA, called respectively the declination and the hour angle. Sun declination and hour angle can be computed with a very large accuracy by astronomical methods. Simplified computation methods without significant error for the present purpose can be found in the literature. An example of such code is given in [13].

If some degrees inaccuracy is acceptable, we can use the following formulae [14]. Sun declination is obtained as

$$\sin\delta(N) = 0.398\sin\left\{\frac{2\pi}{365}\left[N - 82 + 2\sin\frac{2\pi(N-2)}{365}\right]\right\} \tag{2.5}$$

where N is the number of the day of the year

In order to obtain the hour angle H, the following operations are leaded starting with the civil time CT in hours

(a) UT (Universal Time) or GMT (Greenwich Mean Time) is obtained by subtracting from CT the time difference TD (which is defined for each country, with in some countries a seasonal change).

$$\mathrm{UT} = \mathrm{CT} - \mathrm{TD} \tag{2.6}$$

(b) Using the east longitude of the place, we obtain the local mean solar time of the place by

$$\mathrm{MST} = \mathrm{UT} + \frac{(\mathrm{long})}{15} \tag{2.7}$$

where (long) is the east longitude of the place expressed in °, MST and UT being in hours.

(c) Then, the real solar time at the place is obtained using the equation

$$\mathrm{RST} = \mathrm{MST} + E_t \tag{2.8a}$$

where E_t is the time equation, which take into account the fact that the rotation speed of the Earth around Sun is not uniform. We have approximately, in hours,

$$E_t(N) = \frac{1}{60}[9.87 \sin(2N') - 7.53 \cos(N') - 1.5 \sin(N')] \tag{2.8b}$$

with

$$N' = \frac{2\pi(N - 81)}{365} \tag{2.8c}$$

(d) The hour angle is linked to the real solar time by the relation

$$\mathrm{HA} = \frac{\pi}{12}(\mathrm{RST} - 12) \tag{2.9}$$

Once the angles δ and HA are known, we can compute the angles $\theta_{0\ \mathrm{astr}}$ or h_{astr} and φ_0 by a change of coordinates:

$$\sin h_{\mathrm{astr}} = \cos \mathrm{HA} \cos \delta \cos(\mathrm{lat}) + \sin \delta \sin(\mathrm{lat}) \tag{2.10a}$$

$$\cos \varphi_0 \cos h_{\mathrm{astr}} = \cos \mathrm{HA} \cos \delta \sin(\mathrm{lat}) - \sin \delta \cos(\mathrm{lat}) \tag{2.10b}$$

$$\sin \varphi_0 \cos h_{\mathrm{astr}} = \sin \mathrm{HA} \cos(\delta) \tag{2.10c}$$

where (lat) is the north latitude.

It is to be noticed that the determination of φ_0 without ambiguity is possible only using all the two Eqs. 2.10b and 2.10c.

The apparent Sun zenithal angle θ_0 is approximately equal to the astronomical angle $\theta_{0\ \mathrm{astr}}$: a small difference occurs due to the atmospheric refraction. In photovoltaic studies, we assume frequently

$$\theta_0 \approx \theta_{\mathrm{astr}} \text{ or, equivalently, } h \approx h_{\mathrm{astr}} \tag{2.11a}$$

None correction to Eq. 2.11a is useful for small zenithal angle (Sun elevation near to 90°). For larger zenithal angles (Sun elevation near to 0°), the enhanced Saemundsson formula [15] is a little bit better:

$$h \approx h_{\mathrm{astr}} + \frac{p}{1010}\frac{283.15}{273.15+T}\frac{1.02}{60}\,\mathrm{cotg}\left(h_{\mathrm{astr}} + \frac{10.3}{h_{\mathrm{astr}}+5.14}\right) \quad \text{for } -2^\circ < h_{\mathrm{astr}} < 89^\circ \tag{2.11b}$$

where p is the atmospheric pressure in hPa (mbars) and T the temperature in °C, h and h_{astr} being expressed in degrees.

Of course, the formulae are not relevant when $h_{\mathrm{astr}} < -2^\circ$ since it is then certainly the night.

Taking into account the sun radius, sunrise or sunset arrives when

$$h \approx -0.27^\circ. \tag{2.12}$$

2.1.2 Sky and Ground Radiance Modeling

The global irradiance Eq. 2.2 can be split is two parts: direct irradiance and diffuse irradiance

$$G = H_d + H_s \tag{2.13}$$

where the index "d" stands for "direct" and the index "s" for "scattered" (diffuse).

2.1.2.1 Direct Radiation

Direct Sun radiation (also named beam radiation) is assumed coming from a point at the Sun disk center. The direct radiance is thus a Dirac delta on the point of coordinate θ_0 (or h) and φ_0. Then, the corresponding part of Eq. 2.2 reduces to

$$H_d = 0 \quad \text{if } \cos i < 0 \tag{2.14a}$$

$$H_d = H_{\mathrm{dn}} \cos i \quad \text{if } \cos i > 0 \quad (i < 90^\circ) \tag{2.14b}$$

where the index "n" is for "normal" and, thus, H_{dn} is the direct Sun irradiance on a plane perpendicular to Sun beam. i is the incidence angle of direct radiation on the plane. It is easy to compute that incidence angle i, using the particular case of Eq. 2.3 where $\theta = \theta_0$,

$$\begin{aligned} \cos i &= \cos \theta_0 \cos \theta_p + \sin \theta_0 \sin \theta_p \cos(\varphi_0 - \varphi_p) \\ &= \sin h \cos \theta_p + \cos h \sin \theta_p \cos(\varphi_0 - \varphi_p) \end{aligned} \tag{2.15}$$

So, using twice (2.14), it is sufficient to have only one measurement of direct radiation on one irradiated plane to compute H_{dn} and thus the value of H_d on any plane.

When the sky is uniform (clear or uniformly cloudy), H_{dn} is a smooth function of time. However, in case of incomplete cover by thick clouds, H_{dn} experiences quick changes between a value H^*_{dn} for sunshine time and nearly 0 when there is a cloud in front of Sun. In that case, we speak about "bimodal state".

A related notion is the sunshine duration. The sunshine duration is the time for which $H_{\mathrm{dn}} > 120$ W. Thus, if $H^*_{\mathrm{dn}} > 120$ W, the ratio between the sunshine duration and the time of recording is equal to the probability to have $H_{\mathrm{dn}} = H^*_{\mathrm{dn}}$ at a particular time. That probability is also approximately equal to the complement to unity of the cloud cover (in per unit)

2.1.2.2 Circumsolar Diffuse Radiation

Restraining the Eq. 2.2 to the diffuse part, we have

$$H_s = \iint_{i<90^\circ} I_{s\,\mathrm{app}}(\Omega)\mathrm{d}\Omega = \iint_{i<90^\circ} I_{s\,\mathrm{app}}(\cos\theta, \varphi) \cos i \sin\theta \,\mathrm{d}\theta \,\mathrm{d}\varphi \tag{2.16}$$

Searching for a simplified expression of $I_{s\ \mathrm{app}}$ (cos θ, φ), a common approximation [16] is to split that function of two coordinates in a sum of two function of only one coordinate, namely a circumsolar part and an hemispherical part:

$$I_{s\,\mathrm{app}} = I_{s\,\mathrm{app\,circ}}(\Theta) + I_{s\,\mathrm{app\,hem}}(\theta) \tag{2.17}$$

where the circumsolar part is function only of the angle between the Sun direction and the observation direction, and the hemispheric part is only function of θ. That splitting comes from the common observation that the radiance of the part of the sky which is near to the Sun is higher than the other parts of the sky. Angle Θ is simply

$$\cos\Theta = \cos\theta\cos\theta_0 + \sin\theta\sin\theta_0\cos(\varphi - \varphi_0) \tag{2.18}$$

For the irradiance computation, usually, we do not use Eq. 2.18, but we consider that all the circumsolar part comes from the Sun disk center. Then, direct irradiance and circumsolar irradiance can be treated as a whole, which is named "directional" "irradiance". We use the index "d'" for "directional". We obtain

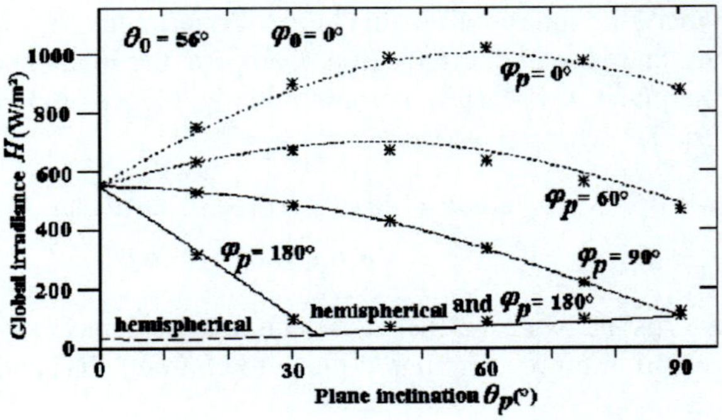

Fig. 2.5 Total irradiance on a surface in function of the inclination for different orientations. * are experimental values [17], curves are computed by Eq. 2.20 with $H_{\mathrm{d'n}}$ and H_{sh} fitted for the best approximation

for the directional component an expression very similar to that obtained for direct component (2.14):

$$H_{d'} = 0 \quad \text{if } \cos i < 0 \tag{2.19a}$$

$$H_{d'} = H_{d'n} \cos i \quad \text{if } \cos i > 0 \quad (i < 90°) \tag{2.19b}$$

As the radiance $I_{s\ \mathrm{app\ hem}}$ depends only of θ, it is clear that the corresponding irradiance depends only of the plane inclination θ_p. We have then

$$G = H_{d'n} \cos i + H_{\mathrm{sh}}(\theta_p) \tag{2.20}$$

That partition can give a good approximation of real irradiance, as it is shown at Fig. 2.5, which uses global irradiance experimental data from the literature [17] versus inclination for different orientations at a fixed time (clear sky, Davis California, $\theta_0 = 90–34° = 56°$).

From Eq. 2.20, it is obvious that, in order to compute the directional part of irradiation, it is sufficient to have two experimental values of global irradiance on two planes of same inclination θ_p but different orientations φ_p provided they are dissymmetric with regard to Sun azimuth φ_0. However, it remains useful to split the directional irradiance into direct and circumsolar irradiances since only the direct component is submitted to bimodal state.

An important particular case of Eq. 2.20 is the irradiance on the horizontal plane, which is often measured in meteorological stations. In that case, following Eq. 2.15 for $\theta_p = 0$, we have $\cos i = \sin h$ and thus Eq. 2.20 becomes:

$$G_{-} = H_{\mathrm{d'n}} \sin h + H_{\mathrm{sh-}} \tag{2.21}$$

2.1.2.3 Ground Diffuse Radiation

The next step in our analysis is to split the hemispherical irradiance into a part coming from the sky and a part coming from ground.

$$H_{\mathrm{sh}} = H_{\mathrm{sh\,sky}} + H_{\mathrm{sg}} \tag{2.22}$$

This ground radiance is often assumed to be isotropic, and thus purely hemispherical. For that reason, we have omitted the index "h " in the last term of Eq. 2.22. Then, reducing the integral Eq. 2.2 to the ground diffuse radiation, we can show that the corresponding irradiance is

$$H_{\text{sg}} = R_{\text{loc}}\, G_{-} \frac{1 - \cos\theta_p}{2} \tag{2.23}$$

where R_{loc} is the local ground albedo (the fraction of the light received by the ground which is reflected).

In the expression Eq. 2.23, the value of G_{-} is given by (2.21). It shall be noticed that the ground diffuse irradiance does not contribute to $H_{\text{s h}-}$.

$$G_{-} = H_{d'-} + H_{\text{sh sky}-} \tag{2.24}$$

2.1.2.4 Sky Hemispherical Diffuse Irradiance

It remains for achieving the analysis to give an expression for the sky hemispherical diffuse irradiance $H_{\text{sh sky}}(\theta_p)$. The simplest assumption is that the sky hemispherical irradiance results from an isotropic radiance. Then, the corresponding part of Eq. 2.2 leads to the expression

$$H_{\text{sh sky}} = H_{\text{si}} = H_{\text{si}-} \frac{1 + \cos(\theta_p)}{2} \tag{2.25}$$

where the index "i" stands for "isotropic sky" and thus implicitly hemispherical.

In that case, the full description of irradiance as a function of the inclination and orientation of a plane is obtained using only three parameters: $H_{\text{d'n}}$, $H_{\text{si}-}$ and R_{loc}. If R_{loc} is known, the two irradiance measurement requested at Sect. 2.1.2.2. are sufficient in order to identify that function.

Unfortunately, the sky isotropy assumption is contrary to the common observation that the part of the sky near to the horizon is often clearer than the other parts of the sky. So, different authors have introduced empirical additional terms, the most known being the horizon circle diffuse irradiance. We shall not consider such terms in that book because we prefer obtain the expression of $H_{\text{sh sky}}$ from a physical analysis of atmosphere, which is achieved below.

2.1.3 Use of an Atmospheric Model

In many cases, the available experimental irradiance data are insufficient, because

- they do not allow computing the irradiance on the considered plane using the methods described in 2.1.2,

- a best approximation of $H_{\text{sh sky}}$ than Eq. 2.25 is required and the number of irradiance measurements is insufficient for that,
- they are given as mean values on too large time intervals,
- information on spectral characteristics is needed but not available.

In particular, we stress the fact that average values of irradiance are not suited for direct use in photovoltaic studies since

- the power generation of photovoltaic modules is not proportional to irradiance,
- for the systems simulations, the energy production timing is of importance.

Then, atmospheric modeling can be used to extend in a realistic way the available experimental data.

2.1.3.1 Air Mass Notion

As the air density increases from 0 at the TOA (top of atmosphere) to its value at the ground level, the air influence on a light ray is more important at low altitude. For that reason, it is useful to use as coordinate not the distance l covered by the light ray but the Air Mass

$$m' \underset{\text{def}}{=} \int \rho \, \mathrm{d}l \tag{2.26}$$

The use of Air Mass in the equation is related to the assumptions that

- the air molecules and particles act on light individually,
- their effect is independent of temperature.

Besides being not totally exact, these assumptions are acceptable for our purpose.

A particular case of Eq. 2.26 is the case of a vertical ray. We define $m'_{z\ \text{ref}}$ as the Air Mass of a standard atmosphere in the vertical direction between TOA and the sea level. Then, we can use instead of absolute Air Mass Eq. 2.26 a relative (without physical dimension) Air Mass

$$m \underset{\text{def}}{=} \frac{m'}{m'_{z\,\text{ref}}} \tag{2.27}$$

Vertical Air Mass m_z is strongly related to the atmospheric pressure. Assuming that the gravitational field g is constant in the entire atmosphere ($g \approx 9.81$ m/s^2), vertical Air Mass is given by

$$m_z = \frac{p}{p_0} \tag{2.28}$$

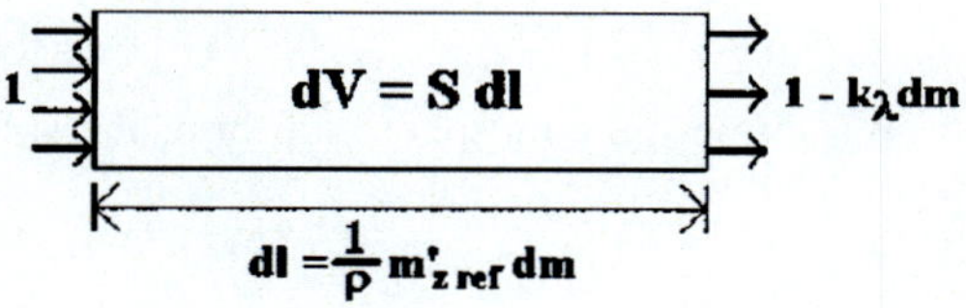

Fig. 2.6 Definition of the extinction coefficient

where p_O is the pressure used in the definition of the standard atmosphere ($p_0 = 1013.25$ hPa).

If one knows the pressure at the sea level, we can find the pressure at an altitude z (in km) using the formula [14]

$$p = p_{\text{sea level}}\,(0.89)^z \text{ for z } < 3 \text{ km} \tag{2.29}$$

Light can be considered as the sum of electromagnetic radiations with different wave length λ. Visible light has a spectrum going from $\lambda = 400$ nm (violet) until $\lambda = 800$ nm (red) (1 nm $= 10^{-9}$ m). Sun radiation includes UV (ultraviolet) of shorter wave lengths and IR (infrared) of longer wave length. Some solar modules are able to get energy even from invisible Sun radiation.

When a beam of light crosses an air volume, its energy is lowered by absorption and diffusion phenomena. The elementary relative lowering is given by the product k_λ dm, where dm is the elementary relative Air Mass in the beam direction and where k_λ is a coefficient named "extinction coefficient". k_λ is function only of λ and of the atmosphere composition. Figure 2.6 gives an illustration of that phenomenon.

2.1.3.2 Direct Radiation

Direct radiation, also called "beam radiation", is considered as coming from the direction computed at 2.2.1.3.

It experiences an Air Mass m which is given, assuming the atmosphere plane and without refraction, by:

$$m \approx \frac{m_z}{\cos(\theta_0)} = \frac{m_z}{\sin h} \approx \frac{m_z}{\sin h_{\text{astr}}} \tag{2.30}$$

If we consider the atmospheric refraction, the way followed by the beam radiation is lightly curved. We can take that fact into account replacing Eq. 2.30 by a best approximation, such as the Kasten formula [14]

$$m \approx m_z \frac{1}{\sin h + 9.40\,10^{-4}(\sin h + 0.0678)^{-1.253}} \tag{2.31}$$

The direct radiation is characterized by its normal irradiance H_{dn}, as in Sect. 2.1.2.1. The extinction mechanism of Fig. 2.6 leads thus to the equation:

$$dH_{dn\lambda} = -k_\lambda H_{dn\lambda} dm \tag{2.32}$$

Assuming the atmosphere uniform, the solution of Eq. 2.32 is simply

$$H_{dn\lambda} = H^e_{dn\lambda}\, e^{-k_\lambda m} \tag{2.33}$$

where $H^e_{dn\lambda}$ is the normal irradiance outside the atmosphere (at the TOA)

Integrating on all the wave length, we obtain the total normal irradiance outside the atmosphere

$$H^e_{dn} = \int H^e_{dn\lambda}\, d\lambda \tag{2.34}$$

When the Sun–Earth distance is equal to its mean value (1 astronomical unit), the value of H^e_{dn} is called the "solar constant" H^e_0 The value of that last one is approximately $H^e_0 = 1361\,\mathrm{W/m^2}$ [18]. Similarly, the mean values of the $H^e_{dn\lambda}$can be found in the literature [19].

However, the Sun irradiation received at TOA varies during a year due to the variation of Sun–Earth distance. So, H^e_{dn} and $H^e_{dn\lambda}$ have to be deduced from their mean values using a distance correction

$$C(N) \approx 1 + 0.034\,\cos\left[\frac{2\pi}{365}(N-2)\right] \tag{2.35}$$

where N is the number of the day of the year.

For each wave length, k_λ can be decomposed in a sum. Each term of the sum take into account the effect of one atmospheric component. In fact, we can consider as only one component the set of all constant gas (N_2, O_2, .. CO_2). Variable gases are the water vapor H_2O, ozone O_3 and sometime NO_2. We have also to take into consideration the aerosol content (dust) and the water condensed in drops or in ice crystals (clouds).

For simplified computation, we prefer to replace the infinite sum Eq. 2.34 by a finite one:

$$H_{dn} = H^e_{dn} \sum_i \alpha_i\, e^{-k_i m} \quad \text{with} \quad \sum_i \alpha_i < 1 \tag{2.36}$$

For terrestrial applications, the sum of the coefficients α_i can be lower than unity since some wave length are so strongly absorbed in the atmosphere that they do not contribute to the irradiance at the ground level.

The decomposition Eq. 2.36 can maintain information on the light spectrum. For that, the spectrum can be divided in bands of wave length, each band corresponding to one or several exponential terms.

On the other hand, for a very simplified computation, we can use only one exponential for the entire spectrum. In that case, it is convenient to admit that the extinction factor k is a function of m. A classical method is to use

$$k(m) = k_0(m)T_{\mathrm{Linke}} \tag{2.37}$$

where k_0 (m) is the value corresponding to the standard atmosphere. T_{Linke} is the Linke turbidity factor, which is a characteristic of the state of the atmosphere. Once $k_0(m)$ has been determined by computation, one measurement of the direct irradiation is sufficient in order to compute T_{Linke}. Historically, several approximations of $k_0(m)$ have been proposed and lead to different value of T_{Linke}. Thus, old experimental values of T_{Linke} can need revision. Kasten [48] has given in 1996 an accurate expression:

$$k_0(m) = \frac{1}{6.6296 + 1.7513\,\mathrm{m} - 0.1202\,\mathrm{m}^2 + 0.0065\,\mathrm{m}^3 - 0.00013\,\mathrm{m}^4} \tag{2.38a}$$

for $1 < \mathrm{m} < 20$

For higher values of m, one can use [49]

$$k_0(m) = \frac{1}{10.4 + 0.718\,\mathrm{m}} \quad \text{for} \quad m > 20 \tag{2.38b}$$

Contrary to the irradiance, the value of T_{Linke} depends only on little of the Sun elevation. Then, assuming that the meteorological conditions are constant during a time period, if one knows the experimental mean value of H_{dn} during that period, it is possible to search the value of T_{Linke} which is constant on that period and which, by computation, leads to the same mean value of H_{dn}. Using that value of T_{Linke}, it is then possible to compute at each time the value of H_{dn}. Of course, if one is in bimodal mode, that computation has to be adapted taking into account the sunshine duration. For photovoltaic computation, it is better to use H^*_{dn} with a sunshine probability in place of a smoothed value of H_{dn} for the above mentioned reasons. For the systems simulations where the energy production timing is of importance, it is possible to generate realistic variations of the direct irradiation at small time scale using the method of Markov chains.

2.1.3.3 Generalities About the Diffuse Radiation

In the following, we use the variables:

$$\mu = \cos\,\theta, \quad \mu_0 = \cos\,\theta_0, \quad \mu_p = \cos\,\theta_p \ldots \tag{2.39}$$

Behind the considered functions are wave length dependent, we shall make abstraction of the subscript "λ", hoping that the formulae could be use for broadband calculations.

The diffuse light is described at each point of the atmosphere by the radiance $I(\theta, \varphi)$ going to the direction (θ, φ). The apparent sky and ground radiances are then related to $I(\theta, \varphi)$ by:

$$I_{s\,\mathrm{app}}(\mu, \varphi) = I(-\mu, \varphi + \pi) \tag{2.40}$$

When an unpolarized parallel beam of normal irradiance E crosses an elementary volume of thickness *dm* (given in relative Air Mass), the increase of scattered radiance is given by the equation

$$\mathrm{d}I(\Theta) = \frac{k\,\omega}{4\,\pi} P(\Theta) E\,\mathrm{d}m \tag{2.41}$$

where ω is the ratio of the scattering coefficient to the sum of the scattering and absorption coefficients, Θ is the angle between the incident and scattered light and $P(\Theta)$ is the so-called phase function. The phase function is normalized in such a way that [20]

$$\frac{1}{4\pi} \int_{4\pi} P(\Theta)\mathrm{d}\Omega = \frac{1}{2} \int_{0}^{\pi} P(\Theta) \sin \Theta \mathrm{d}\Theta = 1 \tag{2.42}$$

where $\mathrm{d}\Omega$ is the elementary solid angle

In presence of aerosols or water drops, the phase function is very complicated and, even when long numerical computations are achieved, we use for radiative transfer computations approximations of this function which depend only of a few parameters, the most important of them being the asymmetry factor defined as

$$g = \frac{1}{4\pi} \int_{4\pi} P(\Theta) \cos(\Theta)\mathrm{d}\Omega = \frac{1}{2} \int_{0}^{\pi} P(\Theta) \cos(\Theta) \sin(\Theta)\mathrm{d}(\Theta) \tag{2.43}$$

From the above definitions, it can be proved that, in a plane atmosphere, the diffuse radiance $I(\mu, \varphi)$ obeys to the equation

$$\mu \frac{\mathrm{dI}(\mu, \varphi)}{\mathrm{d}m_z} = kI(\mu, \varphi) - J(\mu, \varphi) - J_0(\mu, \varphi) \tag{2.44}$$

where $J(\mu, \varphi)$ is the source due to multiple scattering and $J_0(\mu, \varphi)$ is the source due to first scattering.

The source due to multiple scattering is given by the equation

$$J(\mu, \varphi) = \frac{k\omega}{4\pi} \int_{0}^{2\pi} \int_{-1}^{1} I(\mu', \varphi')P(\cos \Theta)\mathrm{d}\mu'\mathrm{d}\varphi' \tag{2.45}$$

with

$$\cos \Theta = \mu\mu' + \sqrt{1 - \mu^2}\sqrt{1 - \mu'^2} \cos(\varphi - \varphi') \tag{2.46}$$

On the other hand, the primary source of diffuse radiation is

$$J_0(\mu, \varphi) = \frac{k\omega}{4\pi} H_0 P(- \cos \Theta_0)\mathrm{e}^{-km_z/\mu_0} \tag{2.47}$$

with

$$\cos(\Theta_0) = \mu\mu_0 + \sqrt{1-\mu^2}\sqrt{1-\mu_0^2}\cos(\varphi - \varphi_0) \tag{2.48}$$

where θ_0 and φ_0 are the coordinates of the Sun and μ_0 is defined by Eq. 2.39.

The equation system Eqs. 2.44–2.48 is very difficult to solve. Currently, an accurate solution of these equations can be obtained only by using time-consuming numerical calculations. Numerous empirical models have been developed. Reviews of these models can be found in the literature [21, 22]. Such models use functions whose numerous parameters are to be extracted from experimental data. They can thus be geography or climate dependent and present unpredictable limitations. In the following, we describe a realistic simplified model using only as data a few atmospheric parameters with well defined physical signification.

2.1.3.4 δ-Approximation and Circumsolar Component

The δ-approximation [35] consist to assume that a fraction f_{circ} of the scattered light keeps the direction of the incident light, let's be

$$P(\Theta) = 2f_{\mathrm{circ}}\delta(1 - \cos\Theta) + (1 - f_{\mathrm{circ}})P_h(\Theta) \tag{2.49}$$

where the first term is a Dirac delta

We consider then that the circumsolar component is due to the first term and has then exactly the same direction as the direct rays. On the other hand, the residual phase function $P_h(\Theta)$ shall be related to hemispherical diffuse radiation. Using the subscript d' for the directional irradiance (direct + circumsolar), we obtain an equation similar to Eq. 2.33:

$$H_{\mathrm{d'n}} = H_{\mathrm{dn}}^{e}\mathrm{e}^{-k'm} \tag{2.50}$$

where k' is the directional extinction coefficient and

$$k' = k(1 - \omega f_{\mathrm{circ}}) \tag{2.51}$$

If necessary, the circumsolar component can be obtained as the difference between Eqs. 2.50 and 2.33.

2.1.3.5 Equations for Hemispherical Component

We obtain equations for the residual diffuse radiance I_h (μ, φ) by replacing in Eqs. 2.44, 2.45 and 2.47 k by k' and ω by:

$$\omega' = \frac{(1 - f_{\mathrm{circ}})\omega}{1 - \omega f_{\mathrm{circ}}} \tag{2.52}$$

Defining I_h (μ) as the mean value of I_h (μ, φ) on $-\pi < \varphi < \pi$, the equation for I_h (μ) is obtained by replacing in the corresponding part of Eq. 2.44 I, J and J_0 by their mean values on φ. We get [24]:

$$\mu \frac{\mathrm{d}I_h(m_z,\mu)}{dm_z} = k' I_h(m_z,\mu) - \frac{k'\omega'}{2}\int_{-1}^{1} I_h(m_z,\mu')P_h(\mu,\mu')\mathrm{d}\mu' - \frac{k'\omega'}{4\pi}P_h(\mu,-\mu_0)H_0 \mathrm{e}^{-k' m_z/\mu_0} \quad (2.53)$$

where

$$P_h(\mu,\ \mu') = \frac{1}{2\pi}\int_0^{2\pi} P_h(\Theta)\,\mathrm{d}\varphi \quad (2.54)$$

The approximation is to consider that I_h (m_z, μ, φ) can be identified with I_h (m_z, μ), and thus considered as a hemispherical radiance.

There are some approximate solutions of the Eq. 2.53. The most known are called two-stream approximations. In those methods, the variables are the upward and downward hemispherical diffuse irradiation:

$$H_h^+(m_z) = 2\pi \int_0^1 I_h(m_z,\mu)\mu\,\mathrm{d}\mu \quad (2.55a)$$

and

$$H_h^-(m_z) = 2\pi \int_0^1 I_h(m_z,-\mu)\mu\,\mathrm{d}\mu \quad (2.55b)$$

H_h^+ and H_h^- are the only variables of the differential system if we use an approximation of $I_h(\mu)$ completely determined by H_h^+ and H_h^-. That approximation can thus have only two freedom degrees.

2.1.3.6 Spherical Harmonics Decomposition

The function $I_h(m_Z, \mu)$ can be decomposed in Legendre polynomial series. The first Legendre polynomials are

$$P_0(\mu) = 1 \quad P_1(\mu) = \mu \quad P_2(\mu) = \frac{1}{2}(3\mu^2 - 1) \quad (2.56)$$

These polynomials obey the orthogonallity condition

$$\frac{2i+1}{2}\int_{-1}^{1} P_i(\mu)P_j(\mu)\mathrm{d}\mu = \delta_{\mathrm{ij}} \tag{2.57}$$

And the recurrence formula [37]

$$(2i+1)\mu P_i(\mu) = (i+1)P_{i+1}(\mu) + i P_{i-1}(\mu) \tag{2.58}$$

We can consider decompositions

$$I_h(m_z,\mu) = I_0(m_z)P_0(\mu) + 3 I_1(m_z)P_1(\mu) + 5 I_2(m_z)P_2(\mu) + \cdots \tag{2.59}$$

and

$$P_h(\mu) = \varpi_0 P_0(\mu) + \varpi_1 P_1(\mu) + \varpi_2 P_2(\mu) + \cdots\cdots \tag{2.60}$$

The Legendre polynomials have the property [25]:

$$\frac{1}{2\pi}\int_{0}^{2\pi} P_i(\cos\Theta)\mathrm{d}\varphi = P_i(\mu)P_i(\mu') \tag{2.61}$$

Using this property and Eqs. 2.56, 2.57, 2.53 can be decomposed as:

$$\begin{aligned}\left[(i+1)\frac{\mathrm{d}}{\mathrm{d}m_z}I_{i+1} + i\frac{\mathrm{d}}{\mathrm{d}m_z}I_{i-1}\right] &= k'(2i+1)I_i - k'\omega'\varpi_i I_i \\ &\quad - \frac{k'\omega'\varpi_i}{4\pi}P_i(-\mu_0)H_0\mathrm{e}^{-k'm_z/\mu_0}\end{aligned} \tag{2.62}$$

Applying the analog of Eqs. 2.42 and 2.43. for $P_h(\mu)$, we have always [24]

$$\varpi_0 = 1 \quad \varpi_1 = 3\,g_h \tag{2.63a}$$

and, by reference to the Henyey-Greenstein phase function, there is a reason [23] to take

$$\varpi_2 = 0 \tag{2.63b}$$

We assume that $\varpi_i = 0$ for all $i\rangle 1$. In the particular case where $g_h = 0$, the hemispherical diffusion is isotropic, i.e. $P_h(\mu) = 1$. In that case, the identification of that I_h (m_z, μ, φ) with I_h (m_z, μ) is exact, and thus also the hemispherical hypothesis.

In order to obtain a system of two differential equations, the decomposition of Eq. 2.59 can only involve two degrees of freedom. The Eddington approximation consists to keep only the first two terms of Eq. 2.59. However, empirical investigations [16] and accurate numerical solutions [26] show that an approximation of the first degree in μ is not sufficient for a correct modeling of the hemispherical radiance.

2.1.3.7 Taking into Account the Second Order Term

The following analysis is equivalent to the method developed in [27]. That approach is to consider the three first terms of the series Eq. 2.59 with an additional condition in order to keep only two freedom degrees. Equation 2.62 yields [28]:

$$\frac{\mathrm{dI}_1}{\mathrm{d}m_z} = a_0 k' I_0 - b_0 k' H_0 \mathrm{e}^{-k' m_z/\mu_0} \tag{2.64a}$$

$$2\frac{\mathrm{dI}_2}{\mathrm{d}m_z} + \frac{\mathrm{dI}_0}{\mathrm{d}m_z} = a_1 k' I_1 - b_1 k' H_0 \mathrm{e}^{-k' m_z/\mu_0} \tag{2.64b}$$

and, assuming $I_3 = 0$,

$$2\frac{\mathrm{dI}_1}{\mathrm{d}m_z} = a_2 k' I_2 - b_2 k' H_0 \mathrm{e}^{-k' m_z/\mu_0} \tag{2.64c}$$

where

$$a_i = (2i+1) - \omega' \varpi_i \tag{2.65a}$$

and

$$b_i = \frac{1}{4\pi}\omega' \varpi_i P_i(-\mu_0) \tag{2.65b}$$

Comparing Eqs. 2.64a and 2.64c, we see that the system is degenerated and can only have a solution if

$$I_2 = c_1 I_0 + c_2 H_0 \mathrm{e}^{-k' m_z/\mu_0} \tag{2.66a}$$

with

$$c_1 = 2\frac{a_0}{a_2} \quad \text{and} \quad c_2 = \frac{b_2 - 2b_0}{a_2} \tag{2.66b}$$

Constraint Eq. 2.66 allows for replacing Eq. 2.64b by

$$\frac{\mathrm{dI}_0}{\mathrm{d}m_z} = a_1' I_1 - b_1' H_0 \mathrm{e}^{-k' m_z/\mu_0} \tag{2.67}$$

where

$$a_1' = \frac{a_1}{1+2c_1} \quad \text{and} \quad b_1' = \frac{b_1 - 2c_2/\mu_0}{1+2c_1} \tag{2.68}$$

The system Eqs. 2.64a and 2.67 is easy to solve. Its solution, completed with Eq. 2.66, is of the form

$$I_0 = \sqrt{a_1'}\gamma_1 \mathrm{e}^{\lambda k' m_z} + \sqrt{a_1'}\gamma_2 \mathrm{e}^{-\lambda k' m_z} + \gamma_3 H_0 \mathrm{e}^{-k' m_z/\mu_0} \tag{2.69a}$$

$$I_1 = \sqrt{a_0}\gamma_1 \mathrm{e}^{\lambda k' m_z} - \sqrt{a_0}\gamma_2 \mathrm{e}^{-\lambda k' m_z} + \gamma_4 H_0 \mathrm{e}^{-k' m_z/\mu_0} \tag{2.69b}$$

$$I_2 = c_1\sqrt{a_1'}\gamma_1 \mathrm{e}^{\lambda\tau'} + c_1\sqrt{a_1'}\gamma_2 \mathrm{e}^{-\lambda\tau'} + (c_1\gamma_3 + c_2)H_0 \mathrm{e}^{-\tau'/\mu_0} \tag{2.69c}$$

where

$$\lambda = \sqrt{a_0 a_1'} \tag{2.70}$$

$$\gamma_3 = \frac{\mu_0(b_1' - a_1' b_0 \mu_0)}{1 - a_0 a_1' \mu_0^2} \tag{2.71a}$$

and

$$\gamma_4 = \frac{\mu_0(b_0 - a_0\, b_1' \mu_0)}{1 - a_0 a_1' \mu_0^2} \tag{2.71b}$$

The coefficients γ_1 and γ_2 are to be determined using boundary conditions. For that, we use the upward and downward horizontal irradiance Eq. 2.55 whose expression is [28], by assuming coefficient I_3 and the next ones equal to 0:

$$H_h^+ = \pi\left(I_0 + 2I_1 + \frac{5}{4}I_2\right) \tag{2.72a}$$

and

$$H_h^- = \pi\left(I_0 - 2I_1 + \frac{5}{4}I_2\right) \tag{2.72b}$$

Thus, introducing Eq. 2.69 into Eq. 2.72,

$$\begin{aligned} H_h^{\pm} = {} & \pi\left[\sqrt{a_1'}\left(1+\frac{5}{4}c_1\right) \pm 2\sqrt{a_0}\right]\mathrm{e}^{\lambda k' m_z}\gamma_1 + \pi\left[\sqrt{a_1'}\left(1+\frac{5}{4}c_1\right) \mp 2\sqrt{a_0}\right]\mathrm{e}^{-\lambda k' m_z}\gamma_2 \\ & + \pi\left[\left(1+\frac{5}{4}c_1\right)\gamma_3 + \frac{5}{4}c_2 \pm 2\gamma_4\right]H_0 \mathrm{e}^{-k' m_z/\mu_0} \end{aligned} \tag{2.73}$$

The boundary conditions are then

$$H_h^- = 0 \quad \text{at the TOA } (m_z = 0) \tag{2.74a}$$

$$H_h^+ = R_{\mathrm{reg}}(H_h^- + H_{d'n}\mu_0) \quad \text{at the ground level} \quad (m_z = m_{z\,\mathrm{sfc}}) \tag{2.74b}$$

where R_{reg} is the regional soil albedo.

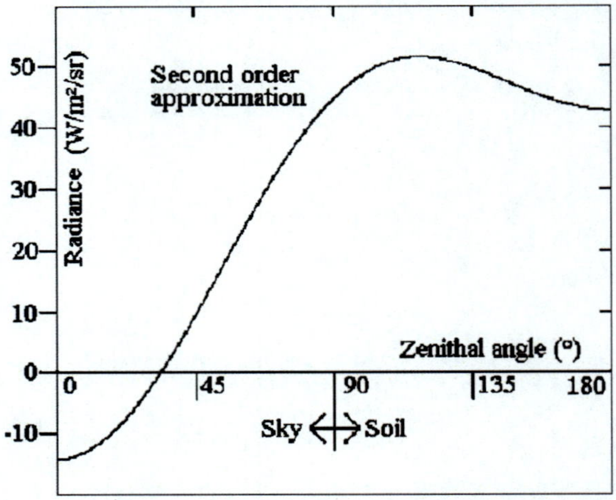

Fig. 2.7 Hemispherical radiance computed by the present model (with $k' m_z = 0.220$ $\omega' = 0.4141$, $g_h = 0$ and $R_{reg} = 0.2779$)

There results a linear system of two equations in γ_1 and γ_2, which can thus be computed. Then, the expression Eq. 2.59 is completely determined. Integrating Eq. 2.59 similarly to Eq. 2.2, we obtain hemispherical irradiance on any plane. By example, considering the case of Fig. 2.5, we obtain good results using $k'\ m_z = 0.220$, $\omega' = 0.4141$ $g_h = 0$ and $R_{\mathrm{reg}} = 0.2779$, behind the fact that the number of freedom degrees is much smaller.

Figure 2.7 shows the corresponding radiance as a function of the zenith angle. We can see that this second order approximation is not realistic since it leads to negative values of the radiance. In fact, the function $I_h(\tau',\mu)$ thus obtained is a smooth function unable to correctly describe the directional properties of radiance near the ground level, because at that boundary the sky and soil radiances do not have the same form.

2.1.3.8 Enhanced Approximation for Hemispherical Radiance

Although the second order model do not lead to realistic values of the radiance, it seems to give an acceptable approximation of the global state of radiation (see Fig. 2.5). Another advantage of that second order model is that the solution includes only three exponentials functions of m_z:

$$\mathrm{e}^{\lambda k' m_z} \quad \mathrm{e}^{-\lambda k' m_z} \quad \mathrm{e}^{-k' m_z/\mu_0} \tag{2.75}$$

Then, if we assume that the second term of Eq. 2.53 can be computed using that model, Eq. 2.53 becomes very simple and can be solved analytically for $\mu < 0$. That enhanced solution contains only one more exponential, that is to say

$$\mathrm{e}^{k' m_z/\mu} \tag{2.76}$$

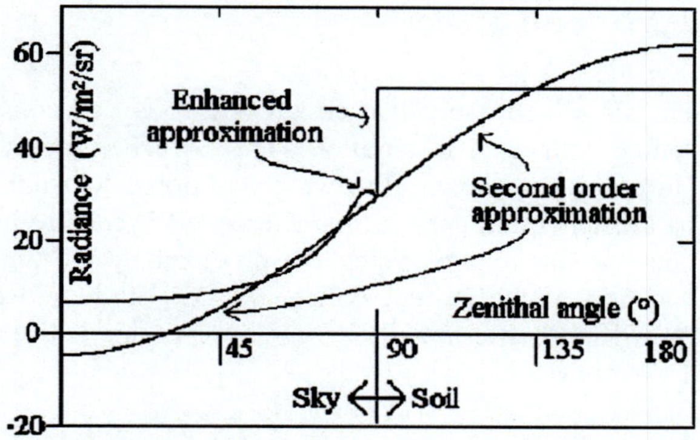

Fig. 2.8 Hemispherical radiance computed by the second ordre model and the enhanced model (with k' $m_z = 0.220$ $\omega' = 0.288$, $g_h = 0$ and $R_{\mathrm{loc}} = R_{\mathrm{reg}} = 0.302$)

with that enhanced model, new parameter identification has been needed in order to keep the good correspondence of Fig. 2.5. We obtain then $\omega' = 0.288$ and $R_{\mathrm{reg}} = 0.302$. The result is shown at Fig. 2.8 assuming that the soil radiance is isotropic and that $R_{\mathrm{loc}} = R_{\mathrm{reg}}$

The present model can take into account the fact that sky brightness is often higher near the horizon. In addition to the soil albedos R_{reg} and R_{loc}, it makes use of only three atmospheric parameters with well defined physical significance: k', ω' and g_h. So, a few number of global irradiance measurements on tilted surfaces is sufficient to determine those parameters. Further investigations are needed to obtain the value of g_h in function of k' and ω', and even a relation between k' and ω', in order to reduce the number of measurements needed for the determination of the model.

2.2 PV Array Modeling

In literature, there are several mathematical models that describe the operation and behavior of the photovoltaic generator. For example, Borowy and Salameh [29] have given a simplified model, with which the maximum power output can be calculated for a module once photovoltaic solar irradiance on the photovoltaic module and the temperature is found, and Jones and Underwood [30] also introduced a simplified model of the maximum power output which has a reciprocal relationship with the temperature module and logarithmic relationship with the solar radiation absorbed by the photovoltaic module. In addition, Jones and Underwood have given the thermal model of the temperature module photovoltaic through the evaluation of many factors. These models differ in the calculation procedure, accuracy and the number of parameters involved in the calculation of the current–voltage characteristic.

2.2.1 Ideal Model

The simplified equivalent circuit of a solar cell consists of a diode and a current source connected in parallel (Fig. 2.9). The current source produces the photo-current I_{ph}, which is directly proportional to solar irradiance G. The two key parameters often used to characterize a PV cell are its short-circuit current and its open-circuit voltage which are provided by the manufacturer's data sheet. The equation of the current voltage I_{pv}–V_{pv} simplified equivalent circuit is derived from Kirchhoff's law.

We have

$$I_{pv} = I_{ph} - I_d \tag{2.77}$$

where

$$I_d = I_0 \left[e^{q\frac{(V_{pv})}{AKT_j}} - 1 \right] \tag{2.78}$$

Thus

$$I_{pv} = I_{ph} - I_0 \left[e^{q\frac{(V_{pv})}{AKT_j}} - 1 \right] \tag{2.79}$$

with I_{ph} (A) is the photocurrent that is equal to short-circuit current, I_0 (A) is the reverse saturation current of the diode, q is the electron charge (1.602×10^{-19} C), K Botzman's constant (1.381×10^{-23} J/K), A is diode ideality factor, Tj is junction temperature of the panels (°K), I_d is the current shunted through the intrinsic diode, V_{pv} is the voltage across the PV cell.

$$I_{pv} = I_{sc} - I_0 \left[e^{q\frac{(V_{pv})}{AKT_j}} - 1 \right] \tag{2.80}$$

We can determine the reverse saturation current I_0 by setting I_{pv}=0 (case when no output current).

$$I_{pv} = 0$$
$$V_{pv} = V_{oc}$$

$$0 = I_{ph} - I_0 \left[e^{q\frac{(V_{oc})}{AKT_j}} - 1 \right] \tag{2.81}$$

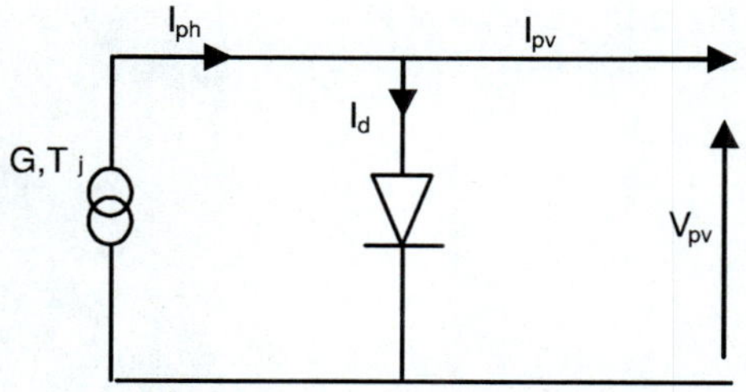

Fig. 2.9 Simplified equivalent circuit of solar cell

Table 2.1 Parameter of the PV panel Siemens SM110-24 [31]

Parameter	Value
P_{mpp}	110 W
I_{mpp}	3.15 A
V_{mpp}	35 V
I_{sc}	3.45 A
V_{oc}	43.5 V
α_{sc}	1.4 mA/°C
β_{oc}	−152 mV/°C

Thus we obtain, taking into account the fact that, with this model, the photocurrent is equal to the short-circuit current:

$$I_0 = \frac{I_{\mathrm{sc}}}{\left[e^{q\frac{(V_{\mathrm{oc}})}{AKT_j}} - 1\right]} \tag{2.82}$$

Application:

The solar cell is modeled and simulated using Matlab software. The simulation is based on the datasheet of Siemens SM110-24 photovoltaic module. The parameters of this solar module are given in Table 2.1 The module is made of 72 solar cells connected in series to give a maximum power output of 110 W (Figs. 2.10, 2.11)

2.2.1.1 Model With Ohmic Losses

To obtain a better representation of the electrical behavior of the cell of the ideal model, the second model takes account of material resistivity and the ohmic losses due to levels of contact. These losses are represented by a series resistance R_s in the equivalent circuit (Fig. 2.12).

The current voltage equation is given as follows:

$$I_{\mathrm{pv}} = I_{\mathrm{ph}} - I_0\left[e^{q\frac{(V_{\mathrm{pv}}+I_{\mathrm{pv}}\cdot R_s)}{AKT_j}} - 1\right] \tag{2.83}$$

or, making the approximation that $I_{\mathrm{ph}} \approx I_{\mathrm{sc}}$,

Fig. 2.10 Example of PV array structure

Fig. 2.11 Bloc diagram of ideal model [195]

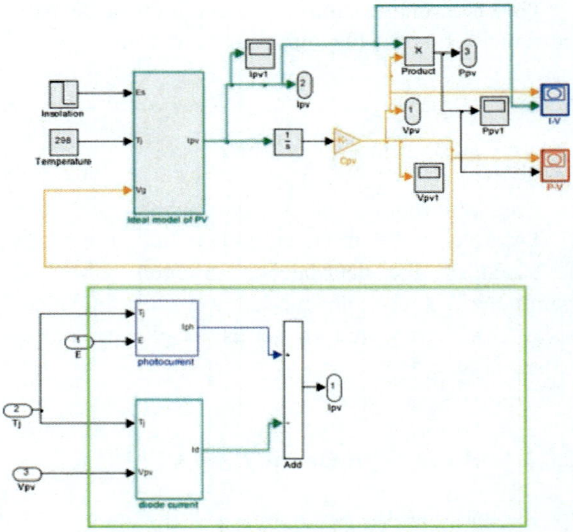

$$I_{\text{pv}} = I_{\text{sc}} - I_0 \left[e^{q\frac{(V_{\text{pv}} + I_{\text{pv}} \cdot R_s)}{AKT_j}} - 1 \right] \tag{2.84}$$

The short-circuit I_{sc} can be calculated at a given temperature T_j:

$$I_{\text{sc-Gref}} = I_{\text{sc-ref}}[1 + \alpha_{\text{sc}} \cdot \Delta T] \tag{2.85}$$

$$\Delta T = T_j - T_{\text{jref}} \tag{2.86}$$

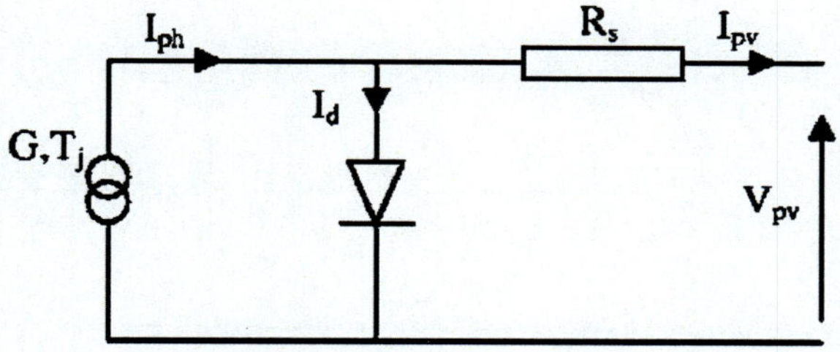

Fig. 2.12 Simplified equivalent circuit of solar cell with R_s

where $I_{\text{sc-ref}}$ is measured under irradiance $G_{\text{ref}} = 1000$ W/m^2 and $T_{\text{j-ref}} = 25°\text{C}$ and is given on the datasheet, $\alpha_{\text{sc}}$is the temperature coefficient of short-current (/°K) and found on the data sheet, T_{jref} is the reference temperature of the PV cell (°K), T_j is the junction temperature (°K).

The short-current generated at any other irradiance G (W/m^2) can be obtained by:

$$I_{\text{sc-G}} = I_{\text{sc-Gref}} \frac{G}{G_{\text{ref}}} \tag{2.87}$$

Applying Eq. 2.84 to the case where $I_{\text{pv}} = 0$ (open-circuit case), one sees that the reverse saturation current at a reference temperature (T_{jref}) is given by:

$$I_{0\text{-Tjref}} = \frac{I_{\text{sc-Tjref}}}{\left[e^{q\frac{(V_{\text{oc-Tjref}})}{AKT_{\text{jref}}}} - 1 \right]} \tag{2.88}$$

Defining

$$V_{\text{th-Tjref}} = \frac{A \cdot K \cdot T_{\text{jref}}}{q} \tag{2.89}$$

one can write Eq. 2.88 in the form:

$$I_{0\text{-Tjref}} = \frac{I_{\text{sc-Tjref}}}{\left[e^{\frac{V_{\text{oc-Tjref}}}{V_{\text{th-Tjref}}}} - 1 \right]} \tag{2.90}$$

The reverse saturation current at any other temperature T_j (°K) can be obtained by:

$$I_0 = I_{0\text{-Tjref}} \cdot \left[\exp\left(\frac{-q \cdot \frac{E_g}{A \cdot K}}{\frac{1}{T_j} - \frac{1}{T_{\text{jref}}}} \right) \right] \cdot \left(\frac{T_j}{T_{\text{jref}}} \right)^{\frac{3}{A}} \tag{2.91}$$

$$I_{pv} = \frac{G}{G_{ref}} \cdot I_{sc-ref} \cdot \left[1 + \alpha_{sc} \cdot \left(T_j - T_{ref}\right)\right] \\ - \left(\frac{I_{sc-ref}}{\exp\left(q \cdot \frac{V_{oc-ref}}{A \cdot K \cdot T_{j-ref}} - 1\right)} \cdot \left[\exp\left(\frac{-q \cdot \frac{E_g}{A \cdot K}}{\frac{1}{T_j} - \frac{1}{T_{jref}}}\right)\right] \cdot \left(\frac{T_j}{T_{jref}}\right)^{\frac{3}{A}}\right) \\ \left[\exp\left(q \cdot \frac{\left(V_{pv} + R_s \cdot I_{pv}\right)}{A \cdot K \cdot T_j}\right) - 1\right] \tag{2.92}$$

where E_g represents the gap energy.

The simple relationship of power for a photovoltaic module is

$$P_{pv} = I_{pv} \cdot V_{pv} \\ = \left\{\frac{G}{G_{ref}} \cdot I_{sc} \cdot \left[1 + \alpha s_c \cdot \left(T_j - T_{ref}\right)\right] \right. \\ \left. - \left(\frac{I_{sc-ref}}{\exp\left(q \cdot \frac{V_{oc-ref}}{A \cdot K \cdot T_{j-ref}} - 1\right)} \cdot \left[\exp\left(\frac{-q \cdot \frac{E_g}{A \cdot K}}{\frac{1}{T_j} - \frac{1}{T_{jref}}}\right)\right] \cdot \left(\frac{T_j}{T_{jref}}\right)^{\frac{3}{A}}\right) \left[\exp\left(q \cdot \frac{\left(V_{pv} + R_s \cdot I_{pv}\right)}{A \cdot K \cdot T_j}\right) - 1\right]\right\} \\ \cdot V_{pv} \tag{2.93}$$

Neglecting the term "-1" added to the exponential in Eq. 2.84, the value of R_s can be obtained by

$$R_s = -\left.\frac{\mathrm{d}V_{pv}}{\mathrm{d}I_{pv}}\right|_{V_{pv} = V_{oc}} - \frac{1}{\mathrm{w}} \\ \mathrm{w} = q \cdot \frac{I_{sc}}{A \cdot K \cdot T_j} \tag{2.94}$$

The first term of Eq. 2.94 ($-\mathrm{d}V_{pv}/\mathrm{d}I_{pv}|_{V_{pv}=V_{oc}}$) can be determined either by experimental or by measures of I_{pv}–V_{pv} characteristic of the manufacturer (datasheets). Eq. 2.84 can be solved using Matlab/simulink or writing a function in Matlab. We can for example Newton's method which is described as [32]:

$$x_{n+1} = x_n - \frac{f(x_n)}{f'(x_n)} \tag{2.95}$$

with $f(x_n)$is the function, $f'(x_n)$ is the derivate of the function.

We have

$$f(x_n) = 0$$

x_n is the first value

x_{n+1} is the next value

We obtain

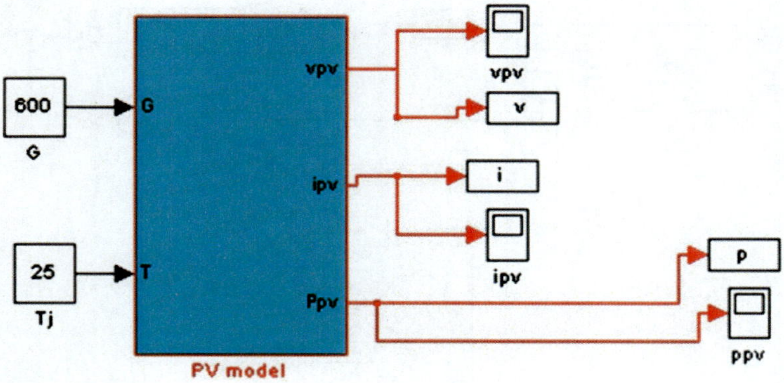

Fig. 2.13 Bloc diagram of model with ohmic losses

$$f(I_{\mathrm{pv}}) = I_{\mathrm{sc}} - I_{\mathrm{pv}} - I_0\left[\mathrm{e}^{q\frac{(V_{\mathrm{pv}}+I_{\mathrm{pv}}\cdot R_s)}{AKT_j}} - 1\right] = 0 \tag{2.96}$$

and then

$$I_{\mathrm{pv\,n+1}} = I_{\mathrm{pv\,n}} - \frac{I_{\mathrm{sc}} - I_{\mathrm{pv}} - I_0\left[\mathrm{e}^{q\frac{(V_{\mathrm{pv}}+I_{\mathrm{pv}}\cdot R_s)}{AKT_j}} - 1\right]}{-1 - I_0 \cdot \frac{q}{A\cdot K\cdot T_j} \cdot R_s \cdot \mathrm{e}^{q(V_{\mathrm{pv}}+I_{\mathrm{pv}}\cdot R_s)/(A\cdot K\cdot T_j)}} \tag{2.97}$$

Defining

$$V_{\mathrm{th}} = \frac{A \cdot K \cdot T_j}{q} \tag{2.98a}$$

we obtain

$$I_{\mathrm{pv}_{n+1}} = I_{\mathrm{pv}_n} - \frac{I_{\mathrm{sc}} - I_{\mathrm{pv}} - I_0\left[\mathrm{e}^{\frac{(V_{\mathrm{pv}}+I_{\mathrm{pv}}\cdot R_s)}{V_{\mathrm{th}}}} - 1\right]}{-1 - I_0 \cdot \frac{R_s}{V_{\mathrm{th}}} \cdot \mathrm{e}^{(V_{\mathrm{pv}}+I_{\mathrm{pv}}\cdot R_s)/(V_{\mathrm{th}})}} \tag{2.98b}$$

The resolution under Matlab/simulink is given in Figs. 2.13 and 2.14.

2.2.1.2 Other One Exponential PV Array Models

In the equations of models above described (Sects. 2.2.1 and 2.2.2), there is only one exponential function. In this section, we present and explain four commonly used models which exhibit the same property.

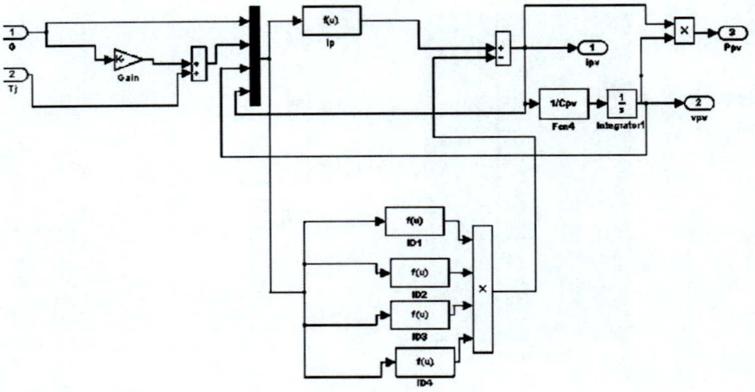

Fig. 2.14 PV subsystem bloc diagram of model with ohmic losses [195]

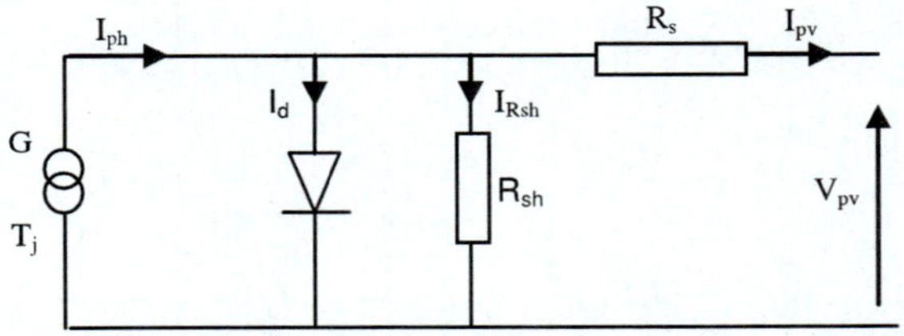

Fig. 2.15 Simplified equivalent circuit of solar cell

2.2.1.3 Model No. 1

The first model studied in this section is defined, as one of these Sects. 2.2.1 and 2.2.2, by an equivalent circuit. This one consists of a single diode for the phenomena of cell polarization and two resistors (series and shunt) for the losses (Fig. 2.15). It can thus be named “one diode model”. This model is used by manufacturers by giving the technical characteristics of their solar cells (data sheets).

$I_{\rm pv}(V_{\rm pv})$ characteristic of this model is given by the following equation [33]:

$$I_{\rm pv} = I_{\rm ph} - I_d - I_{\rm Rsh} \tag{2.99}$$

or, developing the terms $I_{\rm d}$ and $I_{\rm Rsh}$:

$$I_{\rm pv} = I_{\rm ph} - I_0\left[\exp\left(\frac{q(V_{\rm pv} + R_s \cdot I_{\rm pv})}{N_{\rm s-cell}{\rm KT}_j}\right) - 1\right] - \frac{V_{\rm pv} + R_s \cdot I_{\rm pv}}{R_{\rm sh}} \tag{2.100}$$

There are different methods to solve Eq. 2.100 resulting in different approximation mathematical models. The different mathematical models generally include parameters that are provided by photovoltaic modules manufacturers. For this, several methods have been proposed in the literature to determine different parameters. Wolf and Rauschenbach [34] suggest that the current–voltage characteristics of photovoltaic cells can be determined by three different methods. The three characteristics that results are the photovoltaic output characteristic, the pn junction characteristic, and the rectifier forward characteristic. These methods give different results because of the effects of the cell internal series resistance. In Ref. [35], authors propose simple approximate analytical expressions for calculating the values of current and voltage at the maximum power point and the fill factor of a solar cell. While in reference [36], authors use dynamic measurements for integration procedures based on computation of the area under the current–voltage curves.

Recent methods [37] have applied to extract the intrinsic and extrinsic model parameters of illuminated solar cells containing parasitic series resistance and shunt conductance. The approach is based on calculating the Co-content function (CC) from the exact explicit analytical solutions of the illuminated current–voltage (I_{pv}–V_{pv}) characteristics. The resulting CC is expressed as a purely algebraic function of current and voltage from whose coefficients the intrinsic and extrinsic model parameters are then readily determined by bidimensional fitting. Ref [33] proposes an accurate method using Lambert W-function to study different parameters of organic solar cells. The method proposed by [38] uses separate fitting in two different zones in the I_{pv}–V_{pv} curve. In the first one, near short circuit, current fitting is used because the error in current dominates. In the second one, near open-circuit, voltage fitting is used because this is the dominant error. The method overcomes some drawbacks of common procedures: voltage errors are properly managed and no accurate initial guesses for the parameters are needed. This approach is a combination of lateral and vertical optimization to extract the parameters of an illuminated solar cell.

The suggested technique in Ref [35] deals with the extraction of bias independent parameters. It is based on the current–voltage (I_{pv}–V_{pv}) characteristics and the voltage-dependent differential slope $\alpha = \mathrm{d}(\ln I)/\mathrm{d}(\ln V)$ in order to extract the relevant device parameters of the nonideal Schottky barrier, p-n and p-i-n diodes.

In Ref [39], authors propose an approach for photovoltaic (PV) sources modeling based on robust least squares linear regression (LSR) parameter identification method. The parameter extraction is performed starting by the consideration of the temperature and MPPs voltage and current measured values distributions versus solar irradiance. Such distributions show a data placement along a straight line that suggests the possibility to obtain such data by a linear least squares (LSR) [39]. In Ref [40], authors present a method to determine the five solar cell parameters of the single diode lumped circuit model. These parameters are usually the saturation current, the series resistance, the ideality factor, the shunt conductance and the photocurrent. The method includes the presentation of the standard $I_{pv} = f\,(V_{pv})$ function as $V_{pv} = f\,(I_{pv})$ and the determination of the factors of this function that provide the calculation of the illuminated solar cell parameters [40–41].

- The simplified model with four parameters (I_{ph}, I_0, R_s, a)

This model assumes the shunt resistance as infinite and thus neglects it. The model becomes then equivalent to the model of Sect. 2.2.2, but the development presented here is different. Equation 2.100 (≡Eq. 2.83) will be written, using again the definition Eq. 2.98a, as:

$$\begin{aligned} I_{pv} &= I_{ph} - I_0\left[\exp\left(\frac{q(V_{pv} + R_s I_{pv})}{AKT_j}\right) - 1\right] \\ &= I_{ph} - I_0\left[\exp\left(\frac{V_{pv} + R_s I_{pv}}{AV_{th}}\right) - 1\right] \end{aligned} \tag{2.101}$$

To identify the four parameters required for Eq. 2.101, a method is proposed by [42]. They propose to treat the product $A{\cdot}V_{th}$ in Eq. 2.101 as a single parameter denoted a.

$$\begin{aligned} a &= A \cdot V_{th} \\ I_{pv} &= I_{ph} - I_0\left[\exp\left(\frac{V_{pv} + R_s I_{pv}}{a}\right) - 1\right] \end{aligned} \tag{2.102}$$

The following approximations Eqs. 2.103a–2.103b are used to find values of the four parameters under reference conditions

$$I_{ph-ref} = I_{sc-ref} \tag{2.103a}$$

The other parameters are calculated by the following equations:

$$\begin{cases} a_{ref} = \dfrac{\beta_{oc} \cdot T_{jref} - V_{ocref} + E_g}{(T_{jref} \cdot \alpha_{sc}/I_{phref}) - 3} \\ I_{0\,ref} = \dfrac{I_{phref}}{\exp[(V_{ocref}/a_{ref}) - 1]} \\ R_s = \dfrac{a_{ref}\ln\left[1 - (I_{mppref}/I_{phref})\right] - V_{mppref} + V_{ocref}}{I_{mppref}} \end{cases} \tag{2.103b}$$

We can then find the cell parameters at the operating temperature of cells and solar irradiance from:

$$\begin{cases} I_{ph} = \frac{G}{G_{ref}} \cdot \left[I_{phref} + \alpha_{sc} \cdot (T_j - T_{ref})\right] \\ I_0 = I_{0\,ref} \cdot \exp\left[\left(\frac{E_g}{a}\right)\left(1 - \frac{T_{jref}}{T_j}\right)\right] \cdot \left(\frac{T_j}{T_{jref}}\right)^3 \\ R_s = R_{s-ref} \\ a = a_{ref} \cdot (T_j/T_{jref}) \end{cases} \tag{2.104}$$

These four parameters (I_{ph}, I_0, R_s, a) are corrected for environmental conditions using Eq. 2.104 and used in Eq. 2.102.

• The implicit model with five parameters

We can note that Eq. 2.100 is an implicit nonlinear equation, which can be solved with a numerical iterative method such as Levenberg–Marquardt algorithm which requires a close approximation of initial parameter values to attain convergence. Different analytical methods can be used to extract parameters.

1. Proposed analytical method

The five parameters I_L, I_0, R_s, R_{sh}, and A are calculated at a particular temperature and solar-irradiance level from the limiting conditions of V_{oc}, I_{sc}, V_{mpp}, I_{mpp} and using the following definitions of R_{so} and R_{sho}. $I_{pv}(i_{pv})$ characteristic of this model is given by Eq. 2.100. rewritten using again definition Eq. 2.98a:

$$I_{pv} = I_{ph} - I_0\left[\exp\left(\frac{V_{pv} + R_s I_{pv}}{AV_{th}}\right) - 1\right] - \frac{V_{pv} + R_s I_{pv}}{R_{sh}} \tag{2.105}$$

For $V_{pv} = V_{oc}$, $I_{pv} = 0$, we have:

$$0 = I_{ph} - I_0\left[\exp\left(\frac{V_{oc}}{AV_{th}}\right) - 1\right] - \frac{V_{oc}}{R_{sh}} \tag{2.106}$$

For $V_{pv} = 0$, $I_{pv} = I_{sc,}$ we have:

$$I_{sc} = I_{ph} - I_0\left[\exp\left(\frac{I_{sc}R_s}{AV_{th}}\right) - 1\right] - \frac{I_{sc}R_s}{R_{sh}} \tag{2.107}$$

Replacing I_{ph} in Eq. 2.107 by its value extracted from Eq. 2.106, we have:

$$I_{sc} = I_0\left(\exp\frac{V_{oc}}{AV_{th}} - 1\right) + \frac{V_{oc}}{R_{sh}} - I_0\left[\exp\left(\frac{I_{sc}R_s}{AV_{th}}\right) - 1\right] - \frac{I_{sc}R_s}{R_{sh}} \tag{2.108}$$

$$I_{sc} = I_0\left(\exp\frac{V_{oc}}{AV_{th}} - \exp\frac{I_{sc}R_s}{AV_{th}}\right) + \frac{V_{oc}}{R_{sh}} - \frac{I_{sc}R_s}{R_{sh}} \tag{2.109}$$

Thus

$$I_0\left(\exp\frac{V_{oc}}{AV_{th}} - \exp\frac{I_{sc}R_s}{AV_{th}}\right) + \frac{V_{oc}}{R_{sh}} - I_{sc}\left(1 + \frac{R_s}{R_{sh}}\right) = 0 \tag{2.110}$$

We derive Eq. 2.105 with respect to the current:

$$\begin{aligned}1 &= -I_0\left(\frac{dV_{pv}}{dI_{pv}}\frac{1}{AV_{th}} + \frac{R_s}{AV_{th}}\right)\exp\frac{V_{pv} + I_{pv}R_s}{AV_{th}} - \frac{1}{R_{sh}}\frac{dV_{pv}}{dI_{pv}} - \frac{R_s}{R_{sh}}\\ &= -\frac{I_0}{AV_{th}}\exp\frac{V_{pv} + I_{pv}R_s}{AV_{th}}\frac{dV_{pv}}{dI_{pv}} - \frac{I_0R_s}{AV_{th}}\exp\frac{V_{pv} + I_{pv}R_s}{AV_{th}} - \frac{1}{R_{sh}}\frac{dV_{pv}}{dI_{pv}} - \frac{R_s}{R_{sh}}\end{aligned} \tag{2.111}$$

$$-\frac{dV_{pv}}{dI_{pv}}\left(\frac{I_0}{AV_{th}}\exp\frac{V_{pv} + I_{pv}R_s}{AV_{th}} + \frac{1}{R_{sh}}\right) - \frac{I_0R_s}{AV_{th}}\exp\frac{V_{pv} + I_{pv}R_s}{AV_{th}} - \frac{R_s}{R_{sh}} - 1 = 0 \tag{2.112}$$

For $V_{pv} = V_{oc}$ we define

$$\left(\frac{\mathrm{d}V_{\mathrm{pv}}}{\mathrm{d}I_{\mathrm{pv}}}\right)_{V_{\mathrm{pv}}=V_{\mathrm{OC}}} = -R_{s0} \tag{2.113}$$

Replacing Eq. 2.113 in Eq. 2.112, we find

$$\begin{aligned} R_{s0}\left(\frac{I_0}{AV_{\mathrm{th}}}\exp\frac{V_{\mathrm{oc}}}{AV_{\mathrm{th}}}+\frac{1}{R_{\mathrm{sh}}}\right)-\frac{I_0R_s}{AV_{\mathrm{th}}}\exp\frac{V_{\mathrm{oc}}}{AV_{\mathrm{th}}}-\frac{R_s}{R_{\mathrm{sh}}}-1=0 \\ (R_{s0}-R_s)\left(\frac{I_0}{AV_{\mathrm{th}}}\exp\frac{V_{\mathrm{oc}}}{AV_{\mathrm{th}}}+\frac{1}{R_{\mathrm{sh}}}\right)-1=0 \end{aligned} \tag{2.114}$$

For $I_{\mathrm{pv}} = I_{\mathrm{sc}}$, we define:

$$\left(\frac{\mathrm{d}V_{\mathrm{pv}}}{\mathrm{d}I_{\mathrm{pv}}}\right)_{I_{\mathrm{pv}}=I_{\mathrm{SC}}} = -R_{\mathrm{sh0}} \tag{2.115}$$

Replacing Eq. 2.115 in Eq. 2.112 we find:

$$\begin{aligned} R_{\mathrm{sh0}}\left(\frac{I_0}{AV_{\mathrm{th}}}\exp\frac{I_{\mathrm{sc}}R_s}{AV_{\mathrm{th}}}+\frac{1}{R_{\mathrm{sh}}}\right)-\frac{I_0R_s}{AV_{\mathrm{th}}}\exp\frac{I_{\mathrm{sc}}R_s}{AV_{\mathrm{th}}}-\frac{R_s}{R_{\mathrm{sh}}}-1=0 \\ (R_{\mathrm{sh0}}-R_S)\left(\frac{I_0}{AV_{\mathrm{th}}}\exp\frac{I_{\mathrm{sc}}R_s}{AV_{\mathrm{th}}}+\frac{1}{R_{\mathrm{sh}}}\right)-1=0 \end{aligned} \tag{2.116}$$

Dividing Eq. 2.116 by (R_{sh0}-R_s), we find

$$\frac{1}{R_{\mathrm{sh}}}-\frac{1}{R_{\mathrm{sh0}}-R_s}+\frac{I_0}{AV_{\mathrm{th}}}\exp\frac{I_{\mathrm{sc}}R_s}{AV_{\mathrm{th}}}=0 \tag{2.117}$$

At maximum power point, we have

$$I_{\mathrm{mpp}} = I_{\mathrm{ph}} - I_0\left(\exp\frac{V_{\mathrm{mpp}}+I_{\mathrm{mpp}}R_s}{AV_{\mathrm{th}}}-1\right)-\frac{V_{\mathrm{mpp}}+I_{\mathrm{mpp}}R_s}{R_{\mathrm{sh}}} \tag{2.118}$$

From Eq. 2.106, we obtain

$$I_{\mathrm{ph}} = I_0\left(\exp\frac{V_{\mathrm{oc}}}{AV_{\mathrm{th}}}-1\right)+\frac{V_{\mathrm{oc}}}{R_{\mathrm{sh}}} \tag{2.119}$$

Substituting Eq. 2.119 into Eq. 2.118, we get

$$I_{\mathrm{mpp}} = I_0\left(\exp\frac{V_{\mathrm{oc}}}{AV_{\mathrm{th}}}-1\right)+\frac{V_{\mathrm{oc}}}{R_{\mathrm{sh}}}-I_0\left(\exp\frac{V_{\mathrm{mpp}}+I_{\mathrm{mpp}}R_s}{AV_{\mathrm{th}}}-1\right)-\frac{V_{\mathrm{mpp}}+I_{\mathrm{mpp}}R_s}{R_{\mathrm{sh}}} \tag{2.120}$$

$$I_0\exp\frac{V_{\mathrm{oc}}}{AV_{\mathrm{th}}}+\frac{V_{\mathrm{oc}}-V_{\mathrm{mpp}}}{R_{\mathrm{sh}}}-\left(I_0+I_{\mathrm{mpp}}\frac{R_s}{R_{\mathrm{sh}}}\right)-I_0\exp\frac{V_{\mathrm{mpp}}+I_{\mathrm{mpp}}R_s}{AV_{\mathrm{th}}}=0 \tag{2.121}$$

Kennerud and Charles showed that the four parameters A, R_s, I_0, and R_{sh} can be determined by the Newton-Raphson solving simultaneous nonlinear equations

Eqs. 2.110, 2.114, 2.117 and 2.121. However, this method requires long calculations and initial conditions for strict convergence hence, it is difficult to determine these parameters. So it is necessary to have analytical expressions for determining these parameters directly.

As $R_{sh} > > R_s$, we assume for the parameters determination that

$$1 + R_s/R_{sh} \approx 1$$

In Eq. 2.110, we assume also:

$$\exp\frac{V_{oc}}{AV_{th}} \rangle\rangle \exp\frac{I_{sc}R_s}{AV_{th}}$$

In Eq. 2.114, we assume:

$$\frac{I_0}{AV_{th}} \exp\frac{V_{oc}}{AV_{th}} \rangle\rangle \frac{1}{R_{Sh}}$$

Finally, in Eq. 2.117, we assume: $\frac{I_0}{AV_{th}} \exp\frac{V_{pv}+I_{pv}R_{serial}}{AV_{th}} \langle\langle 10\%$ of the remaining terms.

With these simplifications, we get from Eqs. 2.110, 2.114, 2.117 and 2.121:

$$I_0 \exp\frac{V_{oc}}{AV_{th}} - I_{sc} + \frac{V_{oc}}{R_{sh}} = 0 \tag{2.122}$$

$$(R_{s0} - R_S)\frac{I_0}{AV_{th}} \exp\frac{V_{oc}}{AV_{th}} - 1 = 0 \tag{2.123}$$

$$R_{sh} = R_{sh0} \tag{2.124}$$

$$I_0 \exp\frac{V_{oc}}{AV_{th}} + \frac{V_{oc} - V_{mpp}}{R_{sh}} - I_{mpp} - I_0 \exp\frac{V_{mpp} + I_{mpp} \cdot R_s}{AV_{th}} = 0 \tag{2.125}$$

From these last four equations, we obtain an analytic expression of A. For that, from Eq. 2.122, we have:

$$I_0 = \left(I_{sc} - \frac{V_{oc}}{R_{sh}}\right) \exp\left(-\frac{V_{oc}}{AV_{th}}\right) \tag{2.126}$$

From Eq. 2.123, we have

$$\begin{aligned} R_{s0} - R_s &= \frac{1}{\frac{I_0}{AV_{th}} \exp\frac{V_{oc}}{AV_{th}}} \\ R_s = R_{s0} - \frac{1}{\frac{I_0}{AV_{th}} \exp\frac{V_{oc}}{AV_{th}}} &= R_{s0} - \frac{1}{\frac{1}{AV_{th}}\left(I_{sc} - \frac{V_{oc}}{R_{sh}}\right)} \end{aligned} \tag{2.127}$$

Substituting Eq. 2.126 into Eq. 2.125, we obtain:

$$I_{sc} - \frac{V_{oc}}{R_{sh}} + \frac{V_{oc} - V_{mpp}}{R_{sh}} - I_{mpp} = I_0 \exp \frac{V_{mpp} + I_{mpp} \cdot R_s}{AV_{th}}$$

$$I_{sc} - \frac{V_{mpp}}{R_{sh}} - I_{mpp} = \left(I_{sc} - \frac{V_{oc}}{R_{sh}}\right) \exp\left(-\frac{V_{oc}}{AV_{th}} + \frac{V_{mpp} + I_{mpp} \cdot R_s}{AV_{th}}\right)$$

or, substituting Eq. 2.127 in that equation:

$$\frac{I_{sc} - \frac{V_{mpp}}{R_{sh}} - I_{mpp}}{I_{sc} - \frac{V_{oc}}{R_{sh}}} = \exp\left(\frac{-V_{oc} + V_{mpp} + I_{mpp}R_{s0} - \frac{I_{mpp} \cdot A \cdot V_{th}}{I_{sc} - V_{oc}/R_{sh}}}{AV_{th}}\right)$$

$$\ln\left(\frac{I_{sc} - \frac{V_{mpp}}{R_{sh}} - I_{mpp}}{I_{sc} - \frac{V_{oc}}{R_{sh}}}\right) = \frac{-V_{oc} + V_{mpp} + I_{mpp}R_{s0}}{AV_{th}} - \frac{I_{mpp}}{I_{sc} - \frac{V_{oc}}{R_{sh}}}$$

Finally, we obtain the expression of A

$$A = \frac{V_{mpp} + I_{mpp}R_{s0} - V_{oc}}{V_{th}\left[\ln\left(I_{sc} - \frac{V_{mpp}}{R_{sh}} - I_{mpp}\right) - \ln\left(I_{sc} - \frac{V_{oc}}{R_{sh}}\right) + \left(\frac{I_{mpp}}{I_{sc} - \frac{V_{oc}}{R_{sh}}}\right)\right]} \tag{2.128}$$

and I_0, R_s and I_{ph} are obtained by

$$\begin{cases} I_0 = \left(I_{sc} - \frac{V_{oc}}{R_{sh}}\right) \exp\left(-\frac{V_{oc}}{AV_{th}}\right) \\ R_s = R_{s0} - \frac{AV_{th}}{I_0} \exp\left(-\frac{V_{oc}}{AV_{th}}\right) \\ I_{ph} = I_{sc}\left(1 + \frac{R_s}{R_{Sh}}\right) + I_0\left(\exp\frac{I_{sc}R_s}{AV_{th}} - 1\right) \end{cases} \tag{2.129}$$

Once these parameters (A, I_{ph}, R_s, and I_0) are determined, the I_{pv}–V_{pv} characteristic will be calculated by Eq. 2.100 using the Newton Raphson method.

2. Second method (Method of P_i constants)

The model can be enhanced by releasing the constraint of linearity between the photocurrent I_{ph} and the irradiance G. For that, we make choice of an expression for I_{ph} with a term quadratic in G:

$$I_{ph} = P_1 \cdot G \cdot [1 + P_2 \cdot (G - G_{ref}) + P_3 \cdot (T_j - T_{ref})] \tag{2.130}$$

where G is irradiance on the panel plane (W/m^2); G_{ref} corresponds to the reference irradiance of 1000 W/m^2 and T_{jref} to the reference panel temperature of 25°C. P_1, P_2 and P_3 are constant parameters.

The polarization current I_d of junction PN, is given by the expression:

$$I_d = I_0 \cdot \left[\exp\left(\frac{q \cdot (V_{pv} + R_s \cdot I_{pv})}{A \cdot N_{s-cell} \cdot K \cdot T_j}\right) - 1\right] \tag{2.131}$$

$$I_0 = P_4 \cdot T_j^3 \cdot \exp\left(-\frac{E_g}{K \cdot T_j}\right) \tag{2.132}$$

where I_0 (A) is the saturation current, q is the elementary charge, K Botzman's constant, A ideality factor of the junction, T_j junction temperature of the panels (°K), R_s and R_{sh} (Ω) are resistors (series and shunt) and n_s the number of cells in series.

The shunt current is given by:

$$I_{Rsh} = \frac{(V_{pv} + R_s \cdot I_{pv})}{R_{sh}} \tag{2.133}$$

Thus

$$I_{pv} = P_1 \cdot G \cdot \left[1 + P_2 \cdot (G - G_{ref}) + P_3 \cdot (T_j - T_{ref})\right] - P_4 \cdot T_j^3 \cdot \exp\left(-\frac{E_g}{K \cdot T_j}\right) \cdot \left[\exp\left(q \cdot \frac{(V_{pv} + R_s \cdot I_{pv})}{A \cdot n_s \cdot K \cdot T_j}\right) - 1\right] - \frac{(V_{pv} + R_s \cdot I_{pv})}{R_{sh}} \tag{2.134}$$

Different methods exist to determine the constant parameters P_1, P_2, P_3, and P_4.

- First approach

We determine the seven constant parameters P_1, P_2, P_3, P_4, the coefficient A and the resistance R_s and R_{sh} of PV model with a numerical resolution and the use of data sheets PV panels. Parameters commonly provided by module manufacturers are values of short-circuit current I_{sc}, the open-circuit voltage V_{oc} and the point of optimum power (I_{mpp}, V_{mpp}). We determine the system of nonlinear equation as follows:

$$\begin{cases} I_{pv}(V_{oc}) = 0 \\ I_{pv}(0) = I_{sc} \\ I_{pv}(V_{mpp}) = I_{mpp} \\ \left.\frac{dP_{pv}}{dV_{pv}}\right|_{P=P_{mpp}} = I_{mpp} + \left.\frac{dI_{pv}}{dV_{pv}}\right|_{V=V_{mpp}}^{I=I_{mpp}} = 0 \end{cases} \tag{2.135}$$

In order to determine all the seven parameters, three additional equations are needed. In order to determine P_2, points with different irradiance are necessary. Similarly, in order to determine P_3, points with different temperatures are needed.

The system resolution of the nonlinear Eq. 2.135 is done by running the function 'fsolve' contained in toolboxes of Matlab which is based on the least squares method (Fig. 2.16).

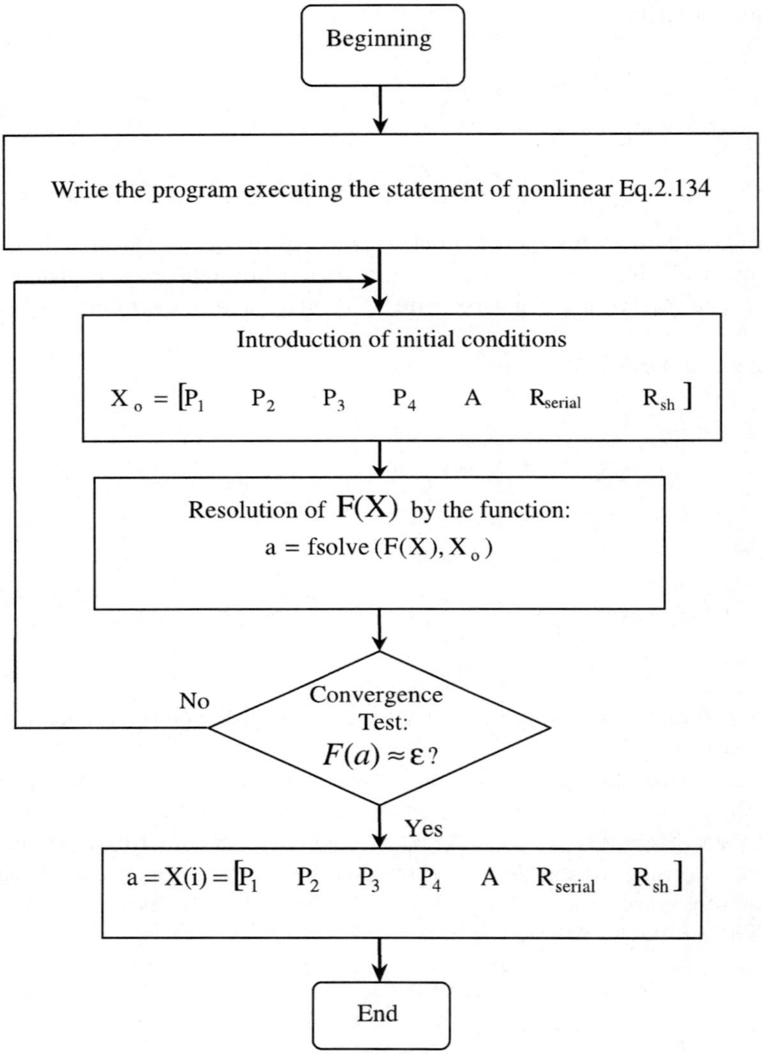

Fig. 2.16 Flowchart of resolution with parameters P_i

- Second approach

To determine the parameters of the panels (P_i, R_s, R_{sh}), we developed a method based on optimization techniques to find the extrema for the function indicated. Its principle is simple; it consists in the first step to define an objective function, called a criterion. This criterion E depends on a set of parameters grouped in a vector p. This objective function is the error between the practical results and those of the simulation. It is given by:

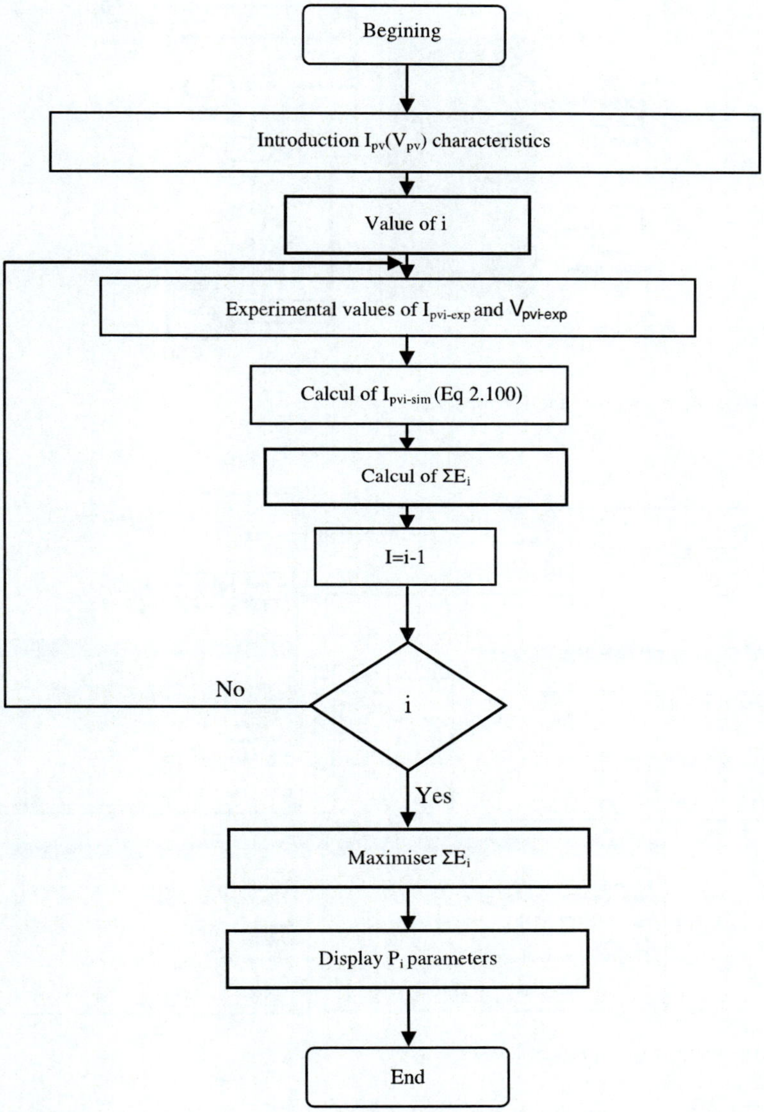

Fig. 2.17 Flowchart to determine the parameters of the panels (P_i, R_s, R_{sh})

$$E(p) = \frac{I_{pv-\exp} - I_{pv-sim}}{I_{pv-\exp}} \quad (2.136)$$

with $I_{\text{pv-sim}}$ is calculated from Eq. 2.134 for the model with a diode, using a wide range of variation of the illumination received by the photovoltaic panel, p contains various parameters to determine (P_1, P_2, P_3, P_4, A, R_s, R_{sh}). We use the least

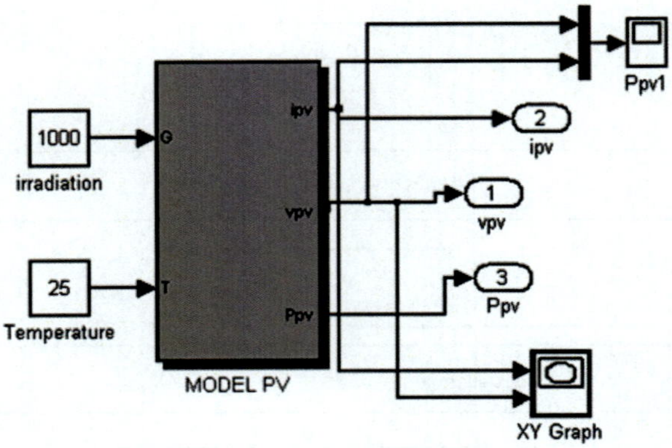

Fig. 2.18 Bloc diagram of the PV model [195]

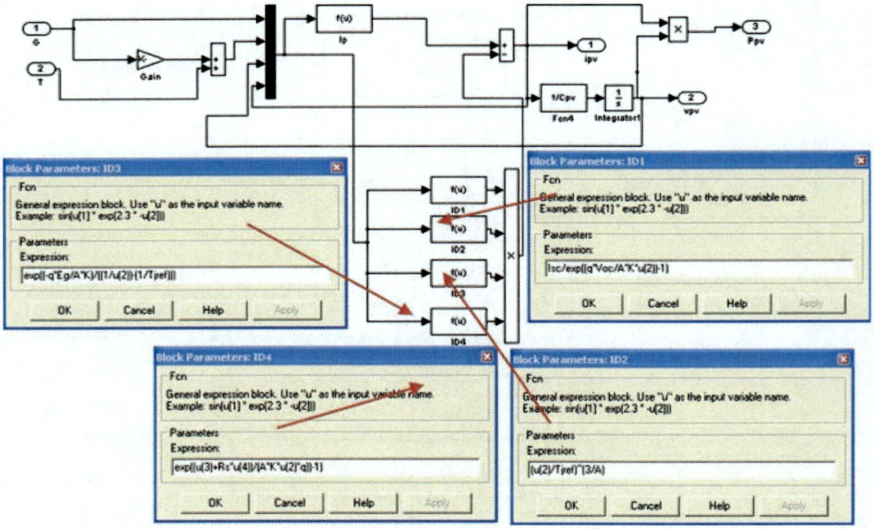

Fig. 2.19 PV subsystem model [195]

squares method to find the value of P that minimizes the function $\Sigma\ (E(p))^2$. We present the proposed flowchart (Figs. 2.17, 2.18, 2.19).

Application in Matlab/simulink (see CD program-model 1)

The simulated current–voltage (I_{pv}–V_{pv}) characteristic and power-voltage (P_{pv}–V_{pv}) of the PV Module is shown in Fig. 2.20. The characteristic is obtained at a constant level of irradiance and by maintaining a constant cell temperature.

The variation in both the I_{pv}–V_{pv} and P_{pv}–V_{pv} characteristics with irradiance level are simulated and the results are shown in Fig. 2.21.

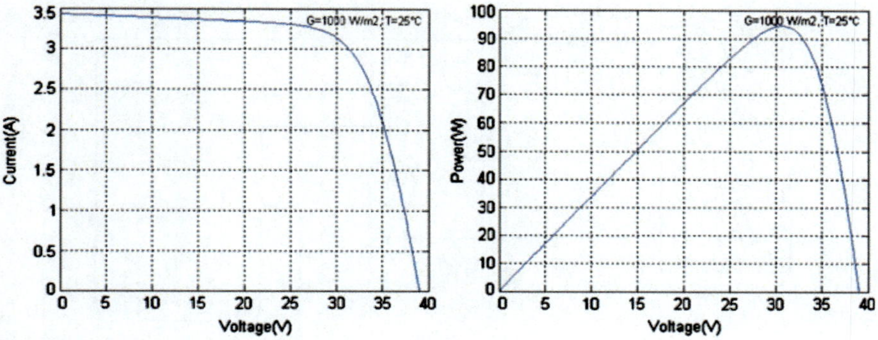

Fig. 2.20 Characteristics $I_{pv} = f\,(V_{pv})$ and $P_{pv} = f\,(V_{pv})$ in STC conditions

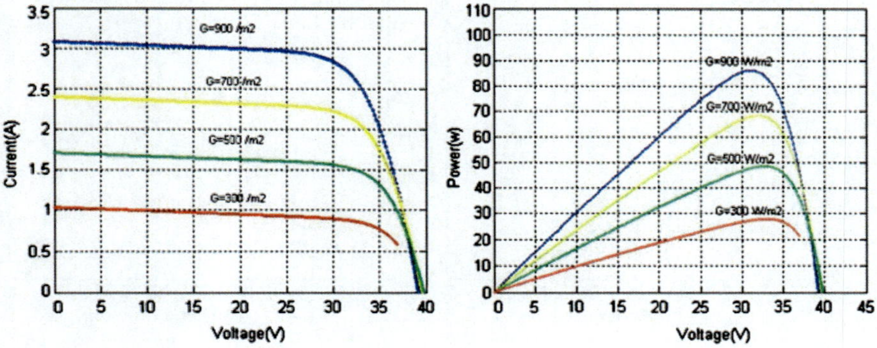

Fig. 2.21 Effects of solar irradiance changing

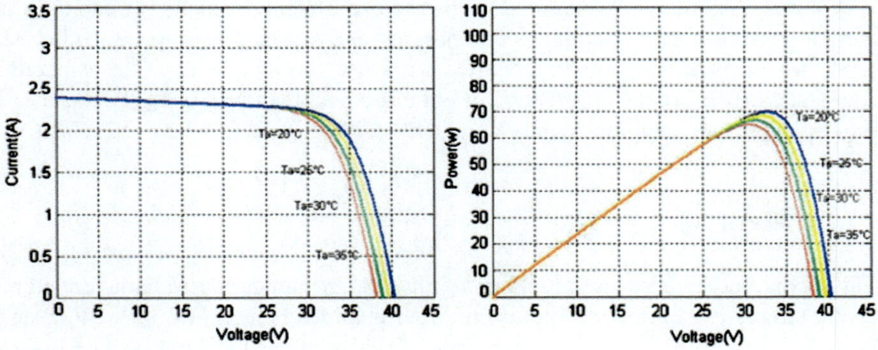

Fig. 2.22 Effects of temperature changing

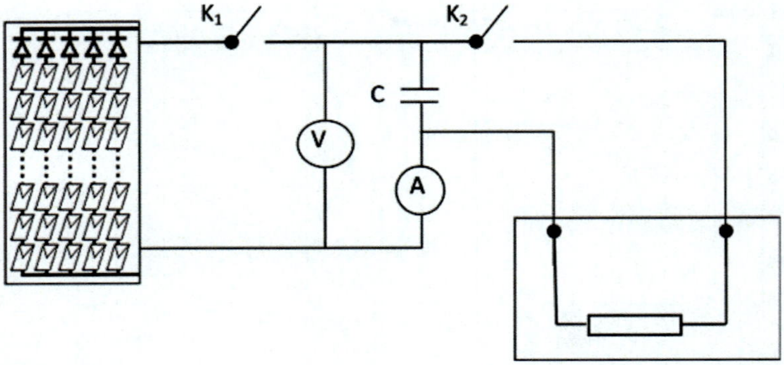

Fig. 2.23 Experimental bench [9]

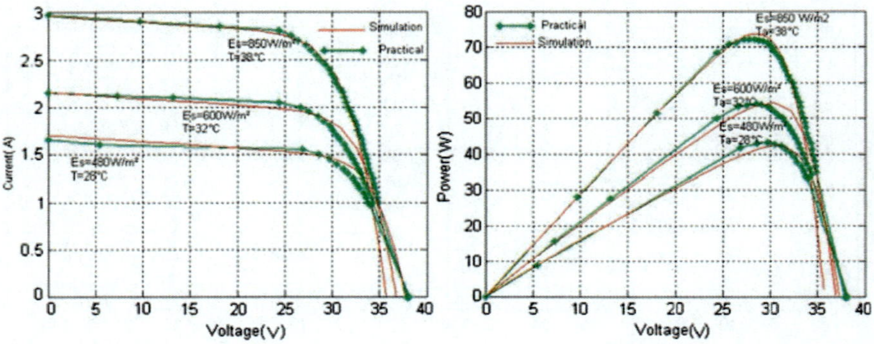

Fig. 2.24 Comparison of experimental results with simulation ones [195]

The simulated I_{pv}–V_{pv} and P_{pv}–V_{pv} characteristics of the solar cell at different cell temperatures are shown in Fig. 2.22.

We make validation through the following experimental bench: We can use the method of resistance variation or the charging and discharging capacitor method. First, we measure the irradiation after we close the switch K_1. The capacitor charges, we measure the different values of current and voltage panel. To discharge the capacitor, we open the switch K_1 and close K_2 (Figs. 2.23, 2.24, 2.25, 2.26).

2.2.1.4 Model No. 2

The second model here presented has no physical meaning, but is characterized by a very simple resolution. It requires only four parameters namely I_{sc}, V_{oc}, V_{mp}, and I_{mp}. The form of the I_{pv}–V_{pv} characteristic of this model is loosely based on that of Model No. 1, that is to say:

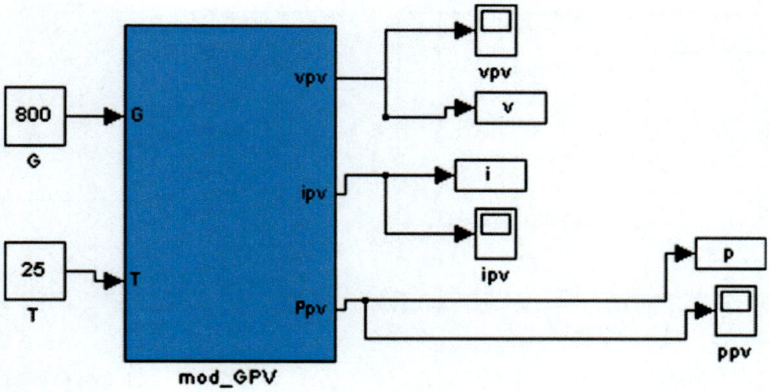

Fig. 2.25 Bloc diagram of Model No. 2 [195]

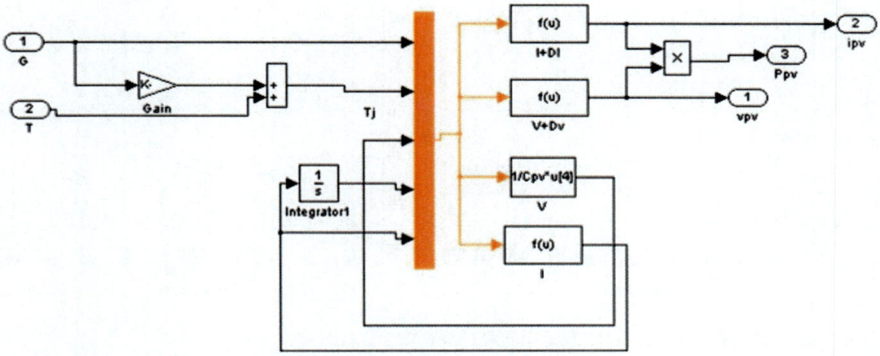

Fig. 2.26 PV subsystem (Model No. 2) [195]

$$I_{pv} = I_{sc}\left\{1 - C_1\left[\exp\left(\frac{V_{pv}}{C_2 \cdot V_{oc}}\right) - 1\right]\right\} \tag{2.137}$$

That equation fits exactly the short-circuit point.
The open-circuit and maximum power points are approximately restituted with:

$$\begin{aligned} C_2 &= \frac{\left(\frac{V_{mpp}}{V_{oc}}\right) - 1}{\ln\left(1 - \frac{I_{mpp}}{I_{sc}}\right)} \\ C_1 &= \left(1 - \frac{I_{mpp}}{I_{sc}}\right)\exp\left(-\frac{V_{mpp}}{C_2 \cdot V_{oc}}\right) \end{aligned} \tag{2.138}$$

The simple relationship of power for a photovoltaic module is

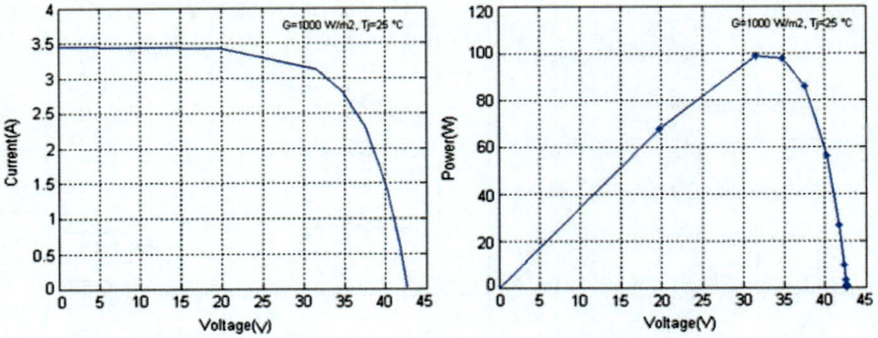

Fig. 2.27 Characteristics $I_{pv} = f(V_{pv})$ and $P_{pv} = f(V_{pv})$ under the STC conditions (Model No. 2)

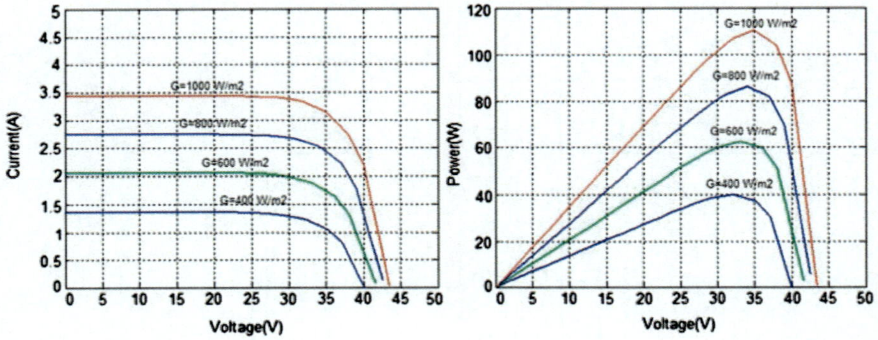

Fig. 2.28 Effects of irradiance changing (Model No. 2)

$$P_{pv} = I_{sc}\left\{1 - C_1\left[\exp\left(\frac{V_{pv}}{C_2 \cdot V_{oc}}\right) - 1\right]\right\} \cdot V_{pv} \tag{2.139}$$

The constant parameters can be determined directly by Eq. 2.138.

Application in Matlab/simulink(see CD program-model2)

Figure 2.27 below shows the characteristic power/voltage and current/voltage obtained under the STC conditions (Standard Test Condition $G = 1000$ W/m^2, $T_j = 25°\text{C}$)

The Figs. 2.28, 2.29 and 2.30 show the influence of irradiance G and temperature T_j on the electrical characteristics.

2.2.1.5 Model No. 3

The third model offers one more freedom degree than Model No. 2. The PV array current I_{pv} obeys to the expression:

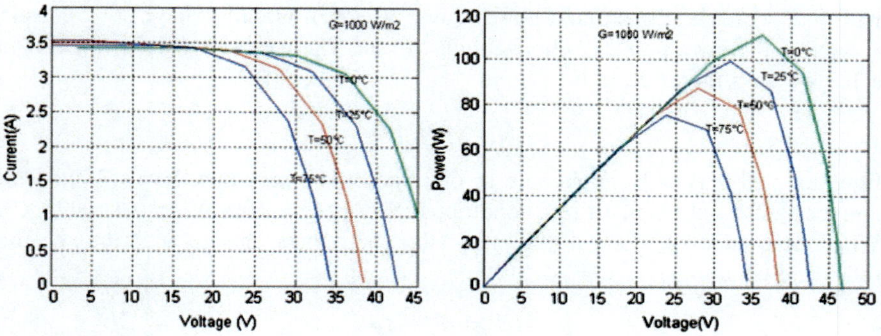

Fig. 2.29 Effect of temperature changing (model No. 2)

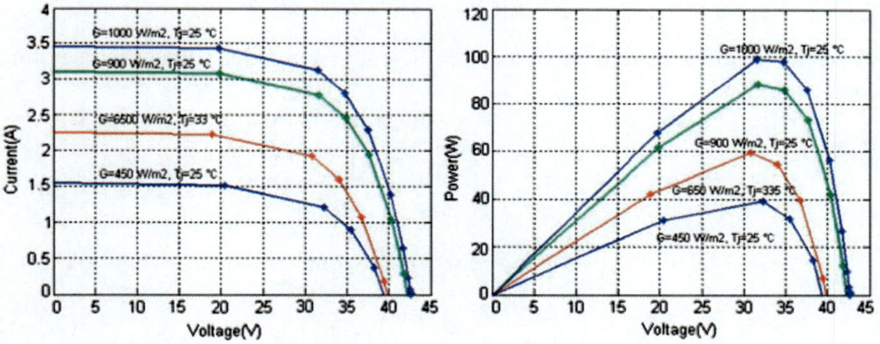

Fig. 2.30 Effects of irradiance and temperature changing (model No. 2)

$$I_{pv} = I_{sc} \cdot \{1 - C_1[\exp(C_2 V_{pv}^m) - 1]\} \tag{2.140}$$

All the three rated points are exactly fitted if the coefficients C_1, C_2 and m verify:

$$C_2 = \frac{C_4}{V_{oc}^m}$$

$$C_3 = \ln\left[\frac{I_{sc}(1 + C_1) - I_{mpp}}{C_1 I_{sc}}\right]$$

$$C_4 = \ln\left[\frac{1 + C_1}{C_1}\right]$$

$$m = \frac{\ln\left[\frac{C_3}{C_4}\right]}{\ln\left[\frac{V_{mpp}}{V_{oc}}\right]}$$

with V_{mpp} voltage at maximum power point; V_{oc} open-circuit voltage; I_{mpp} current at maximum power point; I_{sc} short-circuit current. The parameters determination is achieved with the arbitrary condition

$$C_1 = 0.01175$$

Equation 2.140 is only applicable at one particular irradiance level G and cell temperature T_j, at standard test conditions (STC) ($G = 1000$ W/m^2, $T_j = 25$°C). When irradiance and temperature vary, the parameters change according to the following equations:

$$\Delta T_j = T_j - T_{jref}$$

$$\begin{aligned} \Delta I_{pv} &= \alpha_{sc}\left(\frac{G}{G_{ref}}\right)\Delta T_j + \left(\frac{G}{G_{ref}} - 1\right)I_{sc,ref} \\ \Delta V_{pv} &= -\beta_{oc}\Delta T_j - R_s \Delta I_{pv} \end{aligned} \tag{2.141}$$

where α_{sc} is the current temperature coefficient and β_{oc} the voltage temperature coefficient.

The new values of the photovoltaic voltage and the current are given by

$$\begin{aligned} V_{pv,\,new} &= V_{pv} + \Delta V_{pv} \\ I_{pv,\,new} &= I_{pv} + \Delta I_{pv} \end{aligned} \tag{2.142}$$

The simple relationship of power for a photovoltaic module is (Fig. 2.31)

$$P_{pv} = I_{sc}\{1 - C_1\left[\exp(C_2 V_{pv}^m) - 1\right]\} \cdot V_{pv} \tag{2.143}$$

Application under Matlab/simulink (see CD program-model 3)

The Fig. 2.32 shows the characteristic power/voltage and current/voltage obtained under the STC conditions (Figs. 2.33, 2.34, 2.35).

We make validation through the experimental bench given in Fig. 2.26. We obtain the following results (Fig. 2.36) comparing to the simulation ones.

2.2.1.6 Model No. 4

The fourth model is based on references [39, 40] and [43]. The advantage of this model is that it can be established using only standard data for the module and cells, provided by the manufacturer (data sheets). This model is simple, because it is independent of the saturation current I_0. The current delivered by the solar module (I_{pv}) in any conditions is given by

$$I_{pv} = I_{sc-pv} \cdot \left[1 - \left(\exp\left(\frac{(V_{pv} - V_{oc-pv} + R_{s-pv} \cdot I_{pv})}{A \cdot V_{th-pv}}\right) - \exp\left(\frac{-V_{oc-pv}}{A \cdot V_{th-pv}}\right)\right)\right] \tag{2.144}$$

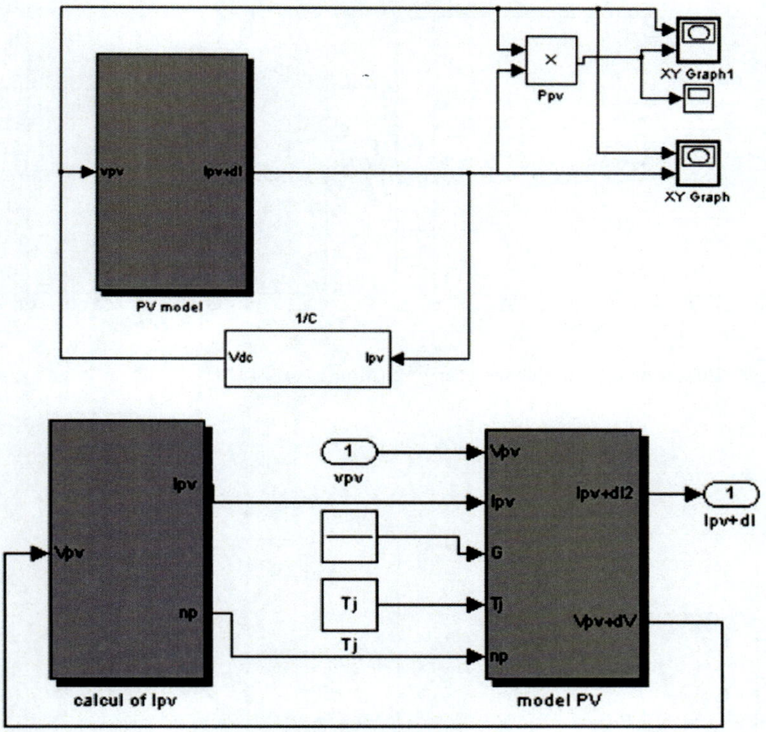

Fig. 2.31 Bloc diagram of Model No. 3 [195]

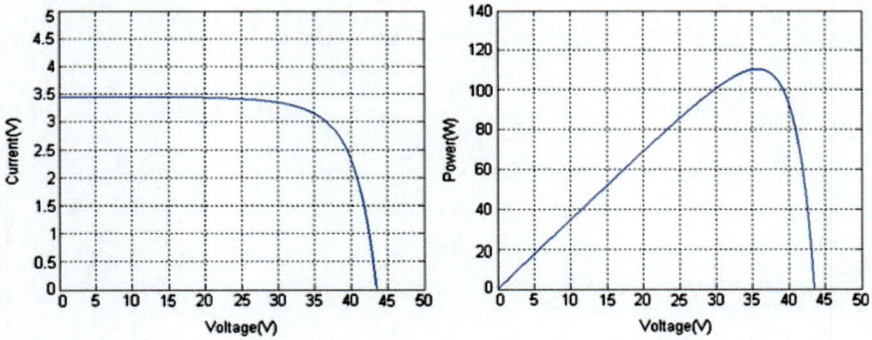

Fig. 2.32 Characteristics $I_{pv} = f(V_{pv})$ and $P_{pv} = f(V_{pv})$ under the STC conditions(Model No. 3) [195]

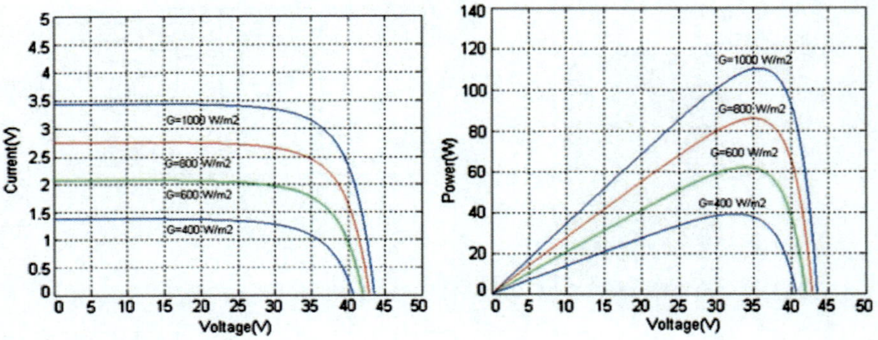

Fig. 2.33 Effects of solar irradiance changing (Model No. 3) [195]

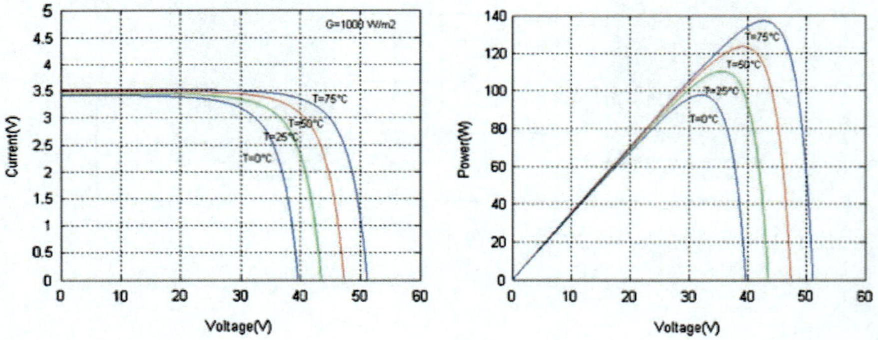

Fig. 2.34 Effects of temperature changing (Model No. 3) [195]

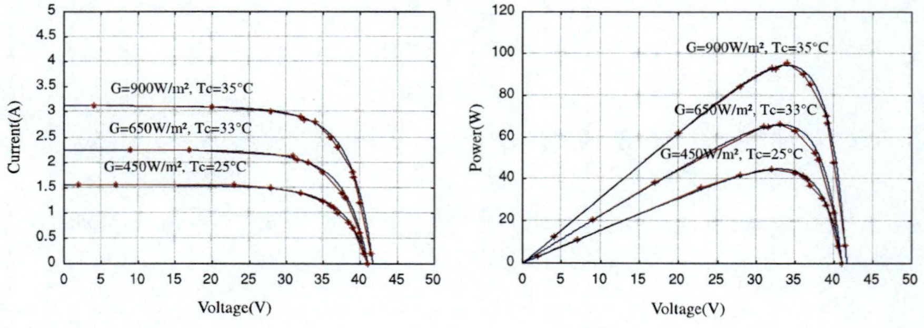

Fig. 2.35 Characteristics $I_{pv} = f(V_{pv})$ and $P_{pv} = f(V_{pv})$ (Model No. 3) [195]

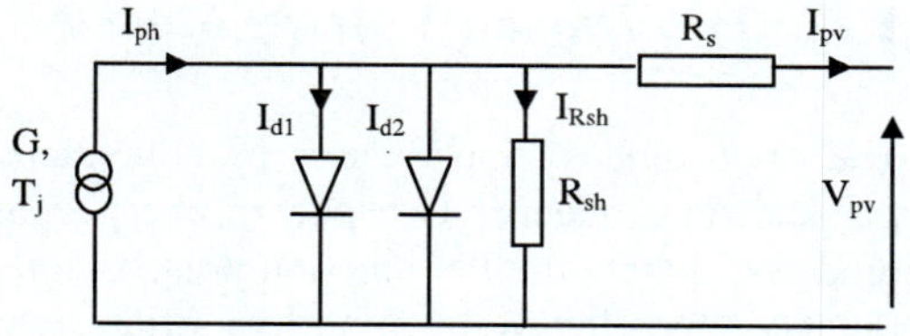

Fig. 2.36 Equivalent circuit for two diode model [44]

with

$$\begin{aligned} I_{sc-pv} &= N_{p-cell} \cdot I_{sc-cell} \\ V_{oc-pv} &= N_{s-cell} \cdot V_{oc-cell} \end{aligned} \quad (2.145)$$

where $N_{s\text{-}cell}$ and $N_{p\text{-}cell}$ are the number of series and parallel cells, respectively.

The open-circuit voltage $V_{oc\text{-}cell}$ is given by

$$V_{oc-cell} = V_{oc-cell-ref} + \beta_{oc} \cdot (T_j - T_{jref}) \quad (2.146)$$

with $V_{oc\text{-}cell\text{-}ref}$ and T_{jref} are the open-circuit voltage and cell temperature at standard conditions, β_{oc} is the temperature coefficient of $V_{oc\text{-}cell\text{-}ref}$

$$V_{oc-cell-ref} = V_{oc-pv-ref}/N_s \quad (2.147)$$

The thermodynamic voltage of the cell $V_{th\text{-}cell}$ is given by:

$$\begin{aligned} V_{th-pv} &= N_{s-cell} \cdot V_{th-cell} \\ V_{th-cell} &= K \cdot T_j/q \end{aligned} \quad (2.148)$$

The resistance of the module is calculated by the following equation:

$$R_{s-pv} = R_{s-cell} \cdot \left(N_{s-cell}/N_{p-cell}\right) \quad (2.149)$$

$R_{s\text{-}cell}$ is calculated by reference to model of Sect. 2.2.2 Eq. 2.94:

$$R_{s-cell} = -\left.\frac{\mathrm{d}V_{pv-cell}}{\mathrm{d}I_{pv-cell}}\right|_{V_{pv}=V_{oc-pv}} - \frac{A \cdot K \cdot T_j}{q \cdot I_{sc-cell}} \quad (2.150)$$

The temperature T_j can be taken equal to T_a, and the thermodynamic voltage $V_{th\text{-}cell}$ can be easily calculated, using the coordinates of the maximum power point of the cell ($V_{mpp\text{-}cell}$ and $I_{mpp\text{-}cell}$).

The expression of $V_{th\text{-}cell}$ can have the following form:

$$V_{th-cell} = \frac{V_{mpp-cell} + R_{serial-cell} \cdot I_{mpp-cell} - V_{oc-cell}}{\ln\left(1 - I_{mpp-cell}/I_{sc-cell}\right)} \quad (2.151)$$

2.2.2 Two Diode PV Array Models

The "two diodes" model uses an equivalent circuit and takes into account the mechanism of electric transport of charges inside the cell. In this model, the two diodes represent the PN junction polarization phenomena. These diodes represent the recombination of the minority carriers, which are located both at the surface of the material and within the volume of the material (Fig. 2.36).

The following equation is then obtained:

$$I_{\mathrm{pv}} = I_{\mathrm{ph}} - (I_{d2} + I_{d2}) - I_{\mathrm{Rsh}} \tag{2.152}$$

with I_{ph} and I_{Rsh} maintaining the same expressions as above Eqs. 2.130 and 2.133. For the recombination currents, we have:

$$\begin{cases} I_{d1} = I_{01} \cdot \left[\exp\left(\frac{q \cdot (V_{\mathrm{pv}} + R_s \cdot I_{\mathrm{pv}})}{A \cdot N_{\mathrm{s-cell}} \cdot K \cdot T_j} \right) - 1 \right] \\ I_{d2} = I_{02} \cdot \left[\exp\left(\frac{q \cdot (V_{\mathrm{pv}} + R_s \cdot I_{\mathrm{pv}})}{2 \cdot A \cdot N_{\mathrm{s-cell}} \cdot K \cdot T_j} \right) - 1 \right] \end{cases} \tag{2.153}$$

The saturation currents are written as:

$$\begin{cases} I_{01} = P_4 \cdot T_{\mathrm{j}}^3 \cdot \exp\left(\frac{-E_g}{k \cdot T_j} \right) \\ I_{02} = P_5 \cdot T_j^3 \cdot \exp\left(\frac{-E_g}{2 \cdot k \cdot T_j} \right) \end{cases} \tag{2.154}$$

with N_{s} is the number of cells in branched series.

The final equation of the model is thereby written as (Figs. 2.37, 2.38):

$$\begin{aligned} I_{\mathrm{pv}} = {} & P_1 \cdot G \cdot \left[1 + P_2 \cdot (G - G_{\mathrm{ref}}) + P_3 \cdot (T_j - T_{\mathrm{ref}}) \right] - \frac{(V_{\mathrm{pv}} + R_s \cdot I_{\mathrm{pv}})}{R_{\mathrm{sh}}} \\ & - P_4 \cdot T_j^3 \cdot \exp\left(\frac{-E_g}{k \cdot T_j} \right) \cdot \left[\exp\left(q \cdot \frac{V_{\mathrm{pv}} + R_s \cdot I_{\mathrm{pv}}}{A \cdot N_{\mathrm{s-cell}} \cdot K \cdot T_j} \right) - 1 \right] \\ & - P_5 \cdot T_j^3 \cdot \exp\left(\frac{-E_g}{2 \cdot k \cdot T_j} \right) \cdot \left[\exp\left(q \cdot \frac{V_{\mathrm{pv}} + R_s \cdot I_{\mathrm{pv}}}{2 \cdot A \cdot N_{\mathrm{s-cell}} \cdot K \cdot T_j} \right) - 1 \right] \end{aligned} \tag{2.155}$$

Application under Matlab/simulink(see CD program-modele2D)

We make validation through the experimental bench given in Fig. 2.25. We obtain the following results (Fig. 2.39) comparing to the simulation ones.

2.2.3 Power Models

2.2.3.1 Model No. 1: Polynomial Model

This model can give the same power of solar modules operating at MPP (Maximum Power Point). It is intended to polycrystalline silicon technology. The maximum power P_{pvmax} can be given by [44] (Figs. 2.40, 2.41, 2.42):

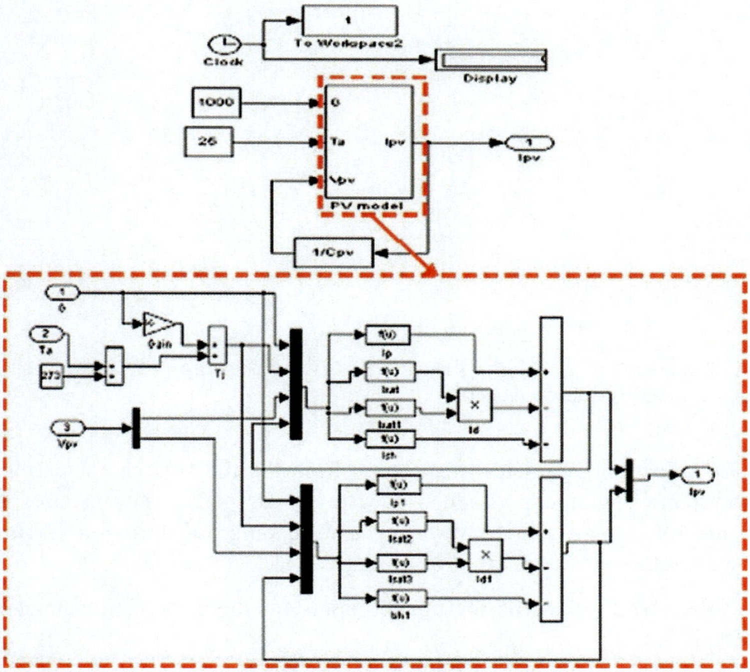

Fig. 2.37 Block diagram of two diode model [195]

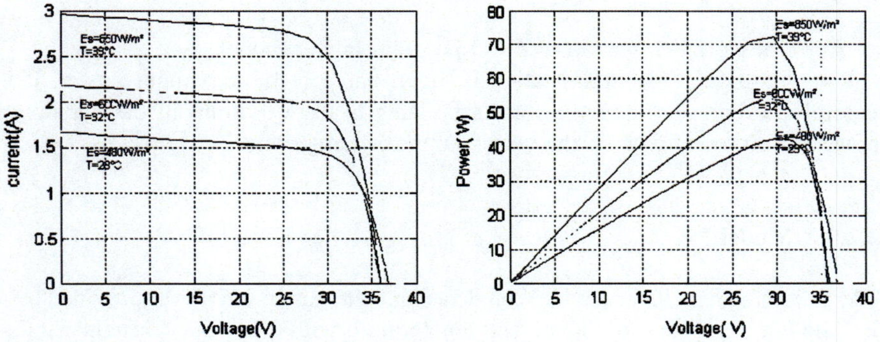

Fig. 2.38 Characteristics $I_{\mathrm{pv}} = f\,(V_{\mathrm{pv}})$ and $P_{\mathrm{pv}} = f\,(V_{\mathrm{pv}})$ simulation results (two diode model) [195]

$$P_{\mathrm{pv-max}} = K_1 \cdot (1 + K_2(T_j - T_{\mathrm{jref}})).(K_3 + G) \tag{2.156}$$

where K_1, K_2, and K_3 are constants to be determined (data sheets)

We can obtain the maximum power for a given irradiance G and panel temperature T_j with only three constant parameters and then solve a simple equation Eq. 2.156.

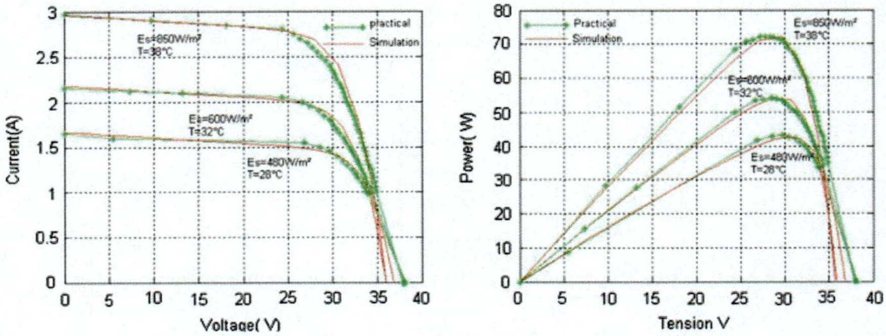

Fig. 2.39 Characteristics $I_{pv} = f\ (V_{pv})$ and $P_{pv} = f\ (V_{pv})$ comparison of experimental results with simulation ones(two diode model)

The identification parameter was carried from the maxima characteristic voltage/power panels with experimental measurements performed on site. Actual values used for several surveys covering a wide range of variation of sunshine actions were taken as:

- K_1 between 0.095 and 0.105 for a panel represents the dispersion characteristics of the panels.
- $K_2 = -0.47\%/°C$ drift in temperature of the panels.
- a parameter (K_3) added to the characteristic of the manufacturer, to obtain results much more satisfying:

Application under Matlab (see CD program Power model 1)

For this model, the optimization is given only on the maximum power. It is possible to represent the curves corresponding to the variations of the maximum power for different days in the year (summer, spring, winter).

2.2.3.2 Model No. 2

The following benchmark model can determine the maximum power provided by a PV module for given irradiance and temperature with only four constant parameters to determine a, b, c, and d. These parameters are obtained solving a simple equation system for a resulting set of measurement points sufficiently extended [30]. We have

$$P_{pv-max} = (a \cdot G + b) \cdot T_j + c \cdot G + d \tag{2.157}$$

where $P_{\text{pv-max}}$ is the maximum power output and where a, b, c, and d are positive constants which can be obtained experimentally.

Application under Matlab (see CD program Power model 2)

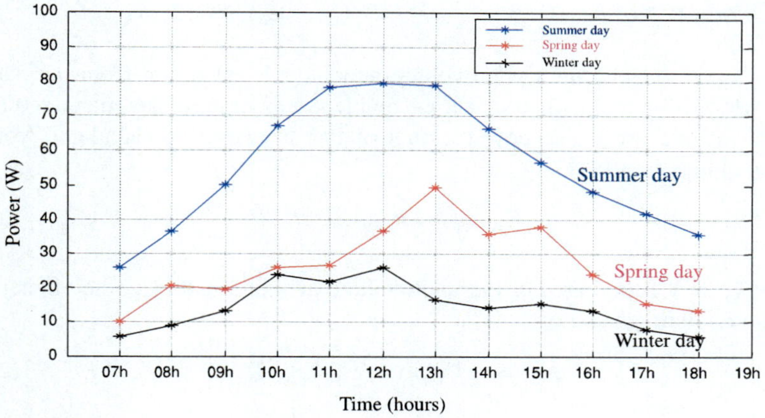

Fig. 2.40 Variations of power for different days of the year (Model No. 1) [9]

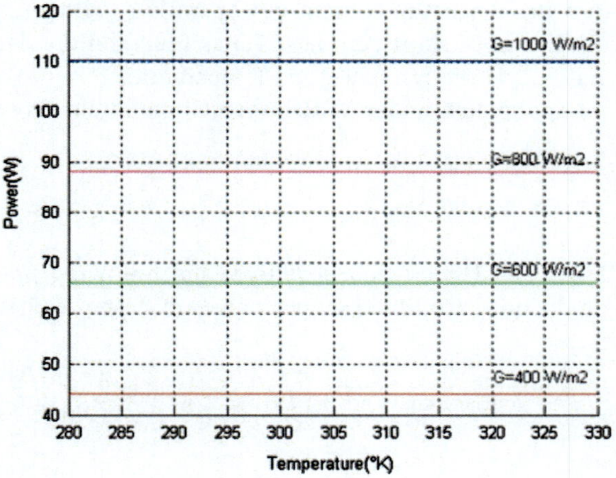

Fig. 2.41 Variations of power (Model No. 3) [195]

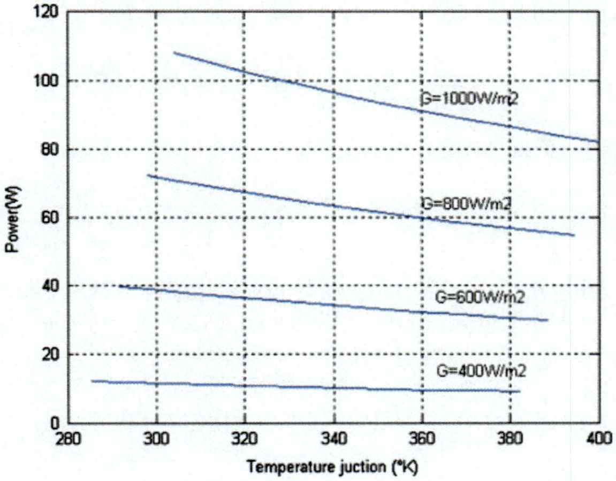

Fig. 2.42 Variations of power (Model No. 4) [195]

2.2.3.3 Model No. 3

The energy produced by a photovoltaic generator is estimated from the data the global irradiation on inclined plane, the ambient temperature data manufacturer for the PV module used. The power output of PV array can be calculated from the following equation [1]:

$$P_{\mathrm{pv-max}} = \eta_{\mathrm{pv}} . A_{\mathrm{pv}} \cdot N_m \cdot G \tag{2.158}$$

where A_{pv} is the total area of the photovoltaic generator and η_{gen} the efficiency of the photovoltaic generator.

$$\eta_{\mathrm{pv}} = \eta_r \cdot \eta_{\mathrm{pc}} \left[1 - \alpha_{\mathrm{sc}} (T_j - T_{\mathrm{jref}})\right] \tag{2.159}$$

where G is a solar radiation on tilted plane module, η_r is the reference efficiency of the photovoltaic generator, η_{pc} is the power conditioning efficiency which is equal to 1 if a perfect maximum power tracker (MPPT) is used, α_{sc} is the temperature coefficient of short-current (/°K) as found on the datasheet, T_j is cell temperature and T_{jref} is the reference cell temperature.

Application under Matlab (see CD program Power model 3)

2.2.3.4 Model No. 4

Jones and Underwood developed the following practical model in 2002 for the production of optimal output power of a photovoltaic module [30, 45]:

$$P_{\mathrm{pv-max}} = \mathrm{FF} \cdot \left(I_{\mathrm{sc}} \cdot \frac{G}{G_{\mathrm{ref}}} \right) \cdot \left(V_{\mathrm{oc}} \cdot \frac{\ln(P_1 \cdot G)}{\ln(P_1 \cdot G_{\mathrm{ref}})} \cdot \frac{T_{\mathrm{jref}}}{T_j} \right) \tag{2.160}$$

where P_1 is a constant coefficient, which can be calculated by the following formula:

$$P_1 = \frac{I_{\mathrm{sc}}}{G} \tag{2.161}$$

FF is the Filling factor.

$$\mathrm{FF} = \frac{P_{\mathrm{pv\,max}}}{V_{\mathrm{oc}} \cdot I_{\mathrm{sc}}} = \frac{V_{\mathrm{mpp}} I_{\mathrm{mpp}}}{V_{\mathrm{oc}} I_{\mathrm{sc}}}$$

with $P_{\text{pv-max}}$ the maximum power under STC conditions.

Application under Matlab (see CD program Power model 4)

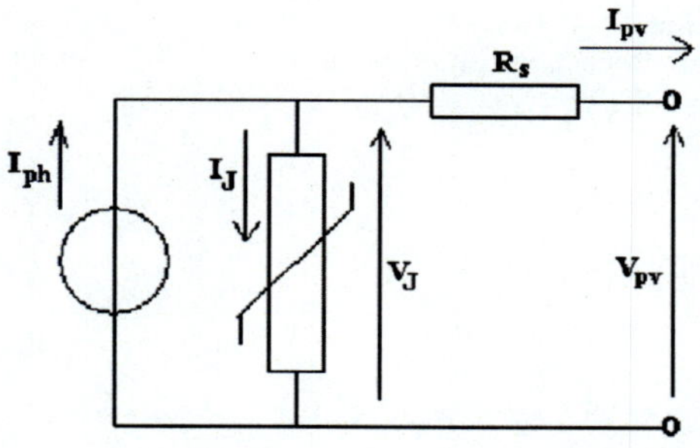

Fig. 2.43 General equivalent circuit for PV cell or module modeling

2.2.4 General Remarks on PV Arrays Models

2.2.4.1 The Forgotten Equation

Some of the above described model use for the parameters identification the values of V_{mpp} and I_{mpp}. However, only one of them (see Eq. 2.135) makes use of the fact that the point (V_{mpp}, I_{mpp}) is the point at maximum power. And yet, for that point, it is easy to prove (see Eq. 1.7) that, whatever the used model may be,

$$\left.\frac{dI_{pv}}{dV_{pv}}\right|_{V_{pv}=V_{mpp}} = -\frac{I_{mpp}}{V_{mpp}} \tag{2.162}$$

That additional equation allows determining one more parameter using the nominal values of a photovoltaic cell or module.

2.2.4.2 General Structure and Consequences

All the models above described which are based on an equivalent circuit have the same general structure shown at Fig. 2.43.

The above described models differ only due to the nature of the nonlinear element J in parallel on the current source I_{ph}. Following the complexity of the model, that element can be formed with a variable number of ideal elements: one or two diodes and sometime a resistance R_{sh}.

The consequence of that general structure is that all the characteristics I_{pv}–V_{pv} corresponding to the same junction temperature but different irradiance can be obtained by translation of one of them [13]. The translation has a vertical component ΔI_{ph} and a horizontal component $-R_s\, \Delta I_{ph}$. It is then possible to deduce the value of R_s from the I_{pv}–V_{pv} characteristics without reference to one particular model. That method is thus an interesting alternative to Eqs. 2.94, 2.103b and 2.129 for determining the series resistance R_s. It has the additional advantage to consider globally the characteristics and not one particular point.

With the general structure of Fig. 2.43, we have, for a variation of V_{pv}

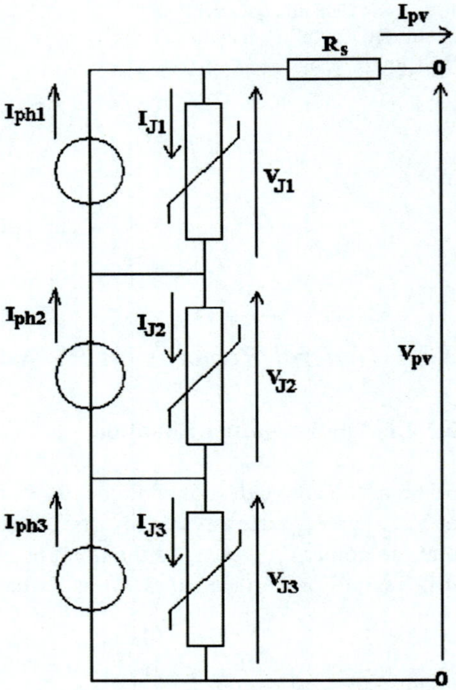

Fig. 2.44 General model for multijunction cell or module

$$dI_{pv} = -dI_J = -\frac{dI_j}{dV_J}[dV_{pv} + R_s dI_{pv}] \tag{2.163}$$

and thus

$$\frac{dI_{pv}}{dV_{pv}} = -\frac{dI_J/dV_J}{1 + R_s(dI_J/dV_J)} \tag{2.164}$$

Reporting (2.164) for the maximum power point into (2.161), we have:

$$[1 + R_s(dI_J/dV_J)|_{V_{pv} = V_{mpp}}]I_{mpp} = [(dI_J/dV_J)|_{V_{pv} = V_{mpp}}]V_{mpp} \tag{2.165}$$

or

$$I_{mpp} = [(dI_J/dV_J|_{V_{pv} = V_{mpp}}](V_{mpp} - R_s I_{mpp}) \tag{2.166}$$

2.2.4.3 Partial Linearity of the Models

Starting with the values of open-circuit voltage, maximum power current and voltage, and short-circuit current, we obtain in all case a system of four equations (three equations for the three nominal points and Eq. 2.165). That system is linear in I_{ph}, $1/R_{\mathrm{sh}}$, I_{d1} and I_{d2}. Then, no approximation formula is necessary for determining that parameters. The only parameters which arrive in the equations system in a nonlinear way are R_s and the non ideality factor A. If one uses the two diodes model, the simplest way is to assume that the factor A is unity. Then, beside R_s which has to be determined by another way, all the parameters of the model are determined using the three nominal points.

2.2.4.4 Remark About the Multijunction Cells

All the above mentioned models are suited for single junction cells. In case of multijunction cells, the model of each cell corresponds to the series connection of two or three circuits with the structure of Fig. 2.43, but with different parameters. Summing all the R_s resistance in only one, we obtain the general model of Fig. 2.44. Of course, that model contains many too parameters. However, if the manufacturer has well equilibrated the junctions for the standard spectrum, we have (for standard spectrum only) equality of the three current sources. In only that case we can combine the three nonlinear elements of Fig. 2.44 and retrieve the simpler structure of Fig. 2.43.

Chapter 3
Power Electronics Modeling

Symbols

$<g>$	Mean value of a quantity g
C	Value of a capacitance
C_{pv}	Input capacitance of a converter connected to photovoltaic array
f	Internal frequency of the converter
i	Instantaneous value of a current
I	Rms value of the current
k	Ratio between input and output voltages of a converter
k_{off}, k_{on}	Constants used for commutation losses estimation
L	Value of an inductance
p	Instantaneous value of a power
P	Active power (mean power on a period)
P_{cap}	Mean power dissipated during commutations due to stray capacitances
P_{ind}	Mean power dissipated during commutations due to stray inductances
p_{loss}	Instantaneous value of a power dissipated as heat
P_{loss}	Mean value of a power dissipated as heat
P_{nom}	Nominal power of the converter
P_{on}	Mean power of commutation losses related to transient from OFF to ON states
P_{off}	Mean power of commutation losses related to transient from ON to OFF states
P_Q	Mean power of commutation losses related to recovery charge of nearby components
Q	Value of an electric charge
R_{diff}	Differential resistance
T	Internal period of the converter
t_{on}	Time of variation of electrical quantities when going from OFF to ON states

D. Rekioua and E. Matagne, *Optimization of Photovoltaic Power Systems*,
Green Energy and Technology, DOI: 10.1007/978-1-4471-2403-0_3,

t_{off}	Time of variation of electrical quantities when going from ON to OFF states
u	Instantaneous value of a voltage
U_{in}	Input voltage of a converter
$U_{in\ nom}$	Nominal input voltage of a converter
U_{thr}	Threshold voltage
α	Duty cycle of a converter
η_{eur}	European efficiency.
η_{x}	efficiency for an operating point at x percent of its nominal power

Photovoltaic generators are almost always associated with some control and power electronics. Even in the case of a direct connection between a solar array and a battery, a non-return diode is needed. Generally, more complex power electronic converters are needed in order to adapt the electrical frequency and voltage level to the planned use. In addition, the rational use of photovoltaic generators is possible only in association with power electronic converters in order to adjust the voltage of the photovoltaic array to the maximum power point independently of the output voltage of the system. Power electronic converters associated to photovoltaic generators have to verify constraints on quality of energy supplied (low level of harmonics), electromagnetic perturbations (low level of EMC) and security. For instance, for photovoltaic generators connected to the AC mains, automatic disconnection from the grid in case of grid islanding is required.

Once those constraints are satisfied, we can look to the properties that have an influence on the energy production of the whole system, i.e. the ability to track the MPPT and the power losses or, equivalently, the efficiency of the power electronic converter.

As windmill systems, photovoltaic systems are fluctuating energy suppliers. The first consequence is that the velocity of the MPPT is important, but above all, these systems work during the most of the time at power levels very lower than their rated power. Thus, the rated efficiency of the power electronic converter is not relevant.

In some cases, the only goal is to maximize the total energy supplied on a large time interval (one year…). Then, we can achieve the design using a weighted average value of the efficiency. However, in many cases, it is necessary to study the time evolution of the working point of the system. It is for example the case when we want to evaluate a risk of energy lack, or if the energy price varies during the time. Then, we have to know the efficiency as a function of the operating point (converted power, input and output voltage …). This is the main subject of this chapter.

3.1 The Origin of Power Losses in Power Electronic Converters

3.1.1 Power Electronics Fundamentals

Power electronic converters are made up of power semiconductors switches (diodes, transistors….) and passive components. Passive components are mainly inductive (inductors, transformers, coupled inductors) or capacitive (capacitors). In non-reactive components (semiconductors, resistors), the power dissipated as heat inside the components is equal to its instantaneous electric power

$$p_{\text{loss}} = ui \tag{3.1}$$

where u is the voltage and i the current of the component.

A semiconductor acting as a switch has only two states, the ON state and the OFF state. The transitions from one state to the other one are achieved in short times named commutations. This use of semiconductors allows the reduction of their losses. Indeed, in the ON state, a significant current i can flow through the component, while the voltage u is low. The power (Eq. 3.1) dissipated during conduction intervals is thus low. It is referred as the "conduction losses". Similarly, in the OFF state, the voltage u across the component can take significant values, while the current which is flowing through it remains very small, so that the power (Eq. 3.1) dissipated in the component is generally negligible in OFF state.

The structure of the power converter must take into account the characteristics of the external circuits connected to its input and output ports. For example, a battery appears as a voltage source, and so it is not acceptable to short circuit it. Moreover, the use of the semiconductors as switches is associated to strong variations of the currents and voltages inside the converter. Generally, these variations are not acceptable for the circuits connected at the input and output ports of the converter. For example, solar modules give all the possible power only at their maximum power point (MPP). So, the converter may not induce significant variations of current and voltage into the solar array. In order to satisfy the above-mentioned constraints, power electronic converter includes, in addition to semiconductors, reactive (inductive and capacitive) components for filtering purpose. These components are not the ideal lossless elements of the circuit theory: they are prone to power losses. For example, the equivalent circuit of the inductive component includes always a "parasitic" series resistance which generates conduction losses. There are also magnetic losses in inductive components and leakage losses in capacitors.

There is also another type of losses due to the fact that the commutations of a semiconductor switch (transition from one state to the other one) are never instantaneous: during a short time interval, u and i take simultaneously significant values, so that a given energy amount is lost at each commutation. There is thus, in addition to the conduction losses, commutation losses which are proportional to the internal frequency of the converter.

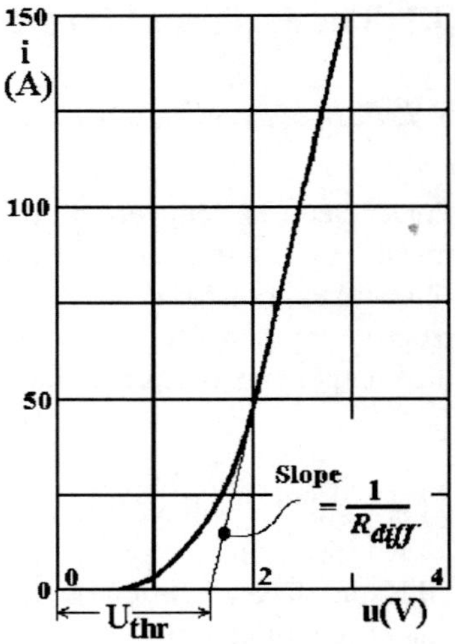

Fig. 3.1 I-U curve for a typical power diode

Today, most of power electronic converters are switching converters, i.e. converters in which the switching frequency of the semiconductor switches is much higher than the frequency found at the input and output ports of the converter (0 Hz for a port connected to a DC circuit, 50 or 60 Hz for a port connected to the AC mains).

According to the power rating of a converter, the switching frequency may range from several kHz for high to medium power converters to more than 1 MHz for very low power converters. The use of a high switching frequency makes easier the filtering of the current and voltage ripples generated at the switching frequency at the input and output of the converter. Moreover, as the energy which has to be stored inside the passive components is lower when the internal frequency is high, the frequency increase allows reducing the size and the cost of the passive components.

Unfortunately, an increase of the switching frequency increases the commutation losses, so the choice of this frequency results from a compromise which is mainly ruled by the switching speed of the semiconductors. As the commutation losses are lower when the components can commute faster, the optimal switching frequencies grow more and more as the semiconductor performances are improved.

Finally, it is necessary to mention the power consumption of auxiliary circuits (drivers of the power switches, sensors, control and communication electronics).

3.1.2 Methods of Elementary Losses Modeling

In the case of a component which includes in its equivalent circuit a constant series resistance R, the power dissipation related to that resistance is simply

$$p = R.i^2 \tag{3.2}$$

And, thus the mean dissipated power during a switching cycle is

$$P = R.I^2 \tag{3.3}$$

where I is the rms value of the current flowing through the semiconductor during that cycle.

Many semiconductors in ON state exhibit a non-linear current–voltage characteristic, as shown in Fig. 3.1.

In that case, we consider often an approximate characteristic of the form

$$u = U_{\text{thr}} + R_{\text{diff}}.I \tag{3.4}$$

where R_{diff} is named "differential resistance" and U_{thr} is named "threshold Voltage". In that case, it is easily found that the mean power related to conduction losses during a cycle is, assuming that the current i can only have the positive sign

$$P = U_{\text{thr}}.\langle i \rangle + R_{\text{diff}} I^2 \tag{3.5}$$

where $< i >$ is the mean value of the current i during that cycle.

In the commutation losses of a semiconductor, we find a term proportional to the time t_{on} (time for going from the OFF-state to the ON-state). Noting u_1 the voltage just before that commutation and i_1 the current just after it, we have:

$$P_{\text{on}} = k_{\text{on}}.u_1.i_1.t_{\text{on}}.f \tag{3.6}$$

where f is the internal frequency of the converter and k_{on} a constant.

Similarly, we have a term proportional to the time t_{off} (time for going from the ON-state to the OFF-state). Noting i_2 the current just before that commutation and u_2 the current just after it, we find:

$$P_{\text{off}} = k_{\text{off}}.u_2.i_2.t_{\text{off}}.f \tag{3.7}$$

Using approximate time dependence of the quantities (linear commutation), one finds for k_{on} and k_{off} values in the range 0.2–0.5.

The commutation losses (Eq. 3.6) and (Eq. 3.7) can be reduced using shorter commutation times but one is limited in that way by the intrinsic limitations of the component or when the maximum values of voltage and current reached during the transient become too high for the components.

There is also commutation losses linked to the stray elements. So, if there is a stray inductance L in series with the commuting element (due to its connections, including the connection inside its packaging), we have power dissipation corresponding to the transfer of the inductance energy to the switch at its turnoff.

$$P_{\text{ind}} = \frac{1}{2}.L.i_2^2.f \tag{3.8}$$

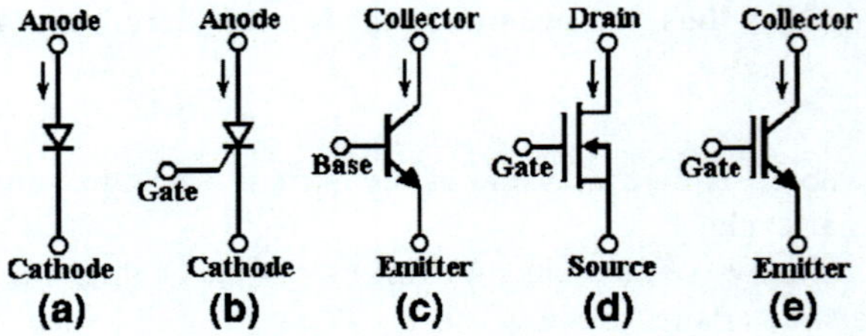

Fig. 3.2 Symbol of some power semiconductors. **a** Diode. **b** Thyristor. **c** bipolar transistor. **d** MosFet transistor. **e** IGBT. Arrow shows the current direction when the semiconductor is conducting

Similarly, if there is a stray capacitance C in parallel on a switch, it causes power dissipation corresponding to the transfer of the capacitance energy to the switch at its turnon.

$$P_{\text{cap}} = \frac{1}{2} C {u_1}^2 f \tag{3.9}$$

Finally, if there is in parallel on the component a source able to supply an electric charge Q under the voltage u_1, it causes power dissipation

$$P_Q = u_1.Q.f \tag{3.10}$$

Magnetic losses in inductive components can be classified into two types.

Hysteresis losses are proportional to the frequency f but they are growing function of the flux variation. The flux variation is approximately equal to u/f, so for a given frequency, they are growing with u. Eddy current losses are proportional to the frequency square f^2 and to the square of the flux, so they depend only of the square of the voltage.

3.1.3 The Most Used Power Semiconductors

The symbols of some power semiconductors with the name of their connections are shown in Fig. 3.2.

Diode (Fig. 3.2a) is not controllable. It becomes conducting when a small voltage is applied in the conduction direction otherwise it becomes blocking.

Thyristor (Fig. 3.2b) is only half controllable. It becomes conducting only if the voltage between anode and cathode is positive and if a current pulse is applied on gate. It becomes blocking when the current tends to zero.

Other semiconductors of the Fig. 3.2 are fully controllable. Bipolar transistor (Fig. 3.2c) is conducting only if a base current is continuously applied. MosFet transistor (Fig. 3.2d) and IGBT (Fig. 3.2e) are in conduction only if a positive gate voltage is applied.

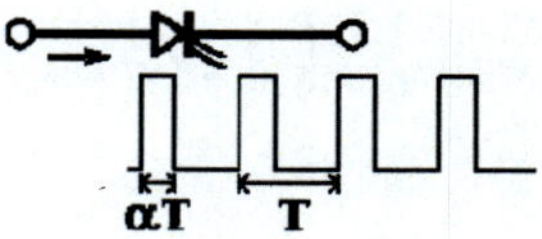

Fig. 3.3 Generic symbol for any of the fully controlled power switches

3.1.4 Particularities of the Semiconductors From the Losses Point of View

Diode (Fig. 3.2a) exhibits a significant threshold voltage. So, in the expression (Eq. 3.5) of losses, the first term must be considered. The drawback is especially penalizing in case of low voltage converters. At low voltage, it can be reduced using Schottky diodes. Another possibility is to replace diodes by MosFet at the price of additional command circuits.

Thyristor (Fig. 3.2b), bipolar transistor (Fig. 3.2c) and IGBT (Fig. 3.2e) also exhibit a threshold voltage, but often lower than the threshold voltage of a diode.

Bipolar transistor (Fig. 3.2c) remains conducting only as long as a base current is applied. This requires a circuit which is able to supply that current. It is thus necessary in the energy balance to take into account the power consumption of that circuit, in addition to conduction losses.

MOSFET transistor (Fig. 3.2d) is controlled by gate voltage. At standstill that voltage can be maintained without current consumption. So, the control circuit consumes energy only during the commutations. The MosFet can commute very fast, then reducing the commutations losses (Eq. 3.6) and (Eq. 3.7). Moreover, it presents no threshold voltage. It is thus possible to reduce the conduction losses by paralleling several MOSFET or, equivalently, using overrated MOSFET. As a rough estimate, IGBT behaves like the bipolar transistor for the main circuit, but as the MosFet for the command circuit.

3.2 The Structures of Converters and the Influence on Their Efficiencies

Inside this section, we shall use the symbol of Fig. 3.3 for any of the fully controlled power switch, i.e. bipolar transistor (Fig. 3.2c), MOSFET transistor (Fig. 3.2d) or IGBT (Fig. 3.2e).

In the following examples, the command of these semiconductors is a periodic digital signal whose period is $T = 1/f$, f being the internal frequency of the converter. The duty cycle α of that signal, as shown in Fig. 3.3, is the ratio between the duration $\alpha\, T$ in the state "on" and the entire period T. We do not give here a detailed description of quantities time evolution: interested readers can find such an analysis in classical power electronics textbooks such as [50, 51].

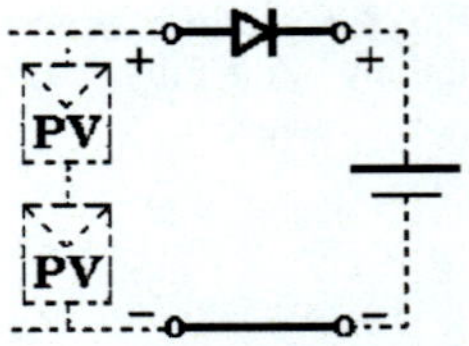

Fig. 3.4 Direct connection of a PV array to a DC bus

The choice of the internal frequency *f* results from a compromise. In one hand, a high frequency *f* allows for an easier filtering of the harmonics at input and output, and thus lowers losses in passive components. On the other hand, commutation losses (and command losses in case of MosFet or IGBT) increase proportionally to that frequency. It thus exists an optimum frequency. This optimum can depend on the working point, but, for practical reasons, designers often opt for a fixed frequency.

3.2.1 Direct Connection to a DC Bus

The simplest interface between a photovoltaic array and a constant voltage DC bus (for example a battery) is to use only one diode (Fig. 3.4).

This circuit has little losses, but, as it is not controlled, the voltage of the PV is fixed. As the optimum voltage value is temperature and irradiation dependent, the use of PV is often far away from the optimum point, and PV cannot supply current when its open-circuit voltage is lower than the DC bus voltage. As that interface has no command possibility, it is often completed by a circuit which can short-circuit or disconnect the PV array when the battery and the load are not able to absorb all the available power.

3.2.2 DC/DC Conversion

3.2.2.1 Transformerless DC/DC Converters

The simplest controllable converters are made of only one transistor, one diode and one inductor as main components. When the input of the converter is the PV array, the best efficiency can be obtained with the boost converter (Fig. 3.5).

For the boost converter, when the duty ratio α goes from 0 to 1, ratio k between input and output voltages goes from 1 to 0. So, it is possible to obtain energy production even for low irradiance (for example at sunrise and sunset). The design must be done in a way that the open-circuit voltage of the PV remains most of the time lower than the DC bus voltage. However, in order to obtain a good efficiency, the input and output voltage levels cannot be too different. Indeed, considering the

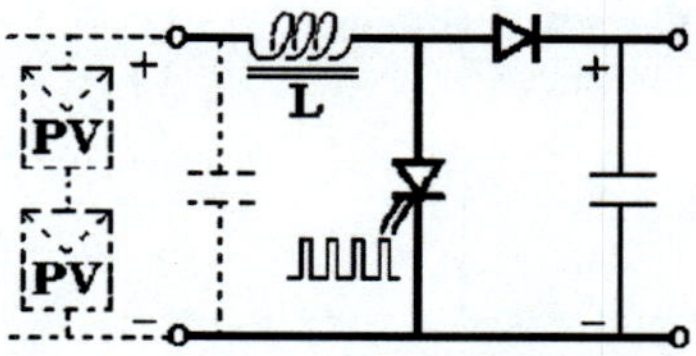

Fig. 3.5 Power components of the boost converter

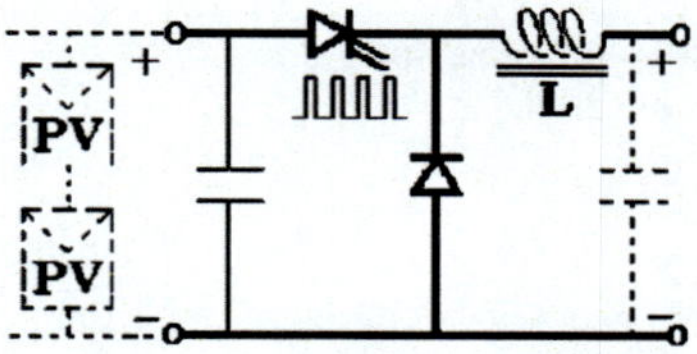

Fig. 3.6 Buck converter

input and output current and voltage, the sizing is governed by the product of the bigger voltage by the bigger current. A big value of the ratio k leads to oversized converter, and then lower efficiency. For that reason, the output voltage is not too large and one can admit that, occasionally, the optimum PV voltage becomes higher than the output voltage. When that case occurs, the converter reduces to direct connection (Fig. 3.4) but it remains possible to short-circuit the PV if necessary by keeping the transistor "on".

The fact that the inductor is in series with input is interesting because it maintains constant the current value during the period T, which is suitable to maintain continuously PV voltage at its optimum value.

In some cases, the PV optimum voltage is most of time higher than the DC bus voltage. It is often the case with a 12 V battery and commercial PV modules. In such a case, the buck converter (Fig. 3.6) can be interesting.

However, with the buck converter, when the duty ratio α goes from 0 to 1, ratio k between input and output voltages goes from 1 to ∞. So, it is not possible to obtain energy production when open voltage of the PV is low (for example at sunrise and at sunset). The buck-boost converter (Fig. 3.7) does not have that drawback because it allows for using all the values of k between 0 and ∞. However, efficiency of the buck-boost converter is slightly lower then the efficiencies of boost or buck converters because, in the buck-boost converter, all the energy carried from input to output is transiently stored inside the inductor. The input current of the buck converter and of the buck-boost converter is chopped, so that in the case of using PV, filtering by a capacitor in parallel on the input is necessary.

3.2.2.2 DC/DC Converters with Transformer

As above mentioned, the efficiency of the DC/DC converters of the Sect. 3.3.2.1 is lowered if the input and output voltage are at very different levels. In that case,

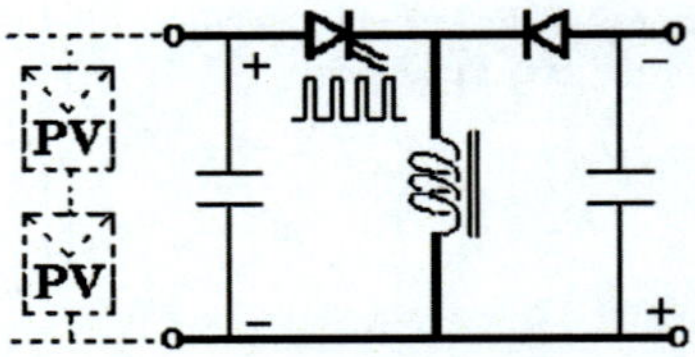

Fig. 3.7 Buck-boost converter

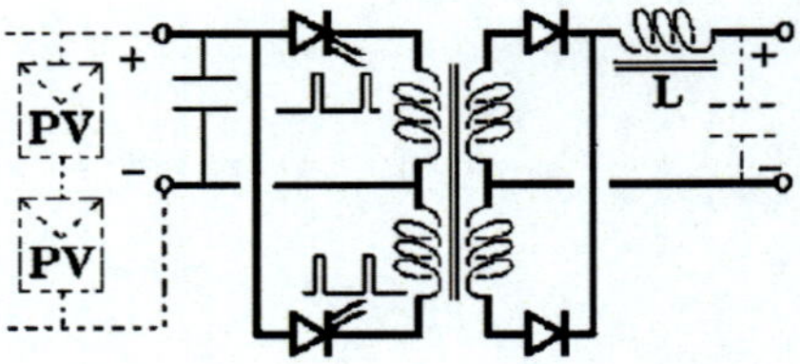

Fig. 3.8 Push-pull forward converter

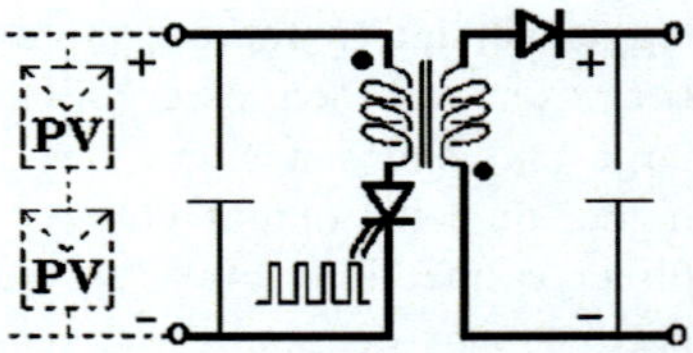

Fig. 3.9 Flyback converter

transformer allows to maintain the efficiency because the rating of each winding is concerned only with the related current and voltage. Transformers are also necessary when a galvanic insulation is needed.

For a given power, transformer size is smaller when the used frequency is larger. Transformer does not work in DC current but it is possible to insert it inside a DC/DC converter with an internal frequency *f*. Then, the duty ratio can be maintained near from 0.5 allowing for a good efficiency. Figure 3.8 shows the schematic of a push–pull forward converter, which operates as a buck converter (thus cannot run with a too low input voltage).

Figure 3.9 shows the schematic of a flyback converter, which derives from a buck-boost converter. In the flyback converter, transformer and inductor are combined into a unique component: a pair of coupled coils.

The efficiency of the flyback converter is lower but it can work with a weak input voltage and it can be realized with multiple outputs, which can be useful in order to charge the elements of a battery in a balanced way.

3.2.2.3 Example of Application of DC–DC Converter with Photovoltaic: a Battery Charger

It is the simplest system where only one converter exists that loads the battery. Then the battery is directly connected to the load (Fig. 3.10).

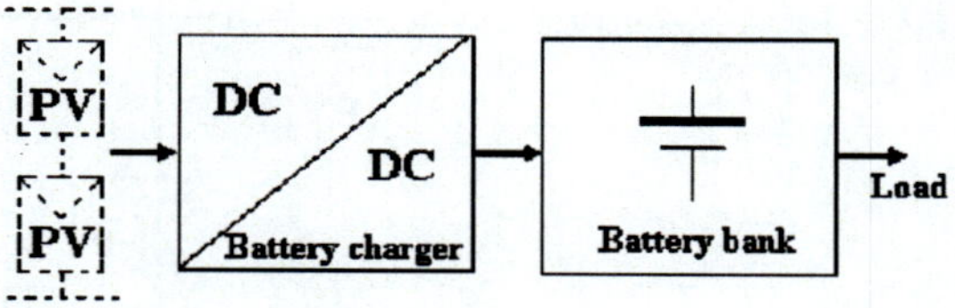

Fig. 3.10 Simple dc system

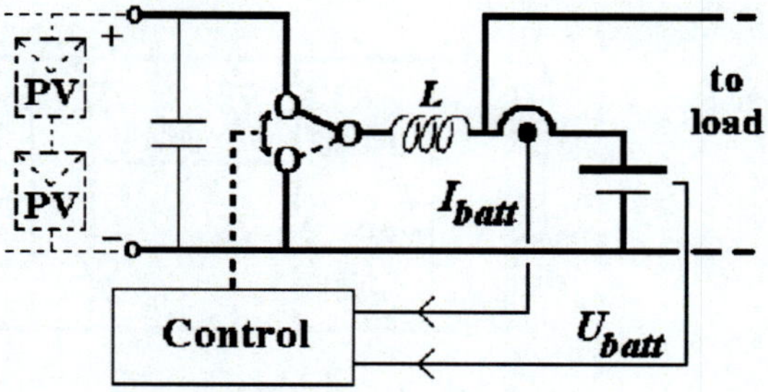

Fig. 3.11 Simplified model of the battery charger

The function of the battery charger is providing energy to a battery bank under controlled voltage and current, in order to improve the service lifetime of the battery bank. We consider as example the circuit of Fig. 3.6 (buck converter). To model the battery charger, we use the equivalent circuit of Fig. 3.11.

As long as the battery is not fully charged, the control circuit act on the switches of DC/DC converter in order to maximize the battery current. However, it can estimate the charge of the battery using measurements of battery voltage, current and often temperature and current integral. It can thus limit the battery current if its value is too large considering the battery state of charge. Often, the control circuit can also acts on a relay in order to disconnect the load when the battery is too discharged. In theory, a model of the battery is needed in order to estimate the battery state of charge. Several control circuits include a fuzzy logic controller which can determinate implicitly such a model after some cycles of charge.

3.2.3 DC/AC Conversion

3.2.3.1 Converters Topologies

DC/AC Converters (Single Stage Inverters)

When the power generated is transmitted to the public network, or used by AC devices, it is necessary to use a DC/AC converter. The most popular is the voltage inverter, whose schematic is given at Fig. 3.12 (single phase inverter) or at Fig. 3.13 (three phase inverter).

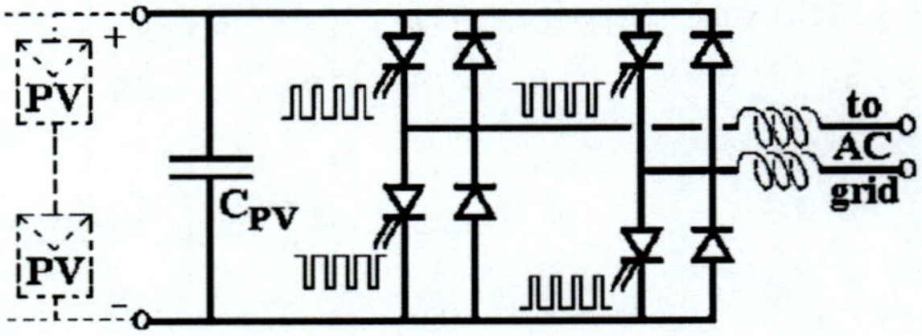

Fig. 3.12 Single phase voltage inverter

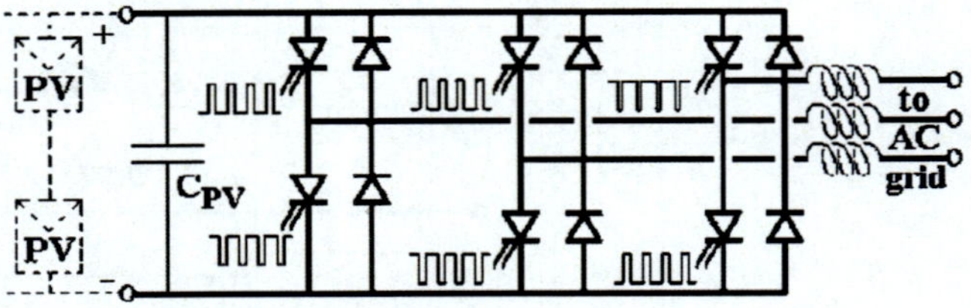

Fig. 3.13 Three phase voltage inverter

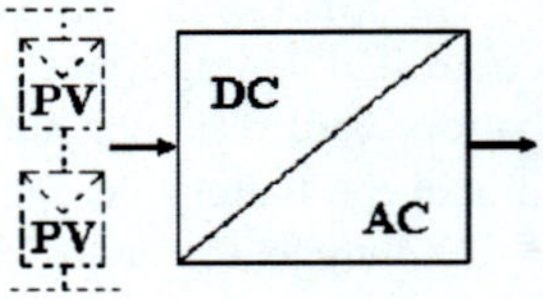

Fig. 3.14 Single stage Inverter

In order to avoid harmonics injection on the AC grid, output filtering is needed. For making that filtering with small reactance, it is necessary to use a high switching frequency, which induces significant commutation losses. On the other hand, the inverter does not provide galvanic insulation. Input voltage of the voltage inverter must be higher than the peak value of the AC voltage, i.e. 315 V in case of single phase 230 V_{rms} and 566 V in case of three phases 400 V_{rms}. Thus, if the output is connected to the public grid and the input to photovoltaic array, this one must deliver a high voltage that is possible only for a series connection of many solar modules and leave only few possibilities MPPT. Moreover, in the case of a single phase inverter, input current is modulated to twice the grid frequency; in that case, if the input is connected to a PV array, use of a capacitor C_{PV} of large value in parallel to the input is needed in order to maintain the PV current at the optimum value of the current. In the following, inverters shall be schematized as in Fig. 3.14, as well as in case of single stage inverter of the Figs. 3.12 and 3.13 than in case of more complex ones.

Complex DC/AC Converters (Multistage Inverters)

The drawbacks mentioned at the DC/AC convertors (Single stage inverters) can be avoided adding to the inverter other conversion stages. In fact, what is named

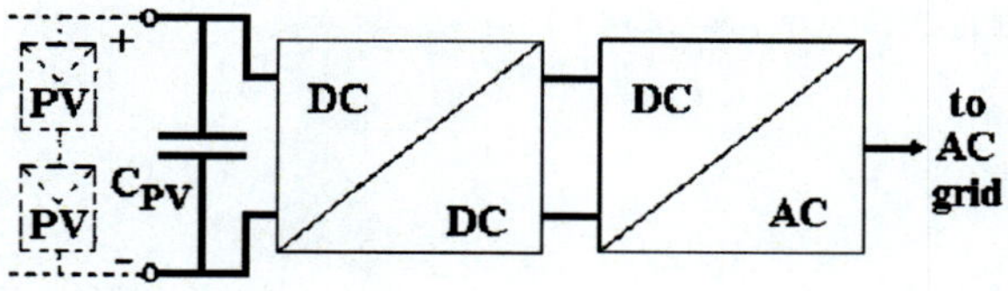

Fig. 3.15 Example of double-stage inverter

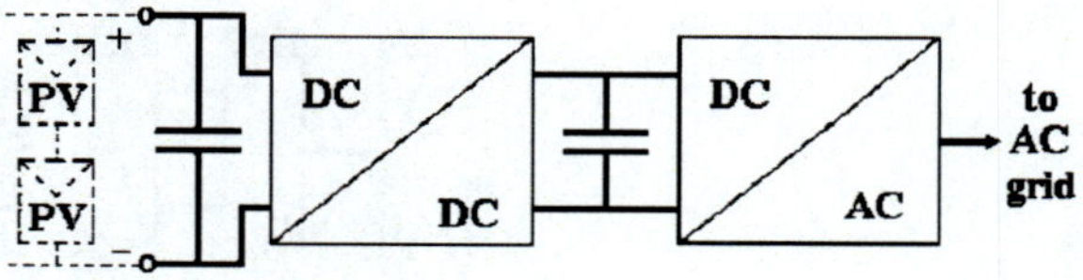

Fig. 3.16 Dual stage inverter

commercially "solar inverter" often includes several power electronic converters. As a first example, we can have a dual-stage inverter (Fig. 3.15). This topology consists of a DC–DC converter which is used for the MPPT, and a DC–AC inverter. The control signals applied to the power switches of the final stage can then be 180 or 120° conduction at the grid frequency (50 or 60 Hz). Then, commutation losses of the inverter are very low. This is possible without strong oversize output filter if the output from the DC–DC converter is modulated to follow a rectified sine wave.

Another example of double stage inverter is given in Fig. 3.16.

In the Fig. 3.16, the DC link between the DC–DC converter which carries out the MPPT and the inverter is fitted with a capacitor C_{link}. Then, it is no longer possible for modulating the DC link voltage to follow a rectified sine wave. But C_{link} is used for energy storage in order to cover the power fluctuations peculiar to single phase AC grid. It is best suited for that use that the input capacitor C_{PV} because larger voltage fluctuations are acceptable on C_{link}. C_{pv} can then be strongly reduced.

Insertion of a Transformer

In some cases, galvanic insulation is required for security reasons, or in order to allow grounding of the photovoltaic array, which is necessary with some photovoltaic technologies. Then, the insertion of a transformer inside the converter is needed. Another case where a transformer is requested is when voltage level of the DC input and AC output are very different. Then, the use of a transformer allows keeping in the converters duty cycles compatible with good efficiency. The simplest topology including a transformer consists to insert a low frequency transformer (LFT) between the output of the converter and the AC grid, as shown in Fig. 3.17.

Unfortunately, at low frequency, transformers are big, heavy and costly. An alternative is to include the transformer inside the DC link of Fig. 3.16. As transformer does not work in DC, it is necessary to insert it between a DC–AC and an AC–DC converter. It is then possible to use a transformer operating at high frequency (HFT). This leads to the general configuration described at Fig. 3.18.

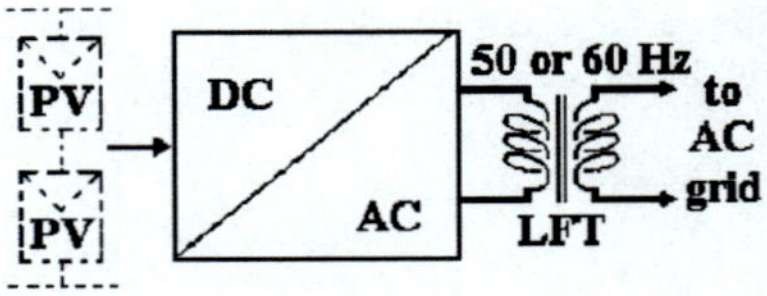

Fig. 3.17 Inverter with LFT

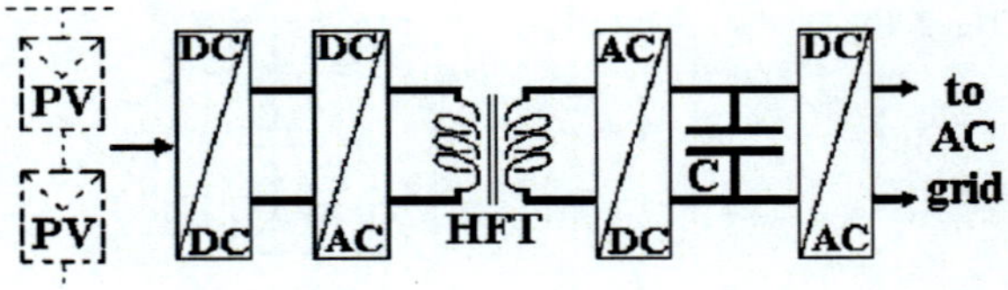

Fig. 3.18 Inverter with HFT

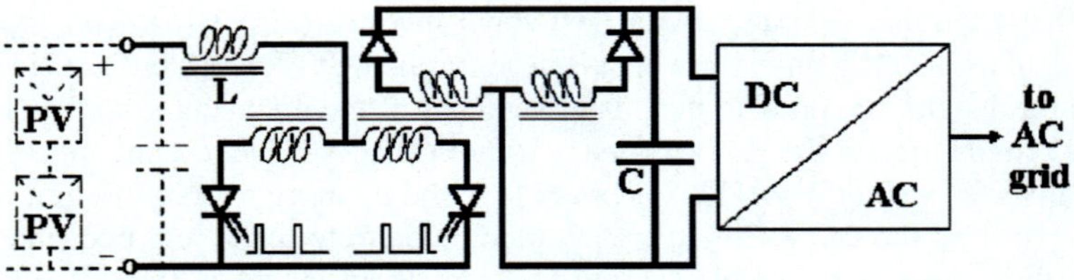

Fig. 3.19 Example of complex "solar inverter" topology

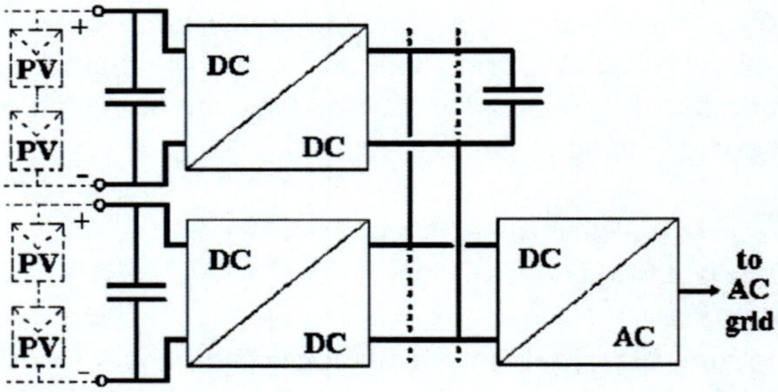

Fig. 3.20 Multi-input inverter

Another solution is to use as input stage a DC/DC converter with transformer, for example the converter of Fig. 3.8 or Fig. 3.9. Figure 3.19 shows a schematic where the input DC/DC converter is boost modified in order to include a transformer.

Multi-Input DC/AC Converters

When the optimum working points of the modules are not all the same ones, for example in case of partial shade, it can be interesting to split the PV array in

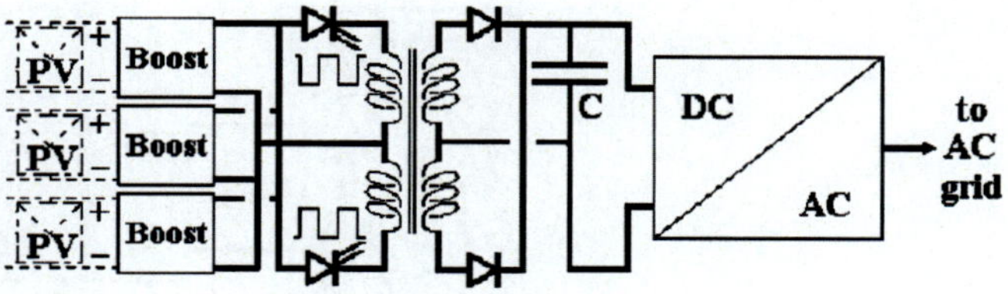

Fig. 3.21 Example of complex "solar inverter" with several independent inputs

several subarrays, each of that subarray being fitted with its own converter and MPPT. We can also have a multi-input inverter (Fig. 3.20). This topology consists of two DC–DC converters connected to the DC link of a common DC–AC inverter.

In that case, we can use several stages of the conversion: only the first stage is own to each subarray and subsequent stages are common. So, we can use at first stage a simple circuit such as a transformerless boost (Fig. 3.5) in order to keep a good efficiency in spite of the reduced power. Figure 3.21 shows a circuit made of several boosts, a common fixed duty factor push–pull converter in order to adapt the voltage level and to provide galvanic insulation, a storage capacitor and finally an inverter.

An important aspect of multi-input inverters is that they can control several different energy sources, for example several photovoltaic subarrays and one wind turbine generator.

3.2.3.2 Systems Configurations

In general, three types of configurations have been identified:

Central Plant Inverter

The past technology is the central plant inverter, and it was based on centralized inverters that interfaced a large number of PV modules to the grid. The PV modules were divided into series connections (called a string).These ones were then connected in parallel, through string diodes, in order to reach high power levels (Fig. 3.22).

String Inverters and AC-Modules

The present technology consists of the string inverters and the AC modules. The string inverter is a reduced version of the centralized inverter, where a single string of PV modules is connected to the inverter (Fig. 3.23). The present solutions use

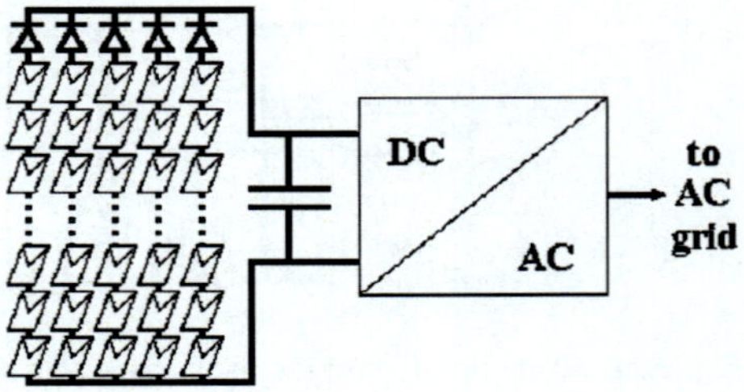

Fig. 3.22 Central plant inverter

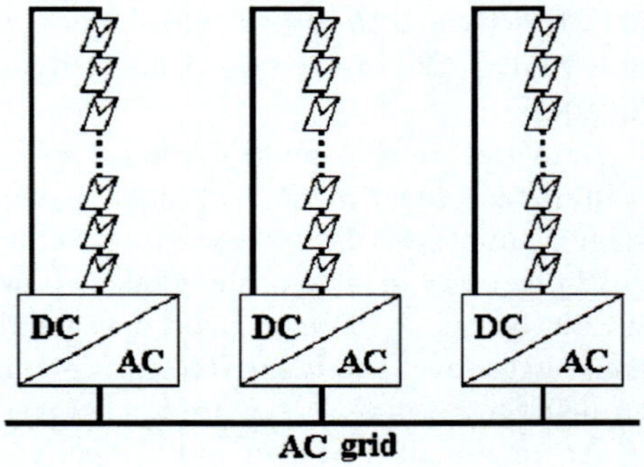

Fig. 3.23 String inverter

self-commutated DC–AC inverters, by means of IGBTs or MOSFETs, involving high power quality in compliance with the standards.

At the limit case, it is possible to associate a DC/AC converter with each module, which can then be called AC-modules (Fig. 3.24). This solution has the advantage of being totally modular. It is an interesting solution when the photovoltaic array is subject to complex shadows, since each module can always be used at its optimal power. However, the DC/AC converters are complex since there is a large difference in the input and output voltage levels, and their nominal power is low since each of them controls only one module. Their efficiency is thus relatively low.

Multi-String Inverter

In the multi-string inverter several strings are interfaced with their own DC–DC converter to a common DC–AC inverter (Fig. 3.25). It is an advantage compared with the centralized system, since every string can be controlled individually. It is also an advantage compared with string inverters and AC modules, since the part of the converter which is critical from the losses point of view is an only DC/AC inverter, which has a higher power and thus a better efficiency.

The choice of appropriate configuration must be motivated by the use conditions imposed by the environment and situation.

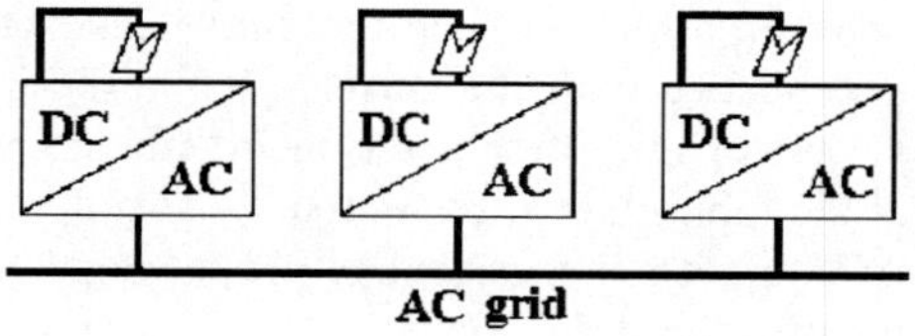

Fig. 3.24 AC modules

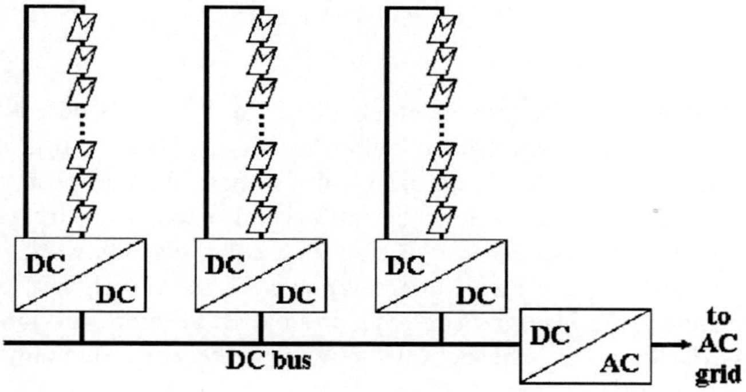

Fig. 3.25 Multi-string inverter

3.3 Empirical Modeling of the Converters

If we admit that the voltage and current on the components are proportional to the voltage and currents at the ports of the converter, we conclude that, at view of the developed expressions above, at fixed internal frequency f, the losses are a sum of terms of the first and second order in the input and output current and voltage, plus a constant term due to the auxiliary circuits. This remark furnishes a help to identify the expression of the losses using data of the manufacturer or other experimental measurements.

The data of the manufacturer are usually expressed in form of efficiency. It is then possible to obtain losses using the relation between efficiency η, loss power P_{loss} and output power P

$$P_{\mathrm{loss}} = \left(\frac{1}{\eta} - 1\right) P \tag{3.11}$$

.

3.3.1 Case of Constant Voltage

In some cases, the converter operates with a constant voltage as well as at the input and the output. This is for instance the case of a DC to AC converter connected between a DC link and the public distribution network. In such a case, only the dependence with the currents has to be identified. As the power electronic converters exhibit often a very good efficiency, we assume that the power is the same at the input than at the output. At fixed voltage (and fixed power factor in case of AC output), the current is proportional to the power, so that it is sufficient to identify the three coefficients of a second-order polynomial

$$P_{\text{loss}} = A + BP + CP^2 \tag{3.12}$$

.

In order to determine the three coefficients of Eq. 3.12, it is necessary to have at one's disposal three data. Manufacturer usually provides value of the efficiency at nominal power. A second data is the so named "European efficiency". It is a mean value of the efficiency at several values of the power. The "European efficiency" is added for the fact that solar systems work most of the time at lower power than their nominal power. It is a weighted average on efficiencies related to a histogram of solar irradiation at temperate latitude on an annual basis. Calling P_{nom} the nominal power of the converter and η_x its efficiency for an operating point at x percent of its nominal power, European efficiency is defined as

$$\eta_{\text{eur}} = 0.03\eta_5 + 0.06\eta_{10} + 0.13\eta_{20} + 0.10\eta_{30} + 0.48\eta_{50} + 0.20\eta_{100} \tag{3.13}$$

Knowledge of the nominal and European efficiencies is not sufficient in order to determine the three coefficients A, B and C of Eq. 3.12. If an additional operating point is not available, we have to put an arbitrary constraints, for example $B = 0$, for computing the values of these coefficients.

3.3.2 Case of Variable Input Voltage

Practically, output voltage of converter is approximately fixed because it is usually either the public grid voltage, either a battery voltage.

However, when the input of the converter is connected to PV array, its voltage can vary because optimal working point of photovoltaic modules depends on temperature and solar irradiation. So, it should be useful to take into consideration the variation of losses in function of input voltage.

For a given power P, a variation of input voltage causes a variation of input current because the power is the product of current and voltage. On that consideration, using also the introductory remark of this section, one could split the terms of (Eq. 3.12) setting

$$P_{\text{loss}} = A_1 + A_2 \frac{U_{\text{in}}}{U_{\text{in nom}}} + A_3 \left(\frac{U_{\text{in}}}{U_{\text{in nom}}} \right)^2 + \left[B_1 + B_2 \frac{U_{\text{in nom}}}{U_{\text{in}}} \right] P$$
$$+ \left[C_1 + C_2 \frac{U_{\text{in nom}}}{U_{\text{in}}} + C_3 \left(\frac{U_{\text{in nom}}}{U_{\text{in}}} \right)^2 \right] P^2 \qquad (3.14)$$

.

Unfortunately, manufacturer data are usually insufficient in order to determine all the coefficients of (Eq. 3.14). When manufacturer gives values of efficiency for several input voltages, it is generally for different configurations of the converter obtained by changing the turn numbers of transformer windings, so obtaining several and different values of the nominal input voltage. Then, there are as many different expressions of losses as different configurations, but often the data for each configuration describe the losses only for the nominal input voltage of that configuration.

3.3.3 Note on Experimental Losses Determination

Even if we have the converter at one's disposal, measuring losses of that converter is difficult because, as the efficiency is always good, obtaining the losses by subtraction between measured input and output powers leads to inaccuracy because it is the difference between two similar values. Thus, it should be necessary to measure input and output powers with a very high accuracy. Another approach is to measure the heat generated by calorimetric methods.

3.4 Circuit Modeling

During the conception of a converter, a circuit model is needed in order to obtain performance evaluation of device which does not exist. Then, it is possible to optimize the device for a particular use, on basis of realistic working conditions. Even if the converter exists, it can be useful to achieve an analysis of its working at view of the difficulties noticed in Sect. 3.1. For obtaining a losses expression, of course, it is possible to complete by this way the manufacturer data only if one knows the structure (see Sect. 3.2) of the converter and the references of all power components used, as well as the command strategy. Analysis of a power electronic circuit can be achieved by specialized software using sophisticated model of each component, and losses can then be accurately computed. Another approach, best fitted for optimization, is to compute the time evolution of quantities analytically, using for that approximated models. Then, an approximate value of losses is obtained as explained in Sect. 3.1.2.

3.5 Note on the Nominal Power Choice

The choice of the converters nominal power is an important aspect of optimization. Indeed, a too large value of the nominal power reduces the ratio between real and nominal powers, and thus the mean efficiency of the converter. On the other hand, for an optimum sizing, it can arrive during limited time that the optimum power of the PV array goes beyond the converter nominal power. In that case, it is important for the protection of the converter to saturate, even if then a part of the available energy is not exploited. This case arrives in particular with amorphous silicon panels, which can produce during the period just after they enter service up to 25% (in the case of single junction) more than their nominal power [52].

3.6 Multi-Agent Systems for the Control of Distributed Energy Systems

In order to handle inherent uncertainties of renewable sources, it is necessary to design controllers able to implement the interface between the grid and the renewable energy systems. The controller must be capable of intelligent and suitable responses to changing environments. In the next generation of grid connected renewable energy systems, the inverters will be the core enabling technologies for the growth of large-scale energy systems. These more intelligent inverters with advanced control features will improve the performance and controllability of future renewable energy systems. The inverter is the intelligent device of the energy conversion system. It has the capability of sensing and storing a wide variety of environmental conditions. By reacting to these changing conditions, the inverter will improve the system's health. Furthermore, the inverter can be used in a distributed control system by taking advantage of the implemented intelligent algorithms and the communication ability of the inverter. This feature will increase the penetration of renewable energy systems by facilitating their connection to the conventional electrical grid. The use of intelligent inverters will increase the reliability of renewable energy systems by improving the quality of the power transferred to the grid. In renewable energy distributed energy systems, the inverters enable renewable energy systems to ride through grid disturbances, operate in islanded or micro-grid modes. Due to environmental awareness, renewable energy penetration in power generation and distribution is continuously increasing. This causes many problems when the distributed renewable systems are connected to the main conventional grid.

Many researches had been conducted on distributed energy systems. The scientific community addressed many issues including technical, environmental and economic issues. Centralized control systems may be able to solve a given problem using powerful computational tools. However, as the complexity of the energy systems increases, computational and communications overheads become a significant problem. To get ride through these issues, decentralized control is an ideal solution because data are locally processed and only the results are transferred, computation time and communications are considerably reduced.

3.6.1 Multi-Agent Systems

Agent-based systems technology has generated lots of interest in recent years because of its promise as a new paradigm for conceptualizing, designing and implementing power control systems. This is particularly attractive for power systems that operate in open and distributed environments.

In order to explore the potential benefits of MAS to power generation and distribution systems, the concepts associated with multi-agent technology need to be described. First, the concept of agent needs to be described. Second, some critical notions in MAS are presented below.

3.6.1.1 Definition of an Agent

The agent concept was defined first from computer engineering. The computer science researchers have proposed several definitions of what an agent is [53–56]. Agents are typically defined to have the following traits:

- Autonomy (they operate without human intervention).
- Cooperation (they interact with other agents).
- Reactivity (they perceive and react to their environment).
- Pro-activeness (they have goal-oriented behavior). [56]: "a software (or hardware) entity that is situated in some environment and is able to autonomously react to changes in that environment". While similar to objects, agents are distinguished from existing software and hardware, thanks to their cooperation and pro-activeness properties.

3.6.1.2 Definition of a Multi-Agent System

If a problem is particularly large or complex, then the best way it can be handled is to use a number of agents that are specialized at solving a specific problem aspect [57, 58]. According to Wooldridge's definition, agents must have the ability to communicate with each other. A multi-agent system is a combination of several agents working in collaboration to achieve the overall assigned goal of the system.

3.6.2 Multi-Agent System in Power Systems

Power system control is currently performed by a central SCADA (Supervisory Control and Data Acquisition System) system in combination with smaller local SCADA systems. The control methods based on SCADA technology are no longer efficient as power systems are increasing in complexity, requiring large amounts of data transfers and computations. However, there is actually a growing trend toward the application of MAS for the control of power systems. The justification of the use of MAS is their ability to be flexible, extensible and fault tolerant [59]. Mc Arthur stated that the MAS systems have been used in two ways: first, as a method to build flexible and extensible power systems, second as a modeling method. Furthermore, four main applications of MAS have been identified [59]:

- Monitoring and diagnostics.
- Protection.
- Modeling and simulation.
- Distributed control.

3.6.3 Distributed Power Systems

Alternative energy production using distributed energy resources attracts growing interest due to their potential benefits to provide reliable, efficient, environmentally friendly and sustainable energy from renewable sources. However, as the degree of penetration increases, the interconnection of distributed energy resources with the main grid involves many problems. The main issues that can affect the quality of supply, include:

- Equipment and public safety.
- Stability.
- Synchronism.
- Reactive power compensation.
- Harmonic injection.
- Central control.
- Market organization.

To address the different issues, the research community has conducted many researchers. It is clear that the distributed nature of this energy production system requires a distributed and autonomous control system.

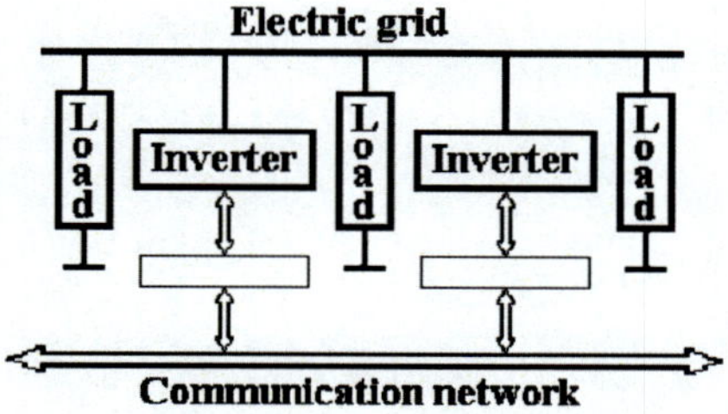

Fig. 3.26 Schematic diagram of distributed control for distributed energy systems

3.6.4 Control Systems for Inverters

Most distributed energy resources are not suitable for direct power transfer to the electrical grid due to the characteristics of the energy produced. Therefore, power electronic interfaces (inverter or converter) and their control systems are required for interfacing the distributed energy resources with the electric grid. The importance of the inverter is increased because its role has two important aspects. First, it extracts and manages the maximum power from the source. Second, it conditions the input power in order to deliver clean and compliant power to the grid.

3.6.5 Application

In this work, the MAS is intended to be used to build a flexible and extensible control system based on control units interconnected with inverters serving as power interfaces. Figure 3.26 illustrates the concept of the control method to be used in this work.

3.7 Conclusion

As power losses inside the electronic converters are dependent from the operation conditions, there is a necessity for evaluating these losses to have more information than for electronic converters classically used. Specific design is suitable in order to keep acceptable efficiency even during the frequent periods where the received power is weak, and more sophisticated modeling is required in order to evaluate the converters influence on solar systems performances. Solar systems design must also take into account the electric grid behavior, and this one is called to become more interactive with the growing of distributed energy resources.

Chapter 4
Optimized Use of PV Arrays

Symbols

a, b, c and d	Coefficients determined by the sampling values of the photovoltaic voltage V_{PV}
BN	Big negative
BP	Big positive
$\frac{dP_{pv}}{dV_{pv}}$	Derivative of PV power by the voltage
I_L	Inductance current
k_1	A constant of proportionality $(0.71 \langle k_1 \langle 0.78)$
k_2	A factor which depends on the current PV $(0.78 \langle k_2 \langle 0.92)$
K_G	A proportional gain
k_c	A positive scaling constant
MN	Means negative
MP	Means positive
R_{pv}	Equivalent load connect to the PV
SN	Small negative
SP	Small positive
V_{opt}	Optimal voltage which gives maximum power
Z	Zero
α	Duty cycle
α_{eq}	Equivalent duty cycle
ΔP_{pv}	Power variation between two operating points
ΔV_{pv}	Voltage variation between two operating points.

The source of photovoltaic electrical energy is the solar cell. Commercial solar cells reach maximum conversion efficiencies of 20–21%, while an efficiency of 25% may be achieved in laboratory [62]. The overall efficiency of a module ranges from 15 to 17% [62]. Under real operating conditions, a lower efficiency than the

D. Rekioua and E. Matagne, *Optimization of Photovoltaic Power Systems*, Green Energy and Technology, DOI: 10.1007/978-1-4471-2403-0_4,

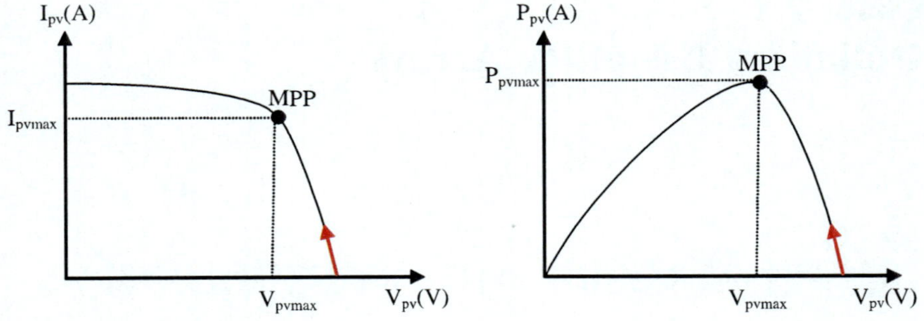

Fig. 4.1 $I_{pv}-V_{pv}$ and $P_{pv}-V_{pv}$ curves [9]

nominal efficiency could be observed [73]. PV arrays must be installed so that they maximize the amount of direct exposure to the sun. That usually means placement in an area clear of shading, in a southward direction and at an angle equal to the latitude of the location. The power provided by the PV array varies with solar irradiance and temperature, since these parameters influence the I–V characteristics of solar cells. In order to optimize the energy transfer from the PV array to the load, it is necessary to force the working point to be at the maximum power point (MPP) [31, 63].

4.1 Introduction to Optimization Algorithms

Photovoltaic energy has increased interest in electrical power applications, since it is considered as an essentially inexhaustible and broadly available energy resource. However, the output power induced in the photovoltaic modules depends on solar irradiance and temperature of the solar cells. Therefore, to maximize the efficiency of the renewable energy system, it is necessary to track the maximum power point of the PV array. The PV array has a unique operating point that can supply maximum power to the load. This point is called the maximum power point (MPP). The locus of this point has a nonlinear variation with solar irradiance and the cell temperature. Thus, in order to operate the PV array at its MPP, the PV system must contain a maximum power point tracking (MPPT) controller. Fig. 4.1

The maximum power point (MPP) is obtained when the derivative of PV power by the voltage (dP_{pv}/dV_{pv}) is zero. Basically, to achieve the maximum power point of operation, the generator voltage V_{pv} is regulated so that it increases when the slope dP_{pv}/dV_{pv} is positive and it decreases when the slope dP_{pv}/dV_{pv} is negative. A control which provides continuous extraction of maximum power point is given by:

$$V_{opt} = K_G \cdot \int \frac{dP_{pv}}{dV_{pv}} \quad dt \approx K_G \cdot \int \frac{\Delta P_{pv}}{\Delta V_{pv}} \quad dt \tag{4.1}$$

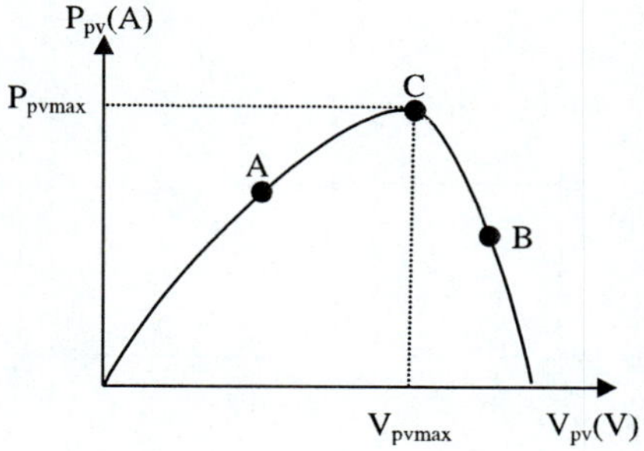

Fig. 4.2 Power–voltage characteristic of a photovoltaic cell

Table 4.1 Control action for different operating points of the power–voltage characteristic of a photovoltaic generator

Operating point	ΔV_{pv}	ΔP_{pv}	$\frac{\Delta P_{pv}}{\Delta V_{pv}}$	Control action
A	⟩0	⟩0	⟩0	Increase V_{pv}
	⟨0	⟨0	⟩0	Increase V_{pv}
B	⟩0	⟨0	⟨0	decrease V_{pv}
	⟨0	⟩0	⟨0	decrease V_{pv}
C	⟩0		0	
	⟨0	No change	0	No change

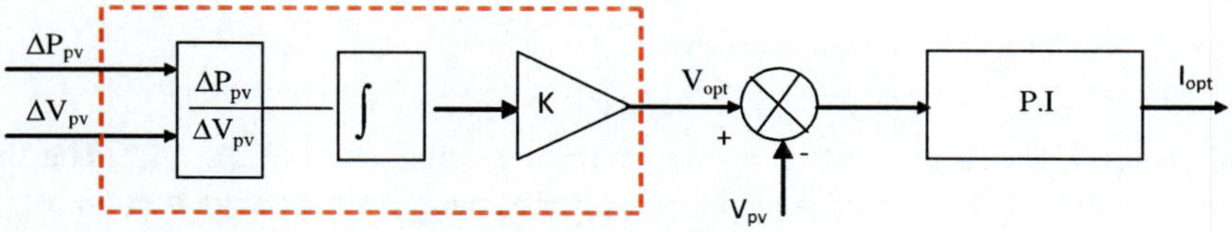

Fig. 4.3 MPPT control scheme and voltage regulation of V_{PV}

where $V_{opt}$is the optimal voltage which gives maximum power, K_G is a proportional gain, $\Delta P_{pv}$is power variation between two operating points and ΔV_{pv} is voltage variation between two operating points.

The following figure shows the curve power–voltage (Fig. 4.2).

Consider the system operating at A, B and C points in Fig. 4.2. Table 4.1 indicates the control signal which will result in each case.

The control block diagram is shown in Figs. 4.3, 4.4.

4.2 Maximum Power Point Tracker Algorithms

By definition [64], an maximum power point tracking (MPPT) control combined to a DC/DC converter allows a photovoltaic generator to produce the maximum continuous power, whatever the metrological conditions (irradiance, temperature).

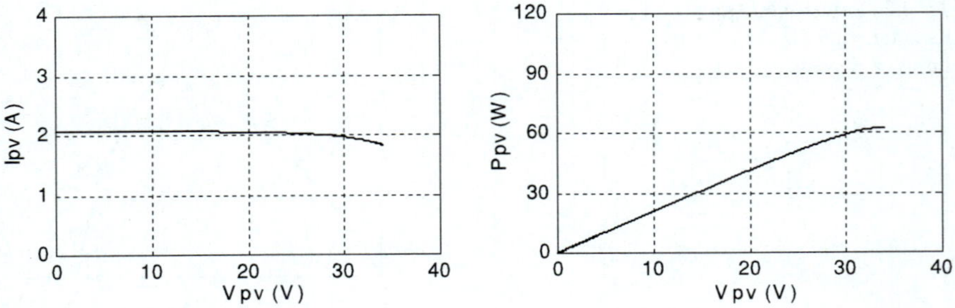

Fig. 4.4 $I_{pv}(V_{pv})$ and $P_{pv}(V_{pv})$ characteristics [9]

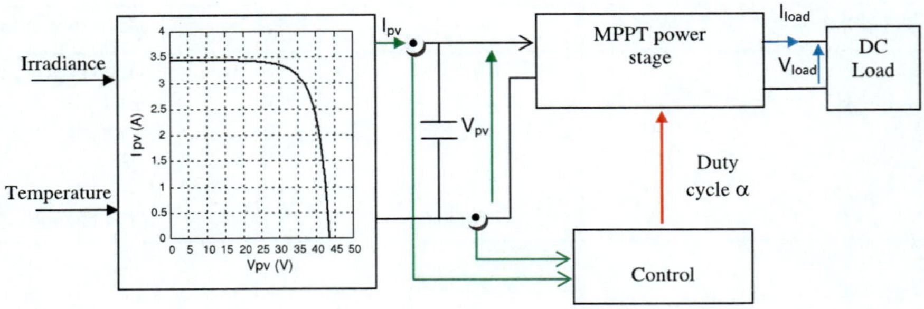

Fig. 4.5 MPPT control via input parameters

This control places the system at maximum operating point (V_{opt}, I_{opt}). The first system with MPPT was introduced in 1968 for a space system [65]. Over the years, several MPPT algorithms have been developed and widely adapted to determine the maximum power point [31, 66, 67]. The control technique the most used consist to act on the duty cycle automatically to place the generator at its optimal value whatever the variations of the metrological conditions or sudden changes in loads which can occur at any time. The main components of the MPPT circuit are its power stage and the controller. The two configurations of MPPT scheme are represented in Figs. 4.5 and 4.6.

In Fig. 4.5 the power stage input voltage V_{pv} and current I_{pv} are used by the controller for the purpose of MPP tracking. In this case, the power stage control parameter α is continuously turned until the PV array is loaded at its MPP.

Many methods have been developed to determine the maximum power point (MPP). In Ref. [68], to track MPP, a look-up table on a microcomputer is employed. It is based on the use of a database that includes parameters and data such as typical curves of the PV generator for different irradiances and temperatures. In Ref. [69], curve-fitting method is used, where the nonlinear characteristic of PV generator is modeled using mathematical equations or numerical approximations. These two algorithms have as disadvantage that they may require a large

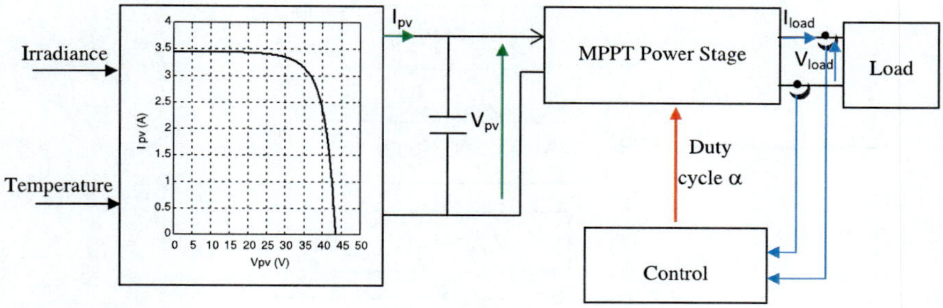

Fig. 4.6 MPPT control via output parameters

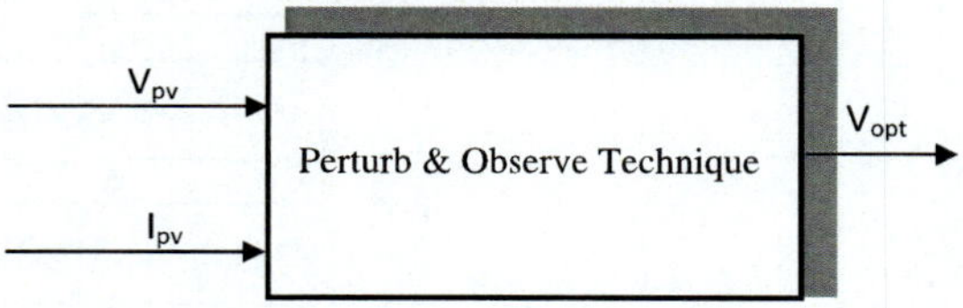

Fig. 4.7 Block diagram of Perturb and Observe method

memory capacity, for calculation of the mathematical formulations and for storage of the data. Open-circuit voltage photovoltaic generator method is employed in Ref. [70], it approximates linearly the voltage of PV generator at the MPP to its open-circuit voltage and a linear dependency between the current at the MPP and the short-circuit current for the short-circuit photovoltaic generator method presented in Ref. [71]. These methods are apparently simple and economical, but they are not able to adapt to changeable environmental conditions. Santos et al. present perturb and observe (P&O) method which is based on iterative algorithms to track continuously the MPP through the current and voltage measurement of the PV module. Most control schemes use the P&O technique because it is easy to implement [73–76] but the oscillation problem is unavoidable. Conductance incremental method presented in Ref. [76] requires complex control circuit. The two last strategies have some disadvantages such as high cost, difficulty, complexity and instability.

Intelligent based control schemes MPPT have been introduced (fuzzy logic, neural network) [76–84]. The fuzzy logic controllers (FLC) are used very successfully in the implementation for MPP searching. The fuzzy controller improves control robustness, it does not need exact mathematical models, it can handle nonlinearity and this control gives robust performance under parameters and load variation. The inputs of a MPPT fuzzy logic controller are usually an error E and an error variation ΔE [76–78]. This method is simple, permits to control the voltage of the PV generator whatever the variation of weather condition in order to obtain the maximum power.

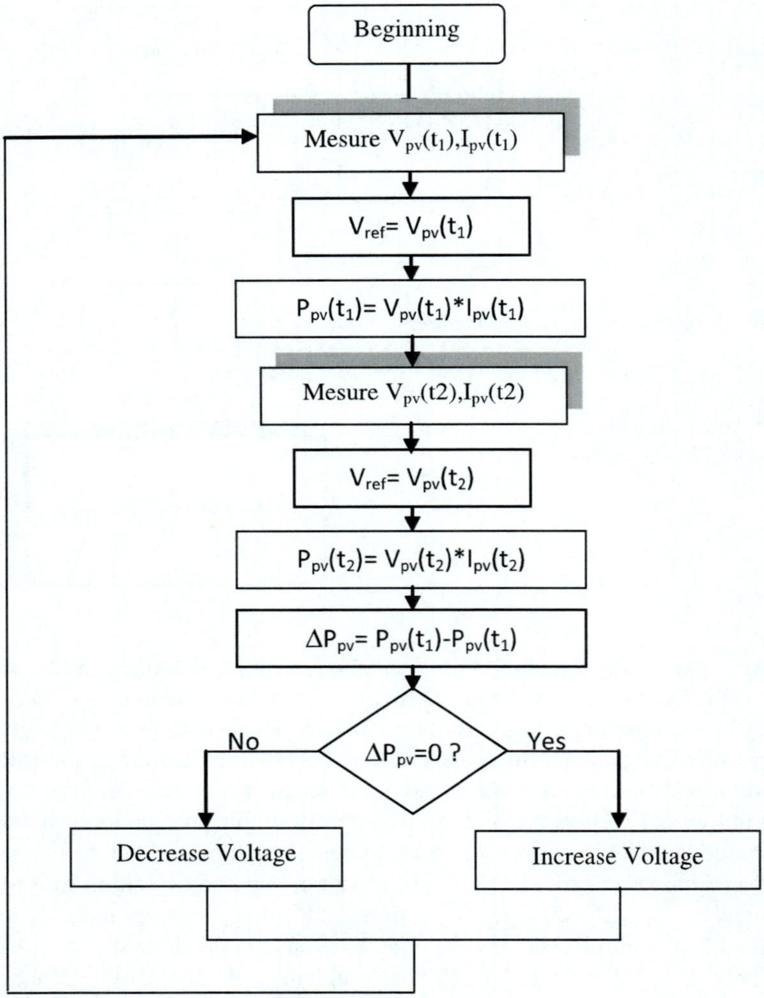

Fig. 4.8 Flowchart of Perturb and Observe method

4.2.1 Perturb and Observe Technique

This is the most widely used method [63, 66, 72, 75, 85]. A feedback loop and few measures are needed. The bloc diagram of the P&O method is given in Fig. 4.7. The panel voltage is deliberately perturbed (increased or decreased) then the power is compared to the power obtained before to disturbance. Specifically, if the power panel is increased due to the disturbance, the following disturbance will be made in the same direction. And if the power decreases, the new perturbation is made in the opposite direction. A flowchart of this method is shown in Fig. 4.8.

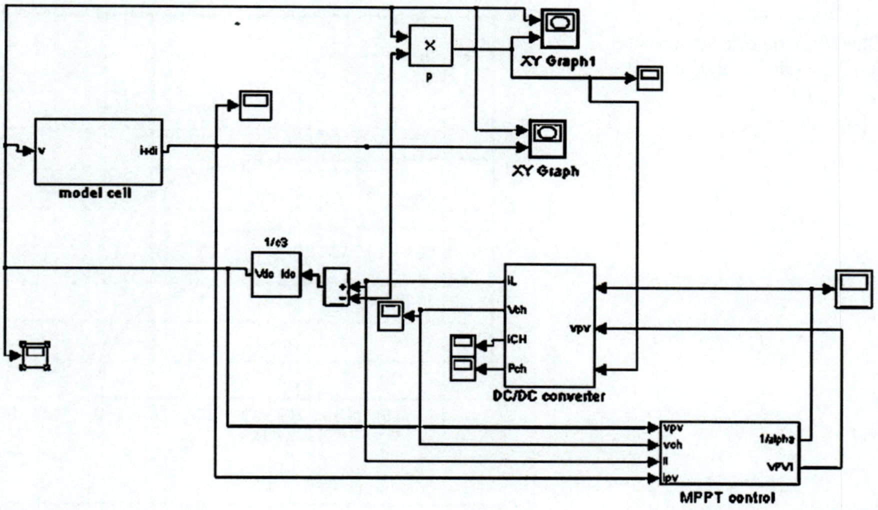

Fig. 4.9 P&O block diagram

The advantages of this method can be summarized as follows: knowledge of the characteristics of the photovoltaic generator is not required, it is relatively simple. Nevertheless, in steady state, the operating point oscillates around the MPP, which causes energy losses. The MPPT is necessary to draw the maximum amount of power from the PV module.

Application:

We make an application under Matlab/simulink (see Fig. 4.9). From the simulations results, it is clear that the system operates closer to a maximum power point for variations in irradiance and temperature (Figs. 4.10 and 4.11)

4.2.2 Modified P&O Method

To overcome the disadvantage of the P&O method in case of atmospheric changes, an improved version of P&O algorithm is proposed. The method is modeled by four input variables ($\Delta P_{pv}(k)$, $\Delta P_{pv}(k-1)$, $\Delta V_{pv-ref}(k)$and $\Delta V_{pv-ref}(k-1)$) and the direction of the disturbance of the next reference voltage $\Delta V_{pv-ref}(k-1)$ is as output variable Fig. 4.12.

We use a rule table (Table 4.2) of the sixteen possibilities. We note that when increasing the panel power on two disturbances in the same direction, there is a new condition which is introduced.

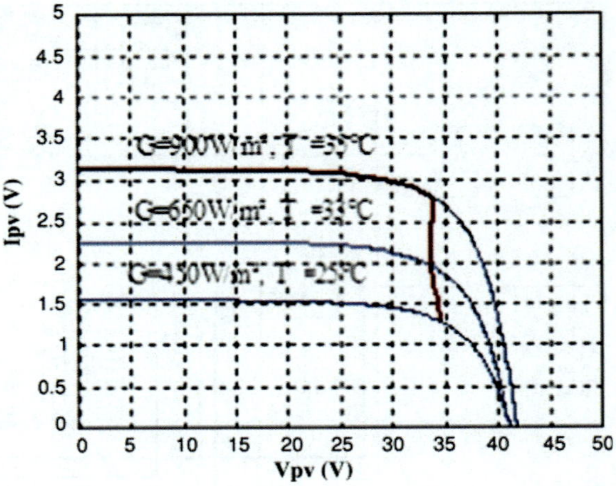

Fig. 4.10 Current–voltage characteristic of a PV module for different irradiances

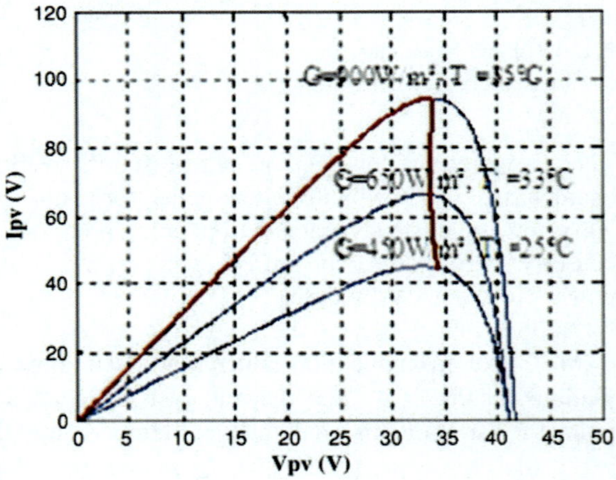

Fig. 4.11 Power-voltage characteristic of a PV module for different irradiances

4.2.3 Incremental Conductance Technique

This method focuses directly on power variations. The output current and voltage of the photovoltaic panel are used to calculate the conductance and the incremental conductance. Its principle is to compare the conductance ($GG = I_{pv}/V_{pv}$) and incremental conductance ($\Delta GG = dI_{pv}/dV_{pv}$) and to decide when to increase or to decrease the PV voltage to reach the MPP where the derivative of the power is equal to zero ($dP_{pv}/dV_{pv} = 0$). The incremental conductance method is often considered effective to search efficiently the maximum power point [66, 86]. However the algorithm to implement is often complex and requires a high calculation capacity, which increases the system control period. The output power of PV array can be given as:

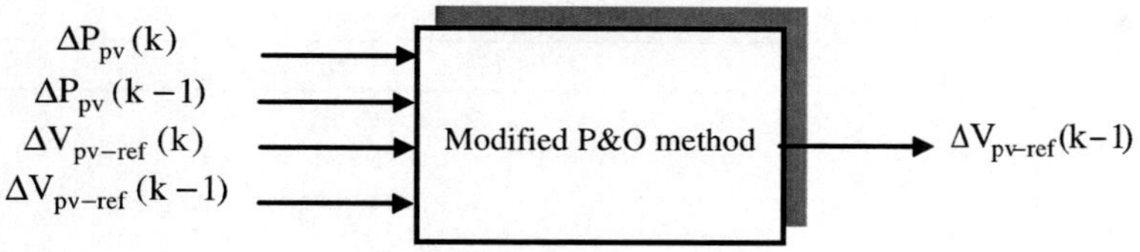

Fig. 4.12 Modified P&O modelisation [88]

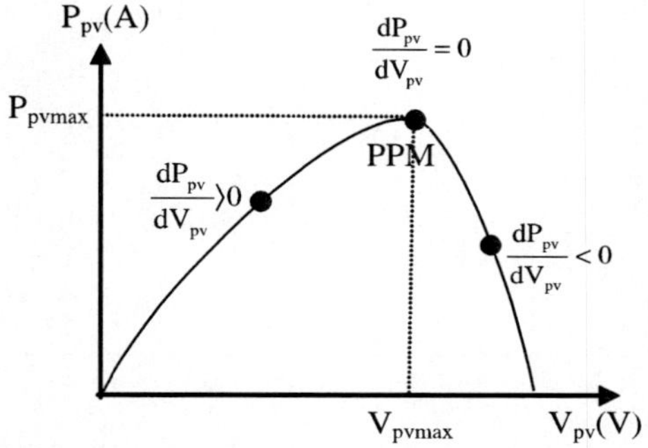

Fig. 4.13 Operating point according to the sign of $\frac{dP_{pv}}{dV_{pv}}$ on the power characteristic [9]

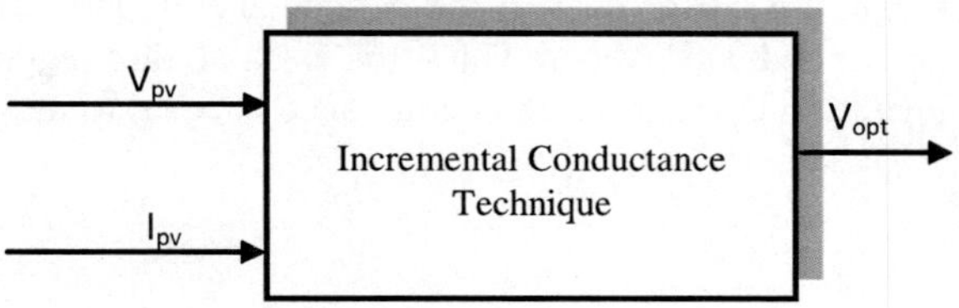

Fig. 4.14 Block diagram of incremental conductance method [9]

$$
\begin{aligned}
&P_{\mathrm{pv}} = V_{\mathrm{pv}} \cdot I_{\mathrm{pv}} \\
&\frac{dP_{\mathrm{pv}}}{dV_{\mathrm{pv}}} = \frac{d\left(V_{\mathrm{pv}} \cdot I_{\mathrm{pv}}\right)}{dV_{\mathrm{pv}}} = Ipv + V_{\mathrm{pv}} \cdot \frac{dI_{\mathrm{pv}}}{dV_{\mathrm{pv}}} \\
&\frac{1}{V_{\mathrm{pv}}} \cdot \frac{dP_{\mathrm{pv}}}{dV_{\mathrm{pv}}} = \frac{I_{\mathrm{pv}}}{V_{\mathrm{pv}}} + \frac{dI_{\mathrm{pv}}}{dV_{\mathrm{pv}}}
\end{aligned}
\tag{4.2}
$$

By defining PV conductance and incremental conductance, yields,

$$
\begin{cases}
GG = \dfrac{I_{\mathrm{pv}}}{V_{\mathrm{pv}}} \\
\Delta GG = -\dfrac{dI_{\mathrm{pv}}}{dV_{\mathrm{pv}}}
\end{cases}
\tag{4.3}
$$

We obtain:

$$
\frac{1}{V_{\mathrm{pv}}} \cdot \frac{dP_{\mathrm{pv}}}{dV_{\mathrm{pv}}} = GG - \Delta GG \tag{4.4}
$$

Table 4.2 Rule table of modified P&O

$\Delta V_{pv-ref}(k-1)$	$\Delta P_{pv}(k-1)$	$\Delta V_{pv}(k)$	$\Delta P_{pv}(k)$	System state	$\Delta V_{pv-ref}(k-1)$
⟨0	⟨0	⟨0	⟨0	Invalid	⟩0
⟨0	⟨0	⟨0	⟩	Invalid	⟩0
⟨0	⟨0	⟩	⟨0	Decreasing of G	⟨0
⟨0	⟨0	⟩0	⟩0	$V_{pv}\langle V_{mpp}$	⟩0
⟨0	⟩0	⟨0	⟨0	$V_{pv} \approx V_{mpp}$	⟩0
⟨0	⟩0	⟨0	⟩0	New condition	⟩0
⟨0	⟩0	⟩0	⟨0	$V_{pv}\rangle V_{mpp}$	⟨0
⟨0	⟩0	⟩0	⟩0	Increasing of G	⟨0
⟩0	⟨0	⟨0	⟨0	Decreasing of G	⟩0
⟩0	⟨0	⟨0	⟩0	$V_{pv}\rangle V_{mpp}$	⟨0
⟩0	⟨0	⟩0	⟨0	Invalid	⟨0
⟩0	⟨0	⟩0	⟩0	Invalid	⟨0
⟩0	⟩0	⟨0	⟨0	$V_{pv}\langle V_{mpp}$	⟩0
⟩0	⟩0	⟨0	⟩0	Increasing of G	⟩0
⟩0	⟩0	⟩0	⟨0	$V_{pv} \approx V_{mpp}$	⟨0
⟩0	⟩0	⟩0	⟩0	New condition	⟨0

Equation 4.4 explains that the operating voltage is below the voltage at the maximum power point if the conductance is larger than the incremental conductance, and vice versa. Thus, the task of this algorithm is to track the voltage operating point at which conductance is equal to incremental conductance. Hence,

$$\frac{dP_{pv}}{dV_{pv}} = 0 \quad \frac{I_{pv}}{V_{pv}} = -\frac{dI_{pv}}{dV_{pv}} \quad GG = \Delta GG \tag{4.5}$$

$$\frac{dP_{pv}}{dV_{pv}} > 0 \quad \frac{I_{pv}}{V_{pv}} > -\frac{dI_{pv}}{dV_{pv}} \quad GG > \Delta GG \tag{4.6}$$

$$\frac{dP_{pv}}{dV_{pv}} < 0 \quad \frac{I_{pv}}{V_{pv}} < -\frac{dI_{pv}}{dV_{pv}} \quad GG < \Delta GG \tag{4.7}$$

In turn, Eqs. 4.6 and 4.7 are used to determine the direction in which a perturbation must occur to shift the operating point toward the MPP, and the perturbation is repeated until Eq. 4.5 is satisfied. Once the MPP is reached, the MPPT continues to operate at this point until a change in current is measured which will correlate to a change in irradiance on the array Fig. 4.13.

The relationship between voltage and current, for the one-exponential model, is given by the following equation:

$$I_{pv} = I_{ph} - I_s\left(\exp\left(\frac{q(V_{pv} + R_s I_{pv})}{AKT_j}\right) - 1\right) - \frac{V_{pv} + R_s I_{pv}}{R_{sh}} \tag{4.8}$$

For an optimal Eq. 4.8, it follows that the derivative of current can be expressed by [87, 88]:

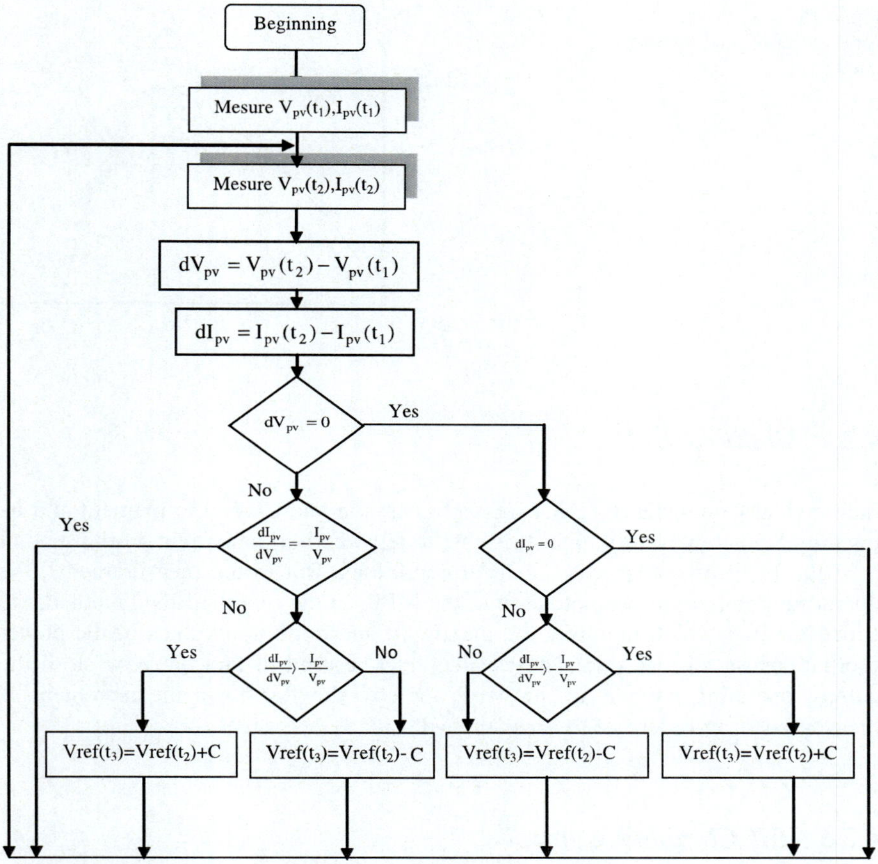

Fig. 4.15 Flowchart of the incremental conductance method

$$\frac{dI_{pv}}{dV_{pv}} = -\left(R_s + \left(\frac{qI_s}{AkT_j}\exp\left(\frac{q(V_{pv}+R_sI_{pv})}{AKT_j}\right) + \frac{1}{R_{sh}}\right)^{-1}\right)^{-1} \tag{4.9}$$

$$\frac{dI_{pv}}{dV_{pv}} = -\frac{1}{R_s + \left(\frac{1}{\frac{qI_s}{AkT_j}\exp\left(\frac{q(V_{pv}+R_sI_{pv})}{AKT_j}\right) + \frac{1}{R_{sh}}}\right)} \tag{4.10}$$

The block diagram of Incremental Conductance method is the same as that of Perturb & Observe method Fig. 4.14.

Figure 4.15 shows the flowchart of this method.

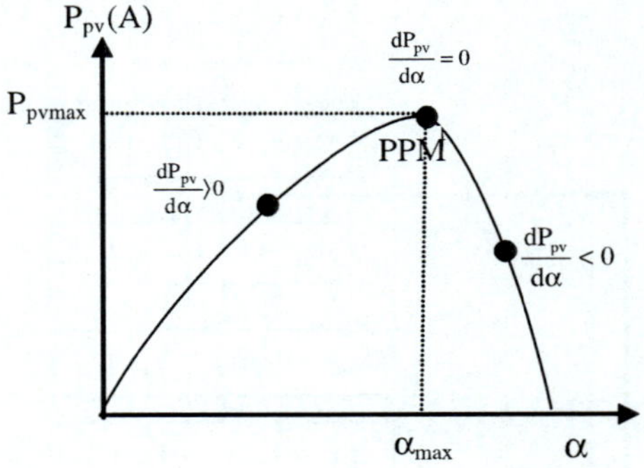

Fig. 4.16 Relationship between power and the duty ratio α [9]

4.2.4 Modified INC

Incremental Conductance (INC) method is simple and easy to implement and its tracking efficiency is very high. It is able to tell whether the current working point is at the MPP or not, in case of ideal conditions. But in the measurements, the operating point could oscillate around the MPP. In the modified INC method, we added the part which monitors the maximum and minimum values of the power oscillations on the PV side. These values can be used to find out how close the current operating point is to the MPP, thereby slowing down the increment of the reference, in order not to cross the MPP [113].

4.2.5 Hill Climbing Control

This method consists to climb the operating point along the generator characteristic to a maximum. It is based on the relationship between the power panel and the value of the duty ratio applied to the static converter. Mathematically the PPM is reached when $\frac{dP_{pv}}{d\alpha}$ is forced to zero by the control Fig. 4.16.

Periodically, the power $P_{pv}(t_2)$ is compared to the previous value $P_{pv}\ (t_1)$. Depending on the outcome of the comparison, we increase or decrease the duty cycle α. Once the PPM point reached, the system oscillates around it indefinitely. The advantage of this method is that it is simple to implement. But there are oscillations around the MPP in steady state and a loss of research of the MPP when climatic conditions change rapidly.

The flowchart of this algorithm is given in Fig. 4.17.

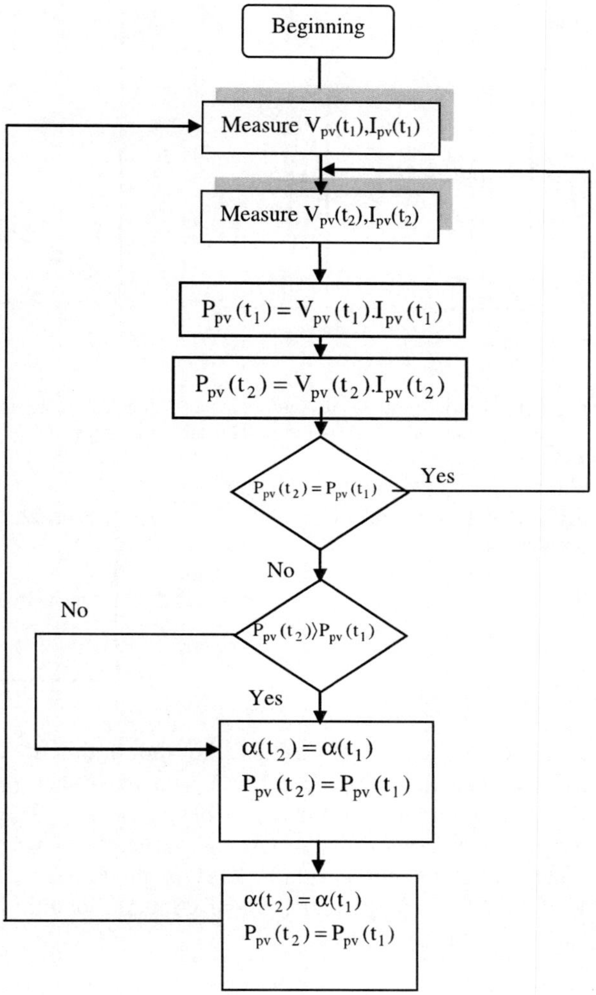

Fig. 4.17 Flowchart of the Hill climbing control

4.2.6 MPPT Controls Based on Relations of Proportionality

These methods are based on proportional relationships between parameters of optimal maximum power point (I_{opt}, V_{opt}) and the characteristic parameters of the panel (I_{sc}, V_{oc}).

4.2.6.1 Constant Reference Voltage Algorithm

The simplest technique to maintain the operation of the PV system near the maximum power point is to control the voltage measured at the PV generator, to its

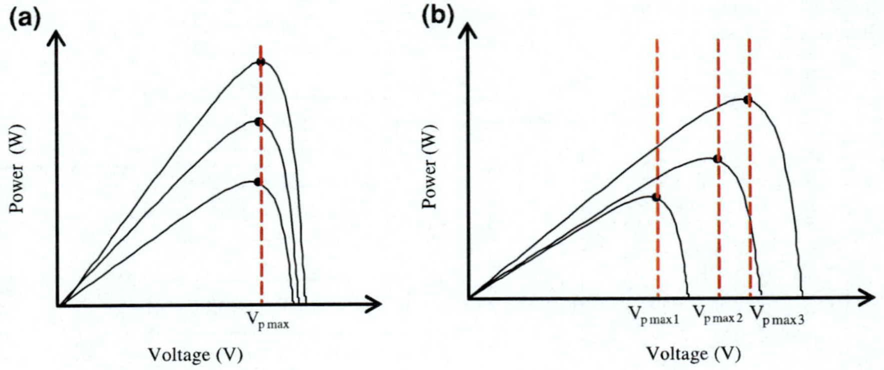

Fig. 4.18 Power–voltage characteristic $P_{pv}(V_{pv})$ of PV generator a-Variable irradiance, constant temperature b-Variable temperature, constant irradiance

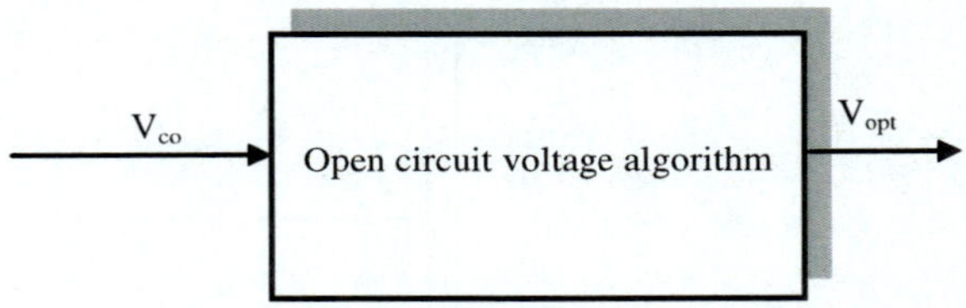

Fig. 4.19 Block diagram of open circuit voltage method [9]

reference voltage corresponding to the optimum voltage [89]. This method assumes that the variation of the optimum voltage to climatic factors variations (irradiance, temperature) is negligible as shown in Fig. 4.18a. However, when the junction temperature of the PV cell varies as shown in Fig. 4.18b, the optimum voltage will be not constant. This method uses a single control loop and is well suited for applications where climatic conditions are stable, such as space satellites [90].

4.2.6.2 Open-Circuit Voltage Photovoltaic Generator Method

It is a very simple method. It consists on comparing the panel voltage V_{pv} with a reference voltage corresponding to an optimal voltage V_{opt}. The voltage error is then used to adjust the duty ratio of the static converter. The reference voltage is obtained by the following equation:

$$V_{opt} = k_1.V_{co} \tag{4.11}$$

with k_1 as a constant of proportionality $(0.71\langle k_1 \langle 0.78)$.

Once k_1 is known, V_{opt} can be computed using Eq. 4.11.

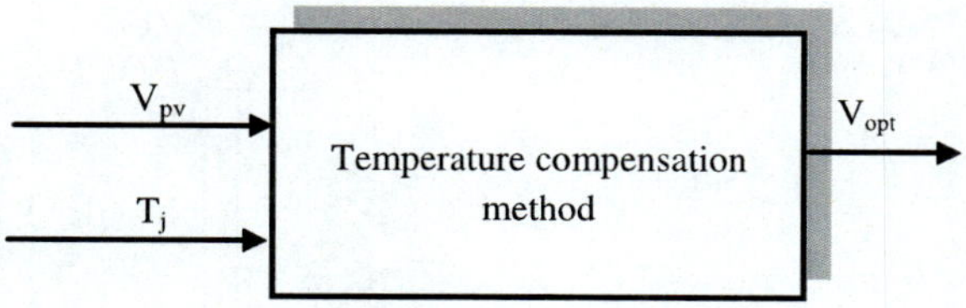

Fig. 4.20 Block diagram of Temperature compensation method [9]

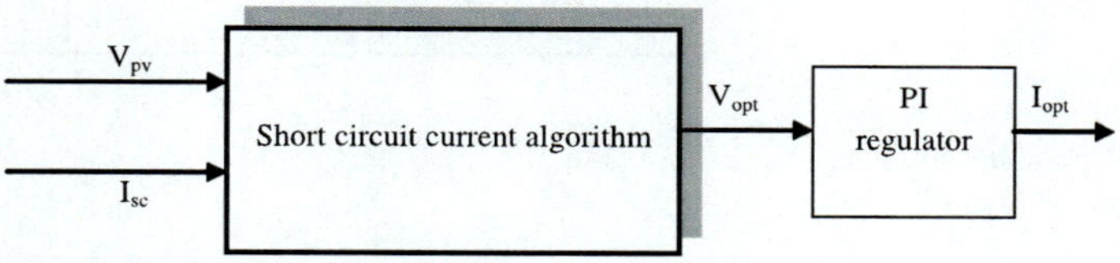

Fig. 4.21 Block diagram of Short circuit current method [9]

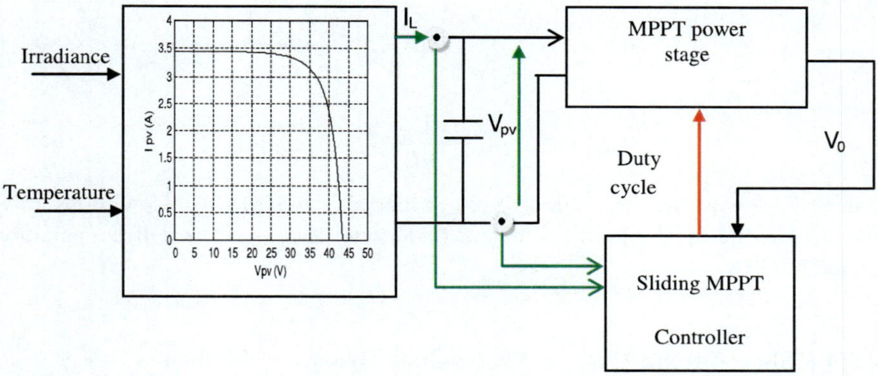

Fig. 4.22 Sliding maximum power point tracker for a photovoltaic system [9]

This requires that the system performs the measurement of the voltage V_{co} for each time period. Then we obtain directly the reference voltage which is the optimum voltage. Although this method is simple, it is difficult to choose an optimal value of k_1 [66, 91] Fig. 4.19.

4.2.6.3 Optimal Voltage with Temperature Compensation

We generate the reference voltage by adding a unique cell junction which is electrically independent of the PV system and having an electrical characteristic identical to the cells of the PV generator [68] Fig. 4.20.

The open-circuit voltage V_{oc} varies with the cell temperature T_j, and the short-circuit current I_{sc} is directly proportional to the irradiance level G. It can be described through the following equation:

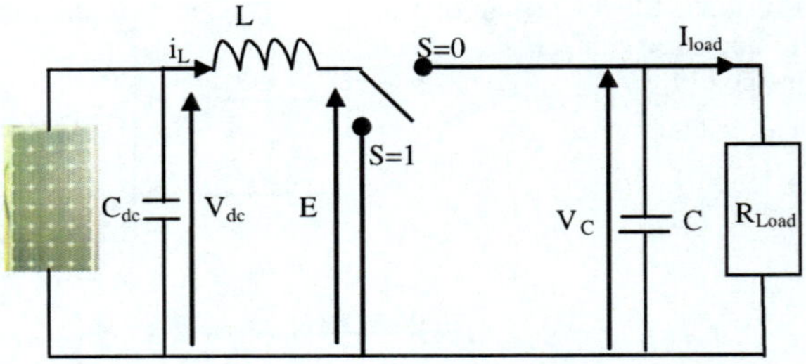

Fig. 4.23 DC–DC converter [97]

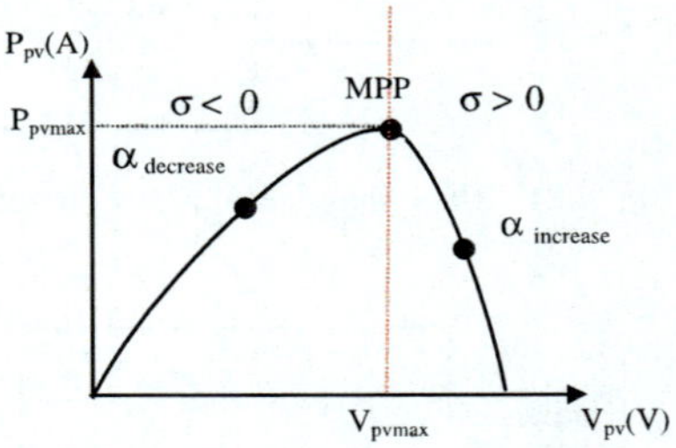

Fig. 4.24 Operating point according to the sign of σ [97]

$$V_{oc} = V_{oc-STC} + \frac{dV_{oc}}{dT}\left(T_j - T_{j-STC}\right) \tag{4.12}$$

where $V_{oc-STC}$is the open-circuit voltage under Standard Test Conditions (V), dV_{oc}/dT is the temperature coefficient (*V/K*), and T_{j-STC} is the cell temperature under STC(°K).

4.2.6.4 Short-Circuit Current Photovoltaic Generator Method

This method is based on knowledge of the linear relationship between current and optimal I_{opt} and the short-circuit current I_{sc}.

$$I_{opt} = k_2 \cdot I_{sc} \tag{4.13}$$

with k_2 a factor which depends on the current PV $(0.78\langle k_2\langle 0.92)$.

According to Eq. 4.12, the current I_{opt} can be determined by measuring the current I_{sc}. In this method the temperature does not affect the coefficient k_2, and we need only a single sensor. The implementation is easy but the accuracy of this control is weak due to the estimation of I_{opt} [71] Fig. 4.21.

4.2.7 Curve-Fitting Method

The nonlinear characteristic of the photovoltaic generator can be modeled using mathematical equations or numerical approximations [69], However, their resolution is very difficult by conventional digital controls. Therefore, their application does not seem appropriate for research of the MPP. Nevertheless, other approaches based on this model are used in [66], [96]. In the last reference, Eq. 4.14 represents the characteristic P_{pv} (V_{PV}) of a photovoltaic generator, where (a, b, c and d) are coefficients determined by the sampling values of the photovoltaic voltage V_{PV}, current i_{pv} and power P_{pv}.The optimum voltage which corresponds to the maximum power is obtained by Eq. 4.15.

$$P_{pv} = a \cdot V_{pv}^{3} + b \cdot V_{PV}^{2} + c_{P} \cdot V_{PV} + d \tag{4.14}$$

$$V_{opt} = \frac{-b + \sqrt{b^2 - 3\,a\,c}}{3\,a} \tag{4.15}$$

The disadvantage of this method is that the numerical approximations used are not valid during the meteorological conditions variations. It requires large memory storage for the calculation of mathematical formulas.

4.2.8 Look-Up Table Method

In this method, the measured values of voltage and photovoltaic current are compared to those stored in the control system, which corresponds to those in the optimal operating conditions under real conditions [68]. This algorithm requires a large memory capacity for data storage and parameters should be adjusted according to the solar panel used. In addition, it is difficult to record and store all possible conditions of system operation.

4.2.9 Sliding Mode Control

4.2.9.1 Control Design of Sliding Mode

The advantages of sliding mode control are various and important: high precision, good stability, simplicity, invariance, robustness etc... This allows it to be particularly suitable for systems with imprecise model. Often it is better to specify the system dynamics during the convergence mode. In this case, the controller structure has two parts, one represents the dynamics during the sliding mode and the other represents the discontinuous system dynamics during the convergence mode.

The design of the control can be obtained in three important steps and each step is dependent on another one:

- The choice of surface.

For a system defined by the following equation, the vector of surface has the same dimension as the control vector (u).

$$\dot{x} = A(x,t) \cdot x + B(x,t) \cdot u \tag{4.16}$$

We find in the literature different forms of the sliding surface and each surface has better performance for a given application. In general, we choose a nonlinear surface. The nonlinear form is a function of the error on the controlled variable (x), it is given by:

$$S(x) = \left(\frac{\partial}{\partial t} + \lambda_x\right)^{r-1} \cdot e(x) \tag{4.17}$$

where $e(x) = \hat{x} - x$ is the difference between the controlled variable x and its reference $\hat{x}$, λ_x is a positive constant, r is the number of times to derive the surface to obtain the control, x is the controlled variable.

The purpose of the control is to maintain the surface to zero. This one is a linear differential equation with a unique solution $e(x) = 0$ for a suitable choice of parameter λ_x with respect to the convergence condition.

- The establishment of the invariance conditions.

The conditions of invariance and convergence criteria have different dynamics that allow the system to converge to the sliding surface and stay there regardless of the disturbance: There are two considerations to ensure the convergence mode.

- The discrete function switching.

This is the first convergence condition. We have to give to the surface a dynamic converging to zero. It is given by:

$$\begin{cases} \dot{S}(x) \succ 0 & si\ S(x) \prec 0 \\ \dot{S}(x) \prec 0 & si\ S(x) \succ 0 \end{cases} \tag{4.18}$$

It can be written as:

$$\dot{S}(x) \cdot S(x) \prec 0 \tag{4.19}$$

- Lyapunov function:

The Lyapunov function is a positive scalar function for the state variables of the system. The idea is to choose a scalar function to ensure the attraction of the variable to be controlled to its reference value.

We define the Lyapunov function as follows:

$$V(x) = \frac{1}{2} \cdot S^2(x) \tag{4.20}$$

The derivative of this function is:

$$\dot{V}(x) = S(x) \cdot \dot{S}(x) \tag{4.21}$$

The function will decrease, if its derivative is negative. This is checked only if the condition (Eq. 4.19) is verified.

- Determination of the control law.

The structure of a sliding mode controller consists of two parts. The first one concerns the exact linearization (u_{eq}) and the second one concerns the stabilization (u_n).

$$u = u_{\mathrm{eq}} + u_n \tag{4.22}$$

where u_{eq} corresponds to the control. It serves to maintain the variable control on the sliding surface. u_n is the discrete control determined to check the convergence condition (Eq.4.19).

We consider a system defined in state space by Eq. 4.16, and we have to find analogical expression of the control (u).

$$\dot{S}(x) = \frac{\partial S}{\partial t} = \frac{\partial S}{\partial x} \cdot \frac{\partial x}{\partial t} \tag{4.23}$$

Substituting Eq. 4.16 and Eq. 4.22 in Eq. 4.23, we obtain:

$$\dot{S}(x) = \frac{\partial S}{\partial x} \cdot \left(A(x,t) + B(x,t) \cdot u_{eq}\right) + \frac{\partial S}{\partial x} \cdot B(x,t) \cdot u_n \tag{4.24}$$

We deduce the expression of the equivalent control:

$$u_{\mathrm{eq}} = -\left(\frac{\partial S}{\partial t} \cdot B(x,t)\right)^{-1} \cdot \frac{\partial S}{\partial t} \cdot A(x,t) \tag{4.25}$$

For the equivalent control can take a finite value, it must:

$$\frac{\partial S}{\partial x} \cdot B(x,t) \neq 0 \tag{4.26}$$

In the convergence mode and replacing the equivalent command by its expression in Eq. 4.24, we find the new expression of the surface derivative:

$$\dot{S}(x,t) = \frac{\partial S}{\partial x} \cdot B(x,t) \cdot u_n \tag{4.27}$$

And the condition expressed by Eq. 4.19 becomes:

$$S(x,t) \cdot \frac{\partial S}{\partial x} \cdot B(x,t) \cdot u_n \prec 0 \tag{4.28}$$

The simplest form that can take the discrete control is as follows:

$$u_n = k_s \cdot sign(S(x,t)) \tag{4.29}$$

where the sign of k_s must be different from that of $\partial S/\partial x \cdot B(x,t)$.

4.2.9.2 A Sliding Maximum Power Point Tracker for a Photovoltaic System

In sliding mode controller, the control circuit adjusts the duty cycle of the switch control waveform for maximum power point tracking as a function of the evolution of the power input at the DC/DC converter. In this control system, it is necessary to measure the PV array output power and to change the duty cycle of the DC/DC converter control signal [97–101] (Figs. 4.22, 4.23).

The system can be written in two sets of state equations depending on the position of switch S.

$$\frac{di_L}{dt} = \frac{1}{L}[V_{dc} - V_C(1-S)] \tag{4.30}$$

$$\frac{dV_C}{dt} = \frac{1}{C}[i_L(1-S) - i_{\text{load}}] \tag{4.31}$$

We introduce the concept of the approaching control [100]. We select the sliding surface as:

$$\frac{dP_{\text{pv}}}{dI_{\text{pv}}} = 0 \tag{4.32}$$

$$\frac{dP_{\text{pv}}}{dI_{\text{pv}}} = \frac{d(R_{\text{pv}}.I_{\text{pv}}^2)}{dI_{\text{pv}}} = 0 \tag{4.33}$$

$$\frac{dP_{\text{pv}}}{dI_{\text{pv}}} = \frac{d(R_{\text{pv}}.I_{\text{pv}}^2)}{dI_{\text{pv}}} = 2.I_{\text{pv}}.R_{\text{pv}} + I_{\text{pv}}^2\frac{dR_{\text{pv}}}{dI_{\text{pv}}} = 0 \tag{4.34}$$

$$\frac{dP_{\text{pv}}}{dI_{\text{pv}}} = I_{\text{pv}}(2.R_{\text{pv}} + I_{\text{pv}}\frac{dR_{\text{pv}}}{dI_{\text{pv}}}) = 0 \tag{4.35}$$

where $R_{\text{pv}} = \frac{V_{\text{pv}}}{I_{\text{pv}}}$ is the equivalent load connect to the PV,
with $I_{\text{pv}} = I_L$.

The non-trivial solution of Eq. 4.35 is:

$$2.R_{pv} + I_{pv}\frac{dR_{pv}}{dI_{pv}} = 0 \tag{4.36}$$

The sliding surface is given as:

$$\sigma \triangleq 2.R_{pv} + I_L\frac{dR_{pv}}{dI_L} \tag{4.37}$$

We have to observe the duty cycleα. Its control can be chosen as:

$$\alpha = \alpha + \Delta\alpha \text{ if } \sigma\rangle 0$$

$$\alpha = \alpha - \Delta\alpha \text{ if } \sigma\langle 0$$

Equations 4.30 and 4.31 can be replaced by an averaged set summing their expressions for S = 0, weighted by $1-\alpha$, and their expressions for S = 1, weighted by α. Then, the result can be written in general form of the nonlinear time invariant system.

$$\overset{\bullet}{\mathbf{X}} = f(\mathbf{X}) + g(\mathbf{X})\cdot\alpha \tag{4.38}$$

The equivalent control α_{eq} is determined from the following condition [100]:

$$\overset{\bullet}{\sigma} = \left[\frac{d\sigma}{d\mathbf{X}}\right]^T.\overset{\bullet}{X} = 0 \tag{4.39}$$

$$\overset{\bullet}{\sigma} = \left[\frac{d\sigma}{d\mathbf{X}}\right]^T\left(f(\mathbf{X}) + g(\mathbf{X}).\alpha_{eq}\right) = 0 \tag{4.40}$$

We obtain the equivalent control α_{eq}:

$$\alpha_{eq} = -\frac{\left[\frac{d\sigma}{d\mathbf{X}}\right]^T\cdot f(\mathbf{X})}{\left[\frac{d\sigma}{d\mathbf{X}}\right]^T\cdot g(\mathbf{X})} = 1 - \frac{V_{pv}}{V_C} \tag{4.41}$$

The equivalent duty cycle must lie in $0\langle\alpha_{eq}\langle 1$.

The real control signal α is proposed as [100]:

$$\alpha = 1 \quad \text{if } \alpha_{eq} + k_c\sigma \geq 1$$

$$\alpha = \alpha_{eq} + k_c\sigma \quad \text{if } 0\langle\alpha_{eq} + k_c\sigma\langle 1$$

$$\alpha = 0 \quad \text{if } \alpha_{eq} + k_c\sigma \leq 0$$

where the control saturate if $\alpha_{eq} + k_c\sigma$ is out of range, where k_c is a positive scaling constant Fig. 4.24.

4.2.10 Method of Parasitic Capacitance Model

The algorithm of the parasitic capacitance (Parasitic Capacitance MPPT) is similar to that of the incremental conductance (INC-MPPT) except that the effect of parasitic capacitance (C_P), which models the storage charges in the p-n solar cells junctions and stray capacitance, is included. A capacitance C_p is added in parallel on the terminals of the previous models. The current in that capacitance is $I_{pc}(t) = C_P.dV_{pv}/dt$, so that the observed current I_{obs} is expressed by [102–103]:

$$\begin{aligned} I_{obs} &= I_{pv} - I_{pc} \\ &= I_{ph} - I_S\left[\exp\left(\frac{(V_{pv} + R_s I_{pv})}{A\,V_{th}}\right) - 1\right] - \left(\frac{V_{pv} + R_s I_{pv}}{R_{Sh}}\right) - C_P\frac{dV_{pv}}{dt} \\ &= F(V_{pv}) - C_P\frac{dV_{pv}}{dt} \end{aligned} \tag{4.42}$$

Equation 4.42 shows the first component (I_{pv}) is function of the voltage $F(V)$ and the second one relates to the current in the parasitic capacitance. Using this notation, the incremental conductance of the solar panel (C_p not included) can be defined as the ratio $dF(V_{pv})/dV_{pv}$ and the instantaneous conductance can be defined as the ratio $F(V_{pv})/V_{pv}$. The MPP is obtained when $dP_{pv}/dV_{pv} = 0$. Multiplying Eq. 4.42 by the panel voltage (V_{pv}) for electric power, and then differentiating the result, the equation of electric power at MPP is obtained and can be expressed as [104]:

$$\frac{dF(V_{pv})}{dV_{pv}} + \frac{F(V_{pv})}{V_{pv}} = \frac{dI_{obs}}{dV_{pv}} + \frac{I_{obs}}{V_{pv}} + C_p\left(\frac{\dot{V}_{pv}}{V_{pv}} + \frac{\ddot{V}_{pv}}{\dot{V}_{pv}}\right) = 0 \tag{4.43}$$

The three terms of Eq. 4.43 represent the negative of the observed incremental conductance, the observed instantaneous conductance and the correction for parasitic capacitance. The first and second derivatives of the voltage of the panel take into account the ripple effect generated by power electronic converter. This ripple can then be used as voltage variation allowing MPPT. Note that if C_P is zero, Eq. 4.43 simplifies and becomes the one used for the incremental conductance algorithm.

4.2.11 Fuzzy Logic Technique

In recent years, fuzzy logic controllers (FLC) are widely used for finding the MPP [76, 77, 79, 80, 83, 85, 92, 93, 94]. The inputs of fuzzy controller are error and its variations; the output is the duty ratio of DC/DC converter or its variation. The fuzzy controller introduced in [77, 78, 92] uses dP_{pv}/dI_{pv} and its variation

$\Delta(dP_{pv}/dI_{pv})$ as inputs, calculates the duty ratio of the MPPT converter in the first reference and the variation of this one in the last two references. While the fuzzy controller in [80] considers the variation of duty cycle as output, it replaces dP_{pv}/dI_{pv} by photovoltaic panel power variations. The fuzzy controller developed in [77] is inadequate when operating conditions vary widely. The adaptive fuzzy controller and hybrid fuzzy controller [92–93] that exploits the characteristics of the fuzzy controller and the theory of neural networks are complex and require a large computing capacity.

The theory of fuzzy logic is used to address the problem of oscillation of the perturbation and observation method (P&O). The proposed fuzzy controller optimizes the amplitude of the disturbance to minimize the oscillations and for a quick response without oscillations. The inputs used to generate the optimal voltage which corresponds to the maximum power are the photovoltaic power variation (ΔP_{pv}) and the variation of the photovoltaic voltage (ΔV_{pv}). As output, the fuzzy controller determines the optimum increment to be added to the operating voltage to extract the maximum power MPP. In contrast to the Perturb and Observe (P & O) method which uses a constant disturbance to determine the operating voltage, it produces the oscillation of the operating point around the MPP in steady state.

4.2.11.1 The Fuzzy MPPT Controller

Fuzzy logic controller is introduced to determine the operating point corresponding to maximum power for irradiance and temperature levels. In this case, inputs of the fuzzy logic controller are power variation (ΔP_{pv}) and voltage variation (ΔV_{pv}). The output is reference voltage variation ($\Delta V_{pv,\ ref}$). In order to converge toward the optimal point, rules are relatively simple to establish. These rules depend on the variations of power ΔP_{pv} and voltage ΔV_{pv}. In accordance with Table 4.3, if the power (P_{pv}) increased, the operating point should be increased as well. However, if the power (P_{pv}) decreased, the voltage ($V_{pv,ref}$) should do the same.

From these linguistic rules, the MPPT algorithm contains measurement of variation of photovoltaic power ΔP_{pv} and variation of photovoltaic voltage ΔV_{pv} and proposes a variation of the voltage reference $\Delta V_{pv,\ ref}$ according to Eq. 4.44.

$$\begin{cases} \Delta P_{\mathrm{pv}} = P_{\mathrm{pv}}(k) - P_{\mathrm{pv}}(k-1) \\ _{\mathrm{pv}} = V_{\mathrm{pv}}(k) - V_{\mathrm{pv}}(k-1) \\ V_{\mathrm{pv-ref}}(k) = V_{\mathrm{pv}}(k-1) + \Delta V_{\mathrm{pv-ref}}(k) \end{cases} \tag{4.44}$$

where $P_{pv}(k)$ and $V_{pv}(k)$ are the power and voltage of the photovoltaic generator at sampled times (k), and $V_{pv,ref}$ (k) the instant of reference voltage. Figure. 4.25 gives an example of a track in the $P_{pv}(V_{pv})$ plan for a constant irradiance and temperature.

Table 4.3 Fuzzy rule table [63]

ΔP_{pv}							
ΔV_{pv}	BN	MN	SN	Z	SP	MP	BP
BN	BP	BP	MP	Z	MN	BN	BN
MN	BP	MP	SP	Z	SN	MN	BN
SN	MP	SP	SP	Z	SN	SN	MN
Z	BN	MN	SN	Z	SP	MP	BP
SP	MN	SN	SN	Z	SP	SP	MP
MP	BN	MN	SN	Z	SP	MP	BP
BP	BN	BN	MN	Z	MP	BP	BP

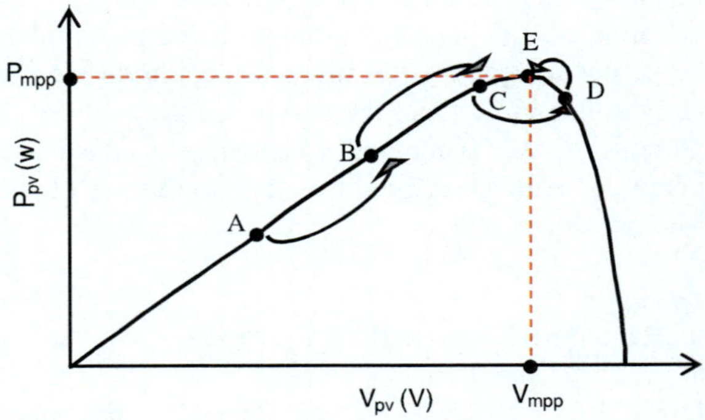

Fig. 4.25 Principle of operation of the MPPT fuzzy controller [9]

The power variation (ΔP_{pv}) is either in the positive direction or in the negative one. The value of ΔP_{pv} can also be small or on the contrary large. This control allows the research of the optimum point while being based on the expert observations. From these judgments, the reference photovoltaic voltage $V_{pv,ref}$ is increased or decreased in a small or large way in the direction which makes it possible to increase the power P_{pv}.

If a great increase in the voltage V_{pv} involves a great increase in the power P_{pv}, we continue to strongly increase the reference voltage $V_{pv,\ ref}$ (point A to point B or point B to point C). If a great increase in the voltage V_{pv} involves a reduction in the power P_{pv} (point C to point D), we decrease the reference voltage $V_{pv,\ ref}$ to obtain a fast increase in the power P_{pv}. If a reduction in the voltage V_{pv} involves a weak increase in the power P_{pv} then we get closer to the optimal reference voltage which is the beginning of stabilization.

When irradiance and temperature vary, the same types of rules are applied to track the maximum power point. The structure of fuzzy logic controller is shown in Fig. 4.26.

As explained previously, the MPPT technique takes measurement of PV voltage and current, and then uses FLC to calculate the reference voltage ($V_{pv,\ ref}$). Then, there is another control loop where proportional and integral (PI) controller

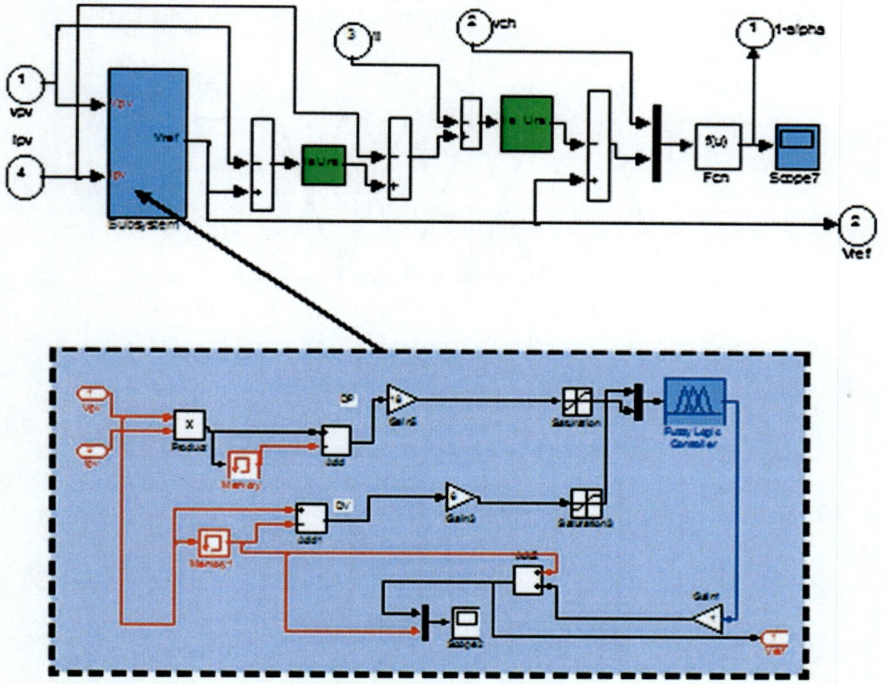

Fig. 4.26 Structure of MPPT fuzzy controller

regulates the input voltage of converter. Its task is to minimize error between $V_{pv,ref}$ and the measured voltage ($V_{pv,m}$). The PI loop operates with a much faster rate and provides fast response and overall system stability [87]. The fuzzy logic controller block includes three functional blocks: fuzzification, fuzzy rule algorithm and defuzzification. Fuzzy logic controller (FLC) inputs can be measured or computed from the voltage and current of solar panel. Figure 4.27 shows the membership functions of input and output variables in which membership functions of input variables ΔP_{pv} and ΔV_{pv} are triangular and has seven fuzzy subsets. Seven fuzzy subsets are considered for membership functions of the output variable $\Delta V_{pv,\ ref}$. These input and output variables are expressed in terms of linguistic variables (such as *BN* (big negative), *MN* (means negative), *SN* (small negative), *Z* (zero), *SP* (small positive), *MP* (means positive), and *BP* (big positive)).

The control rules are indicated in Table 4.3 with ΔP_{pv} and ΔV_{pv} as inputs and $\Delta V_{pv,\ ref}$ as the output. The fuzzy inference is carried out by using Sugeno's method [88], and the defuzzification uses the center of gravity to compute the output of this FLC which is the variation of reference voltage ($\Delta V_{pv,ref}$).

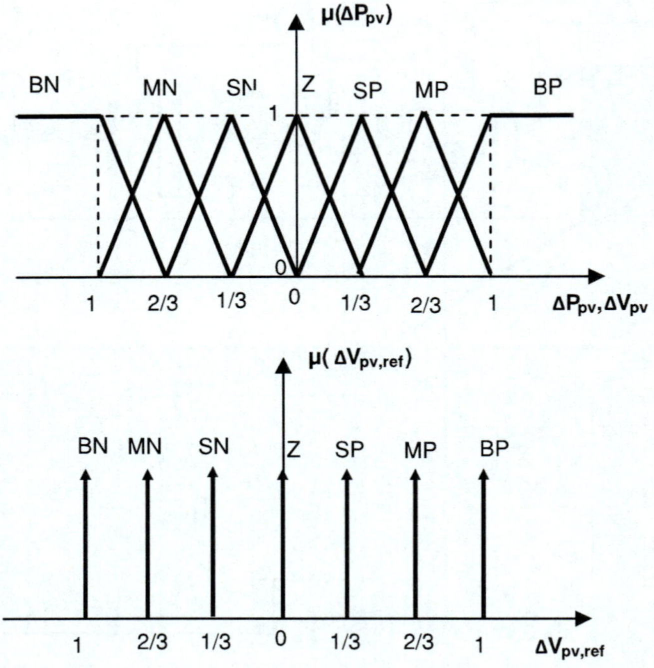

Fig. 4.27 Membership functions of ΔP_{pv}, ΔV_{pv} and of $\Delta V_{\mathrm{pv,\ ref}}$

4.2.11.2 Application Under Matlab/Simulink

We make application under Matlab/Simulink and we present simulation results under the following operating conditions:

- Irradiance level G is changed from 900 to 450 W/m^2 and temperature T_j is changed from 35 to 25°C.
- Temperature (T_j) is changed from 10 to 50°C for constant irradiance of 1000 W/m^2 Fig. 4.28.

The PV characteristics using MPPT control with FLC and the theoretical PV array characteristic are illustrated in Fig. 4.29 for change in irradiance level and temperature. In Fig. 4.29a, b we can see the power-voltage characteristic and power-current characteristic respectively. For $G = 900$ W/m^2 and $T_j = 35$°C the maximum power point corresponds to the values of voltage and current (35.95 V, 2.74 A). The decrease of the irradiance implies a decrease of the PV power, therefore, the maximum power point moves to join the maximum point corresponding to this irradiance, and we can have the corresponding voltage and current value in Fig. 4.29c. In Fig. 4.30, we present PV array characteristics for change in temperature (T_j) at constant irradiance. The open-circuit voltage (V_{oc}) decreases

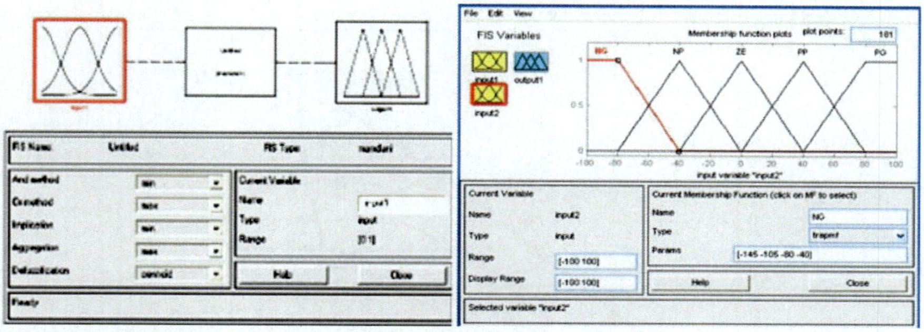

Fig. 4.28 Membership function plots in Matlab [9]

when temperature increases, the maximum power point changes according to the variation of temperature. It is clear that the operating point of this system operates closer to a maximum power point for variations in irradiance and temperature.

The simulation results show the robustness of the FLC for variation in environmental conditions, the PV system is always operating at the maximal power point.

4.2.12 Artificial Neural Networks

4.2.12.1 Definition of ANN

Artificial neural networks (ANN) are electronic models based on the neural structure of the brain. This function permits ANNs to be used in the design of adaptive and intelligent systems since they are able to solve problems from previous examples. ANN models involve the creation of massively paralleled networks composed mostly of nonlinear elements known as neurons. Each model involves the training of the paralleled networks to solve specific problem [105]. ANNs consist of neurons in layers, where the activations of the input layer are set by an external parameter. Generally, networks contain three layers—input, hidden and output. The input layer receives data usually from an external source while the output layer sends information to an external device. There may be one or more hidden layers between the input and output layers. The back-propagation method is the common type of learning algorithm [105].

4.2.12.2 MPPT Using Artificial Neural Networks

In photovoltaic systems, the input variables can be PV array parameters such as open-circuit voltage (V_{OC}) and short-circuit current (I_{SC}), atmospheric data-like irradiance (G) and temperature (T_j), or any combination of these. Ramaprabha R

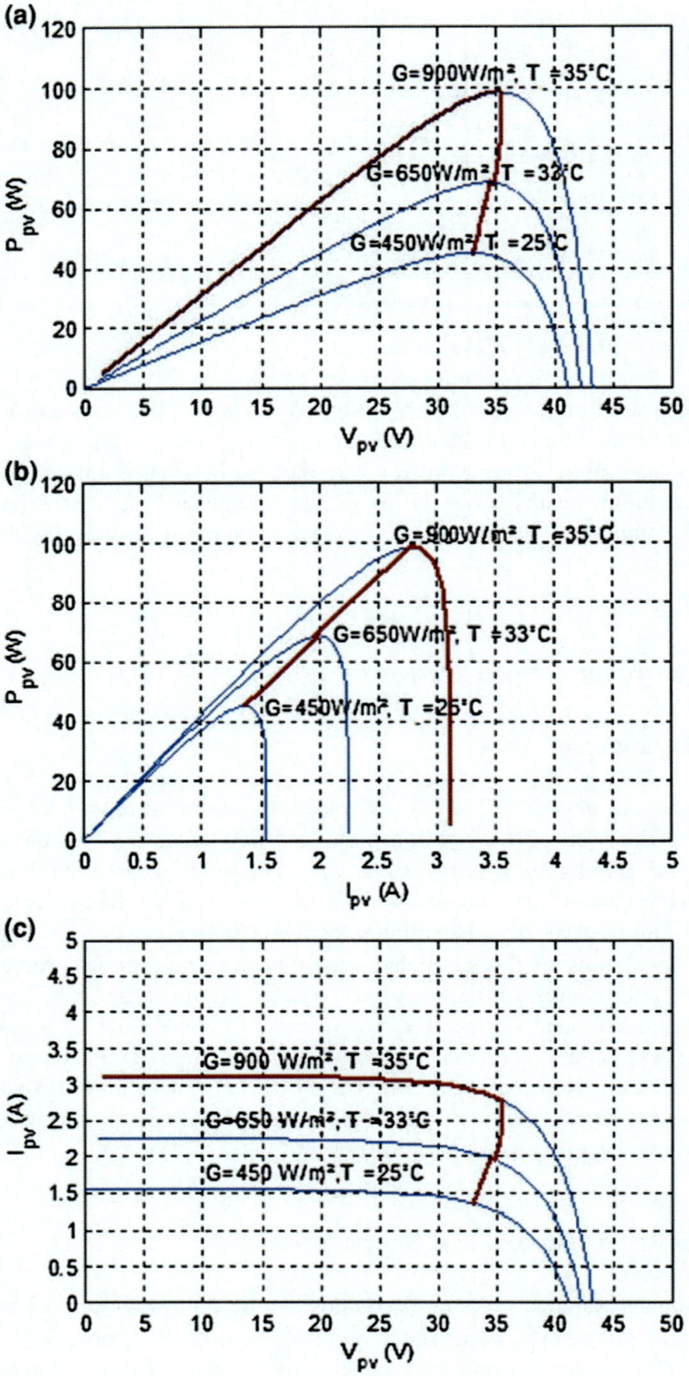
(a)
G=900W/m², T =35°C
G=650W/m², T =33°C
G=450W/m², T =25°C
P_pv (W)
V_pv (V)
(b)
G=900W/m², T =35°C
G=650W/m², T =33°C
G=450W/m², T =25°C
P_pv (W)
I_pv (A)
(c)
G=900 W/m², T =35°C
G=650 W/m², T =33°C
G=450 W/m², T =25°C
I_pv (A)
V_pv (V)

◄**Fig. 4.29** Simulation results of fuzzy logic controller versus theoretical PV array characteristic, irradiance level G is changed from 900 to 450 W/m^2 and temperature Tc is changed from 35 to 25°C. (a): $P_{pv}(V_{pv})$, (b):P_{pv} (I_{pv}), (c):I_{pv} (V_{pv}) a. Power-voltage characteristic of a PV module for different irradiance. b. Power-current characteristic of a PV module for different irradiance. c. Current–voltage characteristic of a PV module for different irradiance

ad Mathur. B. L propose an ANN algorithm with three layers (input layer, hidden layer and output layer). In this case, the block diagram producing maximum power and voltage is given in Fig. 4.31. We have two input neurons, hidden layers and two output neurons. The method used is the back-propagation algorithm (BPA).

In Ref. [107], authors present the ANN block diagram with three input neurons (irradiance, temperature and array voltage). The output neurons are maximum current, maximum voltage and current load. The hidden layer in PBA algorithm consists of four neurons (H_1, H_2, H_3 and H_4) Fig. 4.32.

The proposed architecture of the back-propagation algorithm (BPA) is given as Fig. 4.33.

4.2.13 Neuro-Fuzzy Method

4.2.13.1 Definition

The classic cycle of development of a neuro-fuzzy model can be separated into seven stages [110]:

1. data collection,
2. data analysis,
3. separate databases,
4. the choice of a neural network,
5. formatting data,
6. learning.

A typical architecture of a neuro-fuzzy network is shown in Fig. 4.34 Fig. 4.35. It has a single input variable and five membership functions MF [111].

4.2.13.2 MPPT Using Neuro-Fuzzy Network

The neuro-fuzzy controller has two inputs (e and Δe) and a single output (a), where: e represents the error and Δe the error variation. The two input variables generate action control and adjust the duty cycle α to be applied to the DC–DC converter so as to ensure the adaptation of the power supplied by the GPV. This controller allows automatic generation of fuzzy rules based on the Sugeno inference model [108–109]. The block diagram of a photovoltaic system with a MPPT

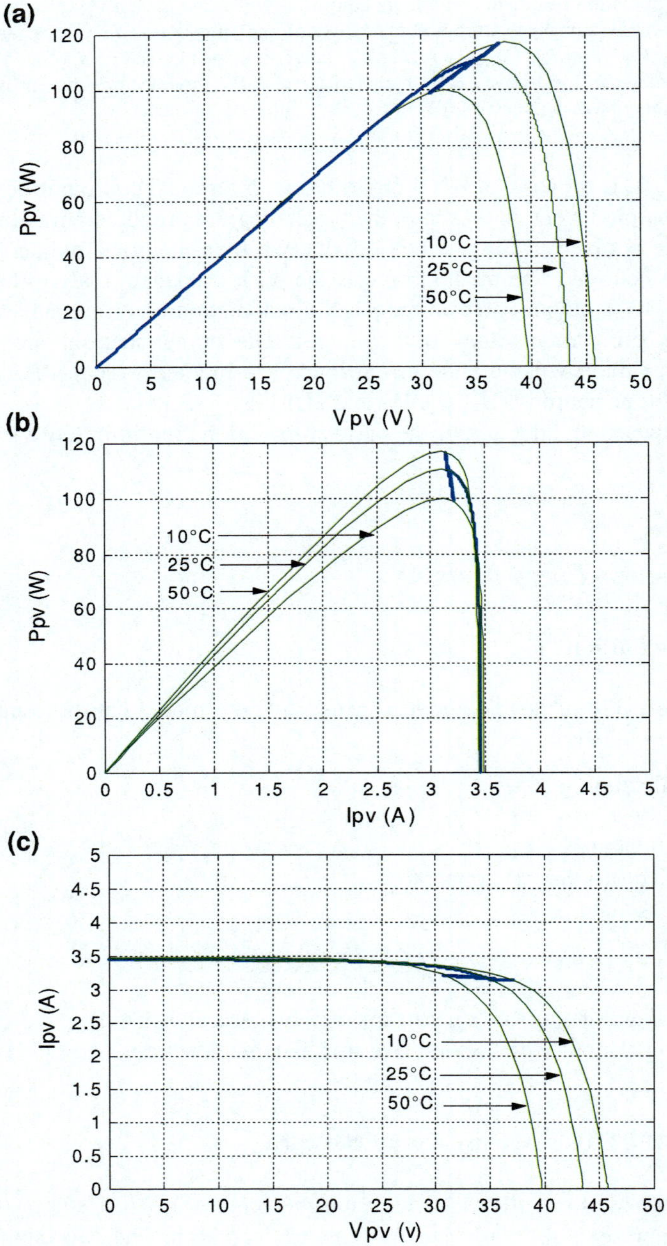
(a)
Ppv (W)
Vpv (V)
10°C
25°C
50°C
(b)
Ppv (W)
Ipv (A)
10°C
25°C
50°C
(c)
Ipv (A)
Vpv (v)
10°C
25°C
50°C

◄**Fig. 4.30** Simulation results of fuzzy logic controller versus theoretical PV array characteristic, temperature T_j is changed from 10 to 50°C at constant irradiance (1000 W/m^2). (a): $P_{pv}(V_{pv})$, (b):P_{pv} (I_{pv}), (c):I_{pv} (V_{pv}) a. Power-voltage characteristic of a PV module for different temperatures. b. Power-current characteristic of a PV module for different temperatures. c. Current-voltage characteristic of a PV module for different temperatures [31]

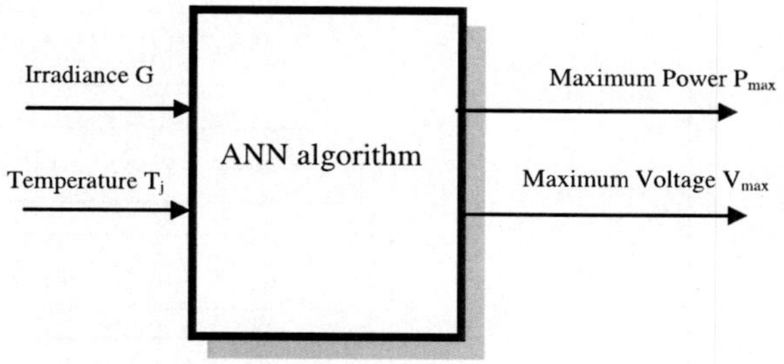

Fig. 4.31 ANN block diagram with two input neurons

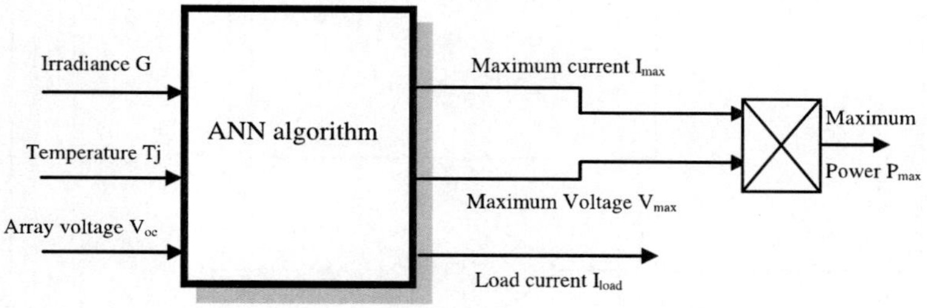

Fig. 4.32 ANN block diagram with three input layers [106]

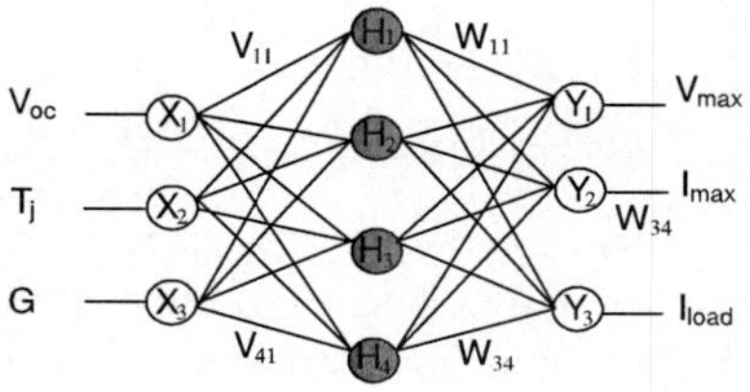

Fig. 4.33 Architecture of the PBA algorithm

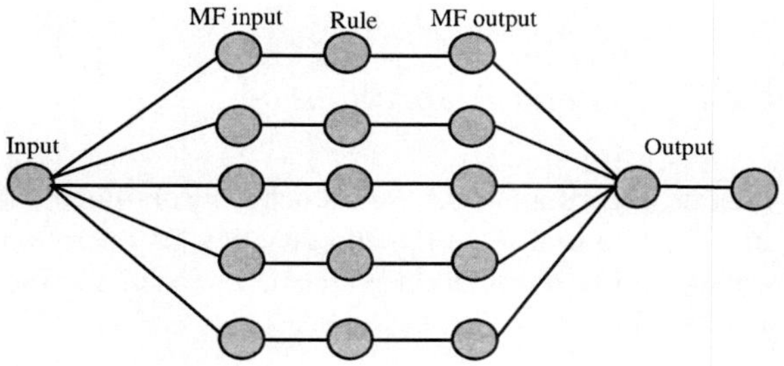

Fig. 4.34 Structure of Neuro-Fuzzy Model [111]

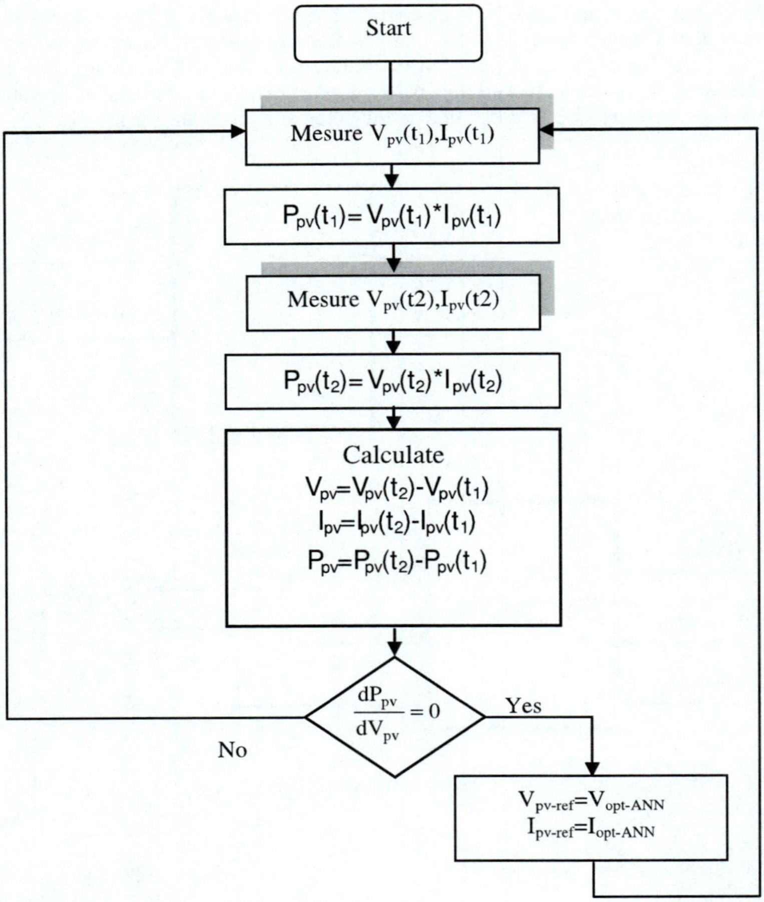

Fig. 4.35 Flowchart of Neuro-Fuzzy Model [111]

control network-based neuro-fuzzy is represented in Fig 4.36.The main advantage of this algorithm comparing to a conventional single ANN algorithm is the distinct generalization ability [110].

4.2.14 Genetic Algorithms

Genetic algorithms (GA) are stochastic optimization algorithms based on mechanisms of natural selection and genetics. Its operation is extremely simple. It starts with an initial population which is encoded for the model of problem by some methods Fig. 4.37.

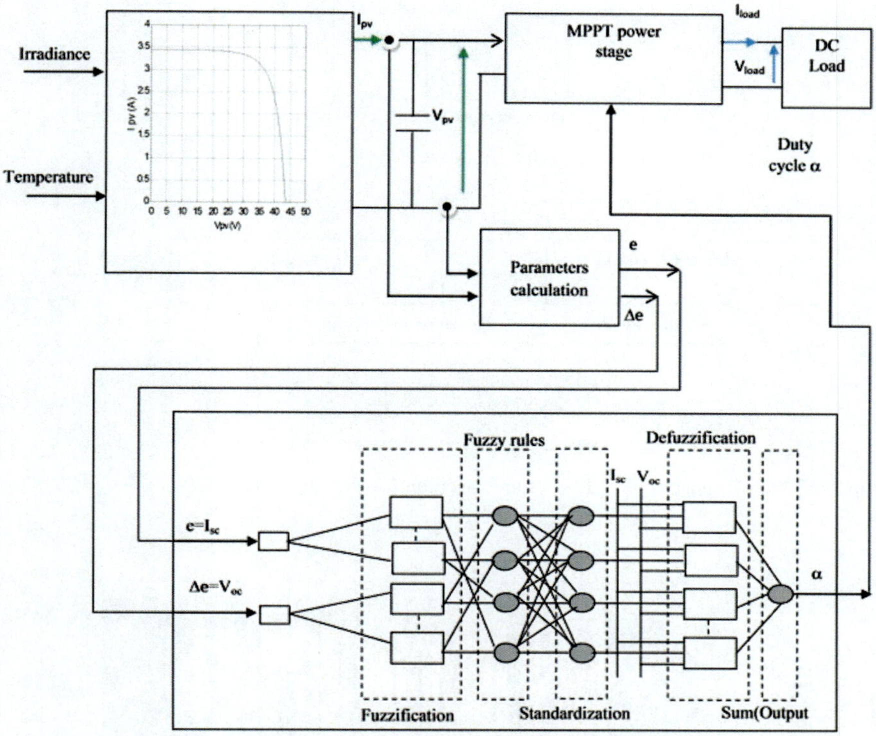

Fig. 4.36 Photovoltaic system with MPPT control by neuro-fuzzy networks

4.3 Efficiency of a MPPT Algorithm

Efficiency $\eta_{\mathrm{MPPT}}$is the most important parameter of an MPPT algorithm. This value is calculated as

$$\eta_{\mathrm{MPPT}} = \frac{\int_0^t P_{\mathrm{pv-max}}(t)dt}{\int_0^t P_{\mathrm{pv-MPPT}}(t)dt} \tag{4.45}$$

where $P_{\mathrm{PV-MPPT}}$ represents the output power of PV system with MPPT, and $P_{\mathrm{pv-max}}$ is the output power at true maximum power point.

4.4 Comparison of Different Algorithms

Hill climbing and P&O methods are different ways but based on the same fundamental method. Hill climbing involves a perturbation in the duty ratio α of the power converter while P&O makes a perturbation in the operating voltage of the

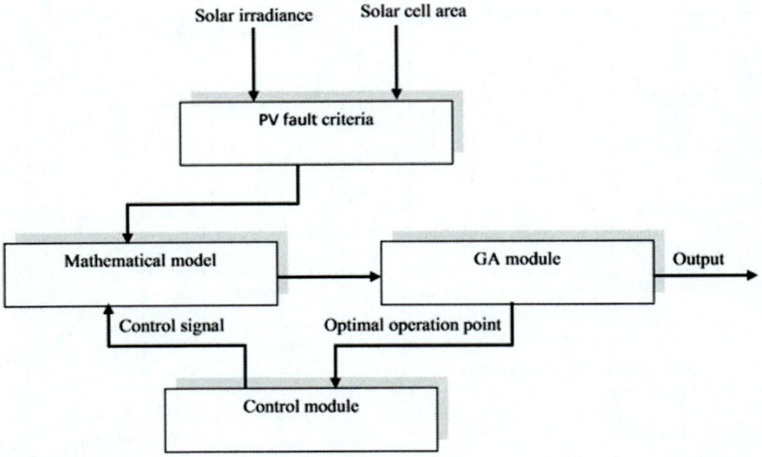

Fig. 4.37 Block diagram of a PV system based on Genetic algorithm [110]

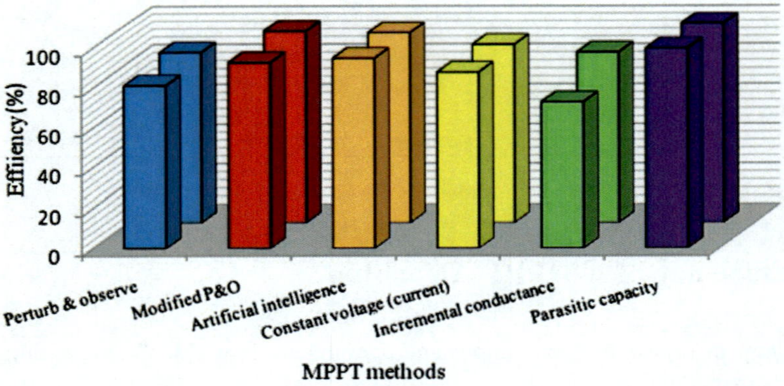

Fig. 4.38 Efficiency comparison of different MPPT methods

PV array. Discrete analog and digital circuitry can be used in the two methods but DSP or microcomputer control is more suitable [110], although the P&O algorithm is easy to implement.The INC method is applicable for DSP and microcontroller which can easily keep track of previous values of voltage and current. The main advantage of this algorithm (INC) comparing to the P&O method is its fast power tracking process. Open-circuit voltage (OCV) method is very easy and simple to implement as it does not necessarily require DSP or microcontroller control but in short-circuit current method we can use a simple current feedback control loop instead using DSP. The disadvantage of the open-circuit voltage or short-circuit current methods is that the online measurement in the two methods causes a reduction in output. The disadvantage of parasitic capacitance algorithm is that the parasitic capacitance is very small in each module. It will be interesting only in large PV arrays, where several module strings are connected in parallel. The disadvantage of intelligent techniques such as fuzzy controller, Neural networks, Neuro-Fuzzy and Genetic algorithms is the high cost of their implementation with complex algorithms that usually need a DSP. Others algorithms of MPPT are used by different authors such as the Delta-adaptive method, the Ripple correlation control (RCC), current sweep method, Load Current or Load Voltage Maximization and *dP*/*dV* or *dP*/*dI* Feedback Control [55]. We make a comparison between some important methods. We represent their minimum and maximum efficiency value in Fig. 4.38.

Chapter 5
Modeling of Storage Systems

Symbols

C	Capacity according to Peukert, at a one-ampere discharge rate (A·h)
C_0	Capacitance of the parallel plates
C_1 and C_2	Capacities of the battery at different discharge-rate states
C_{10}	10-h capacity
D_c and D_d	Diodes which are in series with R_c and R_d, respectively
E_b	Open-circuit voltage of the battery charged
E_m	Main branch electromotive force of the battery
E_t	Battery terminal voltage
I_{batt}	Discharge current
k_{E_b} and k_R	Coefficients that can be calculated experimentally
K_i	Polarization resistance
n	Peukert constant
R_0	Initial battery internal resistance calculated when the battery is full charged
R_{batt}	Internal (ohmic) resistance of the battery
R_c and R_d	Internal resistances associated respectively with the charging and discharging process of the battery
R_0	Nonlinear resistance contributed by the contact resistance of plate to electrolyte
Q	Accumulated ampere-hours divided by full battery capacity
SOC	State of charge
t_i	Time to discharge at current I
U_{batt}	Terminal voltage of the battery
V_{oc}	Open-circuit voltage of a battery cell when fully charged
Z_m	Internal impedance of the battery

Battery is generally needed when PV array cannot be functional as at night or on a cloudy day. The major functions of a storage battery in a PV system are:

D. Rekioua and E. Matagne, *Optimization of Photovoltaic Power Systems*, Green Energy and Technology, DOI: 10.1007/978-1-4471-2403-0_5,

- energy storage capacity and autonomy
- voltage and current stabilization
- supply surge currents

Indeed, battery is necessary to store electrical energy that is produced by the PV array as well as to supply energy to electrical loads. It can supply power to electrical loads at stable voltages and currents, and it can supply surge or high peak operating currents to electrical loads or appliances.

5.1 Description of Different Storage Systems

Cell is the basic electrochemical unit used to generate electrical energy from stored chemical energy or to store electrical energy in the form of chemical energy. It consists of a positive and a negative electrode in a container filled with an electrolyte which provides the essential ionic conductivity between the positive and negative electrodes of a cell. A separator which is an ion-permeable, electronically non-conductive material or spacer prevents short-circuiting of the positive and negative electrodes of the cell.

The desired battery is obtained when two or more cells are connected in an appropriate series/parallel arrangement to obtain the required operating voltage and capacity for a certain load (Fig. 5.1).

5.1.1 Battery Bank systems

In the market, there are many different types of batteries and most of them are subject to further research and development. In PV systems, several types of batteries can be used: nickel-cadmium (NiCd), Nickel-Zinc (Ni-Zn), lead acid etc. Nevertheless it must have some important properties as high charge-discharge efficiency, low self-discharge, and long life under cyclic charging and discharging.

5.1.1.1 Nickel-Cadmium (NiCd) Batteries

The NiCd batteries are commonly known as relatively cheap and robust. The positive nickel electrode is a nickel hydroxide/nickel oxyhydroxide ($Ni(OH)_2$/ NiOOH) compound, while the negative cadmium electrode consists of metallic cadmium (Cd) and cadmium hydroxide ($Cd(OH)_2$). The electrolyte is an aqueous solution of potassium hydroxide (KOH). It is based on the following redox reaction [114]

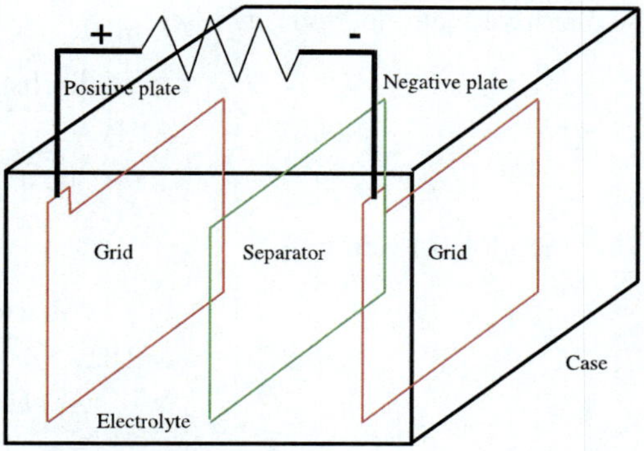

Fig. 5.1 Battery cell composition [10]

$$NiOOH + H_20 + e^- \underset{charge}{\overset{discharge}{\rightleftarrows}} Ni(OH)_2 + OH^- \quad (5.1)$$

$$(1/2)Cd + OH^- \underset{charge}{\overset{discharge}{\rightleftarrows}} (1/2)Cd(OH)_2 + e^- \quad (5.2)$$

The overall reaction is

$$NiOOH + (1/2)Cd + H_20 \underset{charge}{\overset{discharge}{\rightleftarrows}} Ni(OH)_2 + (1/2)Cd(OH)_2 \quad (5.3)$$

Compared to lead acid batteries which are probably the most commonly used batteries in photovoltaic applications, nickel-cadmium batteries can be, in some photovoltaic systems, cost-effective on a life cycle/cost profitable basis.

5.1.1.2 Nickel-Hydrogen Batteries

Nickel-Hydrogen battery has some advantages as long cycle, resistance to overcharge and good energy density, but it has high cost, high cell pressure and low volumetric energy density. It is generally used in space applications and communication satellites [2].

At the nickel electrode, we have:

$$NiOOH + H_2O + e^- \underset{charge}{\overset{discharge}{\rightleftarrows}} Ni(OH)_2 + OH^- \quad (5.4)$$

At the hydrogen electrode, we have:

$$OH^- + (1/2)H_2 \underset{\text{charge}}{\overset{\text{discharge}}{\rightleftarrows}} H_2O + e^- \tag{5.5}$$

And the global reaction is

$$NiOOH + (1/2)H_2 \underset{\text{charge}}{\overset{\text{discharge}}{\rightleftarrows}} Ni(OH)_2 \tag{5.6}$$

5.1.1.3 Nickel-Metal Hydride Batteries

Nickel-Metal Hydride batteries are used generally commercial consumer product. Its disadvantages are high self-discharge and failure leading to high pressure [115].

At the positive electrode, we have

$$NiOOH + H_2O + e^- \underset{\text{charge}}{\overset{\text{discharge}}{\rightleftarrows}} Ni(OH)_2 + OH^- \tag{5.7}$$

At the negative electrode, we have

$$M(H) + OH^- \underset{\text{charge}}{\overset{\text{discharge}}{\rightleftarrows}} M + H_2O + e^- \tag{5.8}$$

And the global reaction is

$$NiOOH + M(H) \underset{\text{charge}}{\overset{\text{discharge}}{\rightleftarrows}} Ni(OH)_2 + M \tag{5.9}$$

5.1.1.4 Nickel-Zinc batteries

The positive electrode is the nickel oxide but the negative electrode is composed of zinc metal. In addition to better environmental quality, this type of battery has a high energy density (25% higher than nickel-cadmium). Its operation is based on the following redox reaction [115]:

Fig. 5.2 Lead acid battery [31]

$$2NiOOH + Zn + 2H_2O \underset{\text{charge}}{\overset{\text{discharge}}{\rightleftarrows}} 2Ni(OH)_2 + Zn(OH)_2 \qquad (5.10)$$

5.1.1.5 Lead Acid Batteries

The lead acid batteries are the most used in PV applications especially in stand-alone power systems because its spill proof and the ease to transport [116]. However, when we use this battery type in PV systems, we can have excessive overcharges mode and loss of electrolyte. In this case, the charge controller must prevent overcharging. The lead acid battery consists of two electrodes immersed in sulfuric acid electrolyte. The negative one is attached to a grid with sponge metallic lead, and the positive one is attached to a porous grid with granules of metallic lead dioxide. During discharge, the lead dioxide on the positive electrode is reduced to lead oxide, which reacts with sulfuric acid to form lead sulfate, and the sponge lead on the negative electrode is oxidized to lead ions that react with sulfuric acid to form lead sulfate. In this manner electricity is generated and during charging this reaction is reversed. There are two types of lead acid batteries (Fig. 5.2) [116–117]:

- Flooded type: in the flooded type battery an aqueous sulfuric acid solution is used.
- Valve regulated lead acid batteries (VRLA): these batteries are closed with a pressure regulating valve to deal with the problem of overpressure risk when overcharging, so that they are sealed. In addition, the acid electrolyte is immobilized.

The active materials are:

1. In the positive plate: lead dioxide (PbO_2),
2. In the negative plate: sponge lead (Pb),
3. In water as the electrolyte: a solution of sulfuric acid (H_2SO_4).

The battery model has two modes of operation: charge and discharge. The battery is in charge mode when the current into the battery is positive, and discharge mode when the current is negative.

It is based on the following reversible chemical reaction:

$$\text{Anode}: \quad Pb + H_2SO_4 \underset{\text{charge}}{\overset{\text{discharge}}{\rightleftarrows}} PbSO_4 + 2H^+ + 2e^- \tag{5.11}$$

$$\text{Cathode} \quad PbO_2 + H_2SO_4 + 2e^- \underset{\text{charge}}{\overset{\text{discharge}}{\rightleftarrows}} PbSO_4 + 2OH^- \tag{5.12}$$

$$\text{Overallreaction} \quad Pb + PbO_2 + 2H_2SO_4 \underset{\text{charge}}{\overset{\text{discharge}}{\rightleftarrows}} 2PbSO_4 + 2H_2O \tag{5.13}$$

In this chemical reaction, we obtain the ideal proportions by weight of the reactants to provide capacity at lowest mass when the quantity of PbO_2, lead and sulfuric acid would be simultaneously reduced to zero. There must be an excess of negative active material (Pb) in the negative plate to extend the life cycle and wet life, and an excess of sulfuric acid in the electrolyte in most cells to maintain adequate acidity for long life and low internal resistance (Fig. 5.3).

5.1.1.6 Sodium-Sulfur (NaS) Batteries

In a sodium-sulfur battery, sodium and sulfur are in liquid form and are the electrodes, sodium being the cathode and sulfur being the anode [118]. They are separated by alumina which has the role of electrolyte. This one allows only the positive sodium ions to go through it and combine with the sulfur to form sodium polysulfide. This type of battery has a high energy density, high efficiency of charge/discharge (89–92%) long cycle life and it is fabricated from inexpensive materials (Fig. 5.4).

The cell reaction is written as:

$$2Na + x.S \underset{\text{charge}}{\overset{\text{discharge}}{\rightleftarrows}} Na_2S_x \tag{5.14}$$

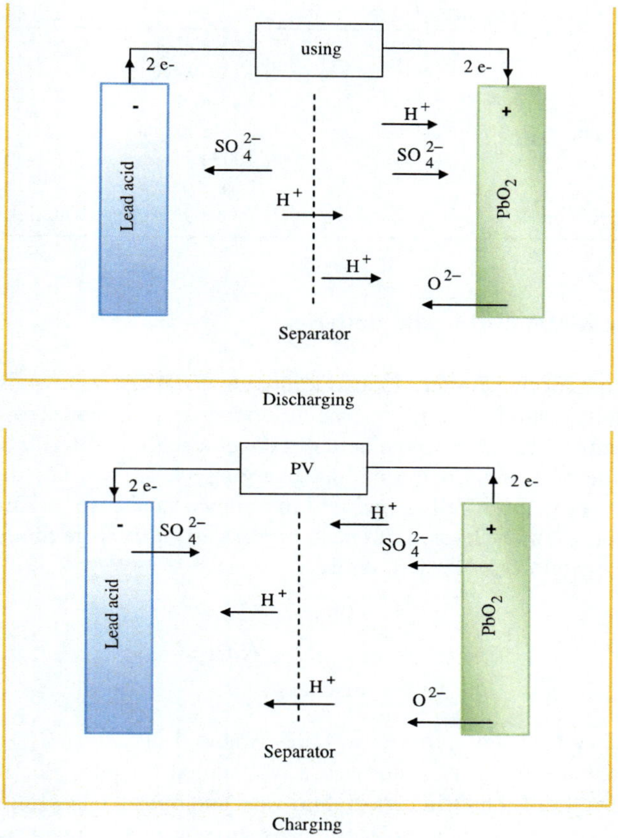

Fig. 5.3 Operation principle of a lead acid battery [117]

Fig. 5.4 Soduim sulfur cell [118]

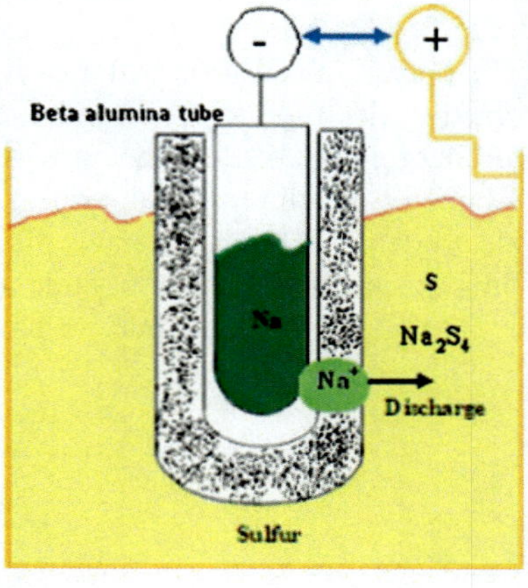

Table 5.1 Zebra battery properties [122]

Specific energy	120 Wh/kg
Specific power	>150 W/kg
Discharge rate	>45 min
Mean discharge rate	2 h or longer
Cycle life	1000 cycles
Cost materials	Low
Technology recommended	for large capacity batteries (>8–40 KWh)

5.1.1.7 Sodium-Metal Chloride Batteries

Sodium-Metal Chloride Battery is also known as ZEBRA (zero emission battery research activity) battery and it is a system operating at around 270–350°C [119, 120]. The chemical reaction in the battery converts sodium chloride and nickel to nickel chloride and sodium during the charging phase. During discharge, the reaction is reversed. Each cell is enclosed in a robust steel case. A ZEBRA battery is designed for a 2 h discharge with peak power capability as required [119–122]. The normal operation can be written as:

$$2NaCl + Ni \underset{discharge}{\overset{charge}{\rightleftarrows}} NiCl_2 + 2Na \tag{5.15}$$

Zebra battery has some proprieties [122] (Table 5.1):

For applications, ZEBRA batteries are well suited for pure electric cars, vans and buses as well as for hybrid cars, hybrid vans and hybrid buses, and can be used in stationary systems such as telecom back-up power and regenerative energy supply by photovoltaic or wind generators [122].

5.1.1.8 Lithium-ion (Li-ion) batteries

The operation of Li-ion batteries is based on the transfer of lithium ions from the positive electrode to the negative electrode while charging and vice versa while discharging [123, 124]. The positive electrode of a Li-ion battery consists of one of a number of lithium metal oxides, which can store lithium ions and the negative electrode of a Li-ion battery is a carbon electrode. The electrolyte is made up of lithium salts dissolved in organic carbonates.

At the positive electrode we have:

$$XLi^+ + Li_{1-X}CoO_2 + Xe^- \underset{charge}{\overset{discharge}{\rightleftarrows}} LiCoO_2 \tag{5.16}$$

At the negative electrode:

$$Li_XC \underset{\text{charge}}{\overset{\text{discharge}}{\rightleftarrows}} XLi^+ + C + Xe^- \tag{5.17}$$

5.1.1.9 Lithium-Ion Polymer Batteries

The major difference with Li-ion batteries is that the electrolyte consists of a solid ion-conducting polymer material [125]. The polymer electrolyte also serves as a separator. The advantage of this architecture is related to the absence of liquid in the battery which increases the energy density, the safety and life.

The overall reaction during charging and discharging of the considered Li-ion battery, in which Li^+ ions are involved, is given by [126]:

$$LiCoO_2 + C_6 \underset{\text{discharge}}{\overset{\text{charge}}{\rightleftarrows}} CoO_2 + LiC_6 \tag{5.18}$$

The Li^+ ions move from the LiCoO2 electrode to the graphite electrode during charging. They move in the other direction during discharging. Lithium-ion Polymer batteries are being developed as energy storage system for different applications (satellites, space shuttle…).

5.1.1.10 Li-Metal Batteries

The biggest advantage of storing lithium in its metallic form, instead of its ionic form surrounded by carbon atoms in the maximum ratio of 1:6, is a gain in energy density and specific energy [127–129]. However, the use of metallic lithium introduces the severe problem of its very high reactivity (Table 5.2).

5.1.1.11 Comparison of Different Batteries

We give some typical numbers of characteristics of different types of batteries (Table 5.3).

Sodium nickel chloride battery or ZEBRA battery has a relatively high energy density and no electric self-discharge but belongs to the high-temperature batteries class. It can provide new levels of performance, capability, safety and cost-effectiveness and it is ideal for surface ships, submersibles submarines applications, and electric vehicles [118, 130].

NiCd batteries are among the hardest batteries to charge. While with lithium ion and lead acid batteries you can control overcharge by just setting a maximum

Table 5.2 Comparison of different Li-ion batteries [114]

Cathode materials	Lithium cobalt oxide ($LiCoO_2$)	Lithium manganese oxide ($LiMn_2O_4$)	New material ($Li(NiCoMn)O_2$)	Lithium iron phosphate ($LiFePO_4$)
Working voltage (V)	3.7	3.8	3.6	3.2
Charge termination voltage (V)	4.25	4.35	4.3	4.2
Overcharge tolerance (V)	0.1	0.1	0.2	0.7
Cycle life (cycles)	400	300	400	1000
Energy density (Wh.kg)	180	100	170	130

charge voltage, the nickel-based batteries do not have a "float charge" voltage. So the charging is based on forcing current through the battery. We can also remark that NiCd batteries are lighter and as efficient and reliable as lead acid batteries.

NaS battery has fast response in charging and discharging performances, and then can be applicable for absorbing fluctuations in photovoltaic systems [6].

Lithium-ion (Li-ion) batteries have become very popular due to their high energy density, longevity and their good performance even when temperatures are low. Their disadvantage is the relatively expensive surveillance measures which have to control the cell in order to avoid explosion if they get too hot.

The best batteries used in PV systems are lead acid batteries, captive electrolyte lead acid (VRLA) and nickel-cadmium (NiCd). These batteries must have certain essential properties [131]:

- Low self-discharge
- High charge-discharge efficiency
- Long life under cyclic charging and discharging
- Ease of transport
- Low maintenance

5.1.2 Battery Bank Model

Different mathematical models exist to predict the performance of batteries. None of these models is completely accurate and the factors that affect battery performance are [132–133]:

- State of charge
- Battery storage capacity

Table 5.3 Comparison of different batteries [118]

Batteries	Lead acid	NiCd	NiZn	Li/ion	Li/polymer	Sodium nickel chloride ($Na\text{-}NiCl_2$)	NAS
Energy/weight (Wh/kg)	40–50	60–75	50–60	150–200	150–200	125	150–240
Cell voltage (V)	2	1.25	1.5	3.7	3.7	2.58	≈ 2
Voltage at the end of discharge slow/fast (V)	1.8–1.65	1.09/0.7	1.6	2.75/1.45	/	1.58	1.74–2.076
Cycle number	600–1500	800–2500	500–1000	500–1000	100–10.000	1000–2500	2500
Operating temperature (°C)	0–45	0–50	50	Ambient	Ambient	270–350	300
Charging time (h)	5–20	10–15	1.5	1–2	1–5	2	8
Efficiency (%)	65–85	70–80	65–80	85–95	90–95	90	89–92

Table 5.4 Peukert constant

AGM batteries	Gel batteries	Flooded batteries	Typical lead acid batteries	Lithium-ion batteries
1.05–1.15	1.1–1.25	1.2–1.6	1.35	1.1

- Rate of charge/discharge
- Environmental temperature
- Age/Shelf life

5.1.2.1 Electrochemical Battery Models

The simplest models are based solely on electrochemistry. These models can predict energy storage but they are neither able to model phenomena such as the time rate of change of voltage under load nor do they include temperature and age effects.

5.1.2.2 Peukert Equation

A cell is characterized by its capacity. It is an amount of electricity, expressed in Ah, and that it is able to return after a full charge, and discharged at a constant current. This capacity varies depending on several factors, such as the intensity of discharge, temperature and electrolyte concentration.

The Peukert equation (5.19) is an empirical formula which approximates how the available capacity of a battery changes according to the rate of discharge [129–134].

$$I_{\text{batt}}^{n}.t = C \tag{5.19}$$

where I_{batt} is the discharge current, n the Peukert constant, t the time to discharge at current I_{batt} and C is the capacity according to Peukert, at a one-ampere discharge rate, expressed in A·h.

The Peukert constant increases with age for any of the battery types (Table 5.4).

Equation 5.19 shows that at higher currents, there is less available capacity in the battery. The Peukert constant is directly related to the internal resistance of the battery, and indicates how well a battery performs under continuous heavy currents.

We can relate the capacity at one discharge rate to another combination of capacity and discharge rate. Then we obtain:

$$C_1 = C_2 . \left(\frac{I_{\text{batt2}}}{I_{\text{batt1}}}\right)^{n-1} \tag{5.20}$$

where C_1 and C_2 are capacities of the battery at different discharge-rate states.

The state of charge (SOC) at a constant discharge rate can be obtained by the following equation:

$$\mathrm{SOC}(t) = 1 - \left(\frac{I_{\mathrm{batt}}}{C}\right) \cdot t \tag{5.21}$$

The current is continuously variable over time. We consider the constant current between two calculation steps. Then we can determine the expression of the change in charge state of the cell at time t_k:

$$\Delta\mathrm{SOC}(t_k) = \frac{I_{\mathrm{batt}\ k}}{C_1} \cdot \left(\frac{I_{\mathrm{batt}\ k}}{I_{\mathrm{batt}\ 1}}\right)^{n-1} \cdot \Delta t \tag{5.22}$$

This approach also takes into account the phases of recharging the battery. Indeed, if the current in the cell becomes negative, its state of charge increases. Ultimately, cell state of charge is expressed by:

$$\mathrm{SOC}(t_k) = \mathrm{SOC}(t_{k-1}) + \Delta\mathrm{SOC}(t_k) \tag{5.23}$$

5.1.2.3 Shepherd Model Equation

The model describes the electrochemical behavior of the battery directly in terms of voltage and current. It is often used in conjunction with the Peukert equation to obtain battery voltage and state of charge given power draw variations [135]:

$$E_t = E_c - K_i Q - R_{\mathrm{batt}} I_{\mathrm{batt}} \tag{5.24}$$

where E_t is the battery terminal voltage, E_c a constant potential, K_i a polarization coefficient and R_{batt} the internal (ohmic) resistance of the battery. We have

$$R_{\mathrm{batt}} = R' + K_R \frac{1}{1-Q} \tag{5.25}$$

with K_R the electrolyte resistance at full charge and Q the accumulated ampere-hours divided by full battery capacity. The residual resistance R' is thus given by

$$R' = R_0 + K_R$$

with R_0 the initial battery resistance at full charge ($Q = 0$).

The open-circuit voltage or no-load battery terminal voltage for this model is simply:

$$E_{\mathrm{oc}} = E_0 - K_i Q \tag{5.26}$$

The determination of the Shepherd model parameters is based on constant current discharges at low current levels.

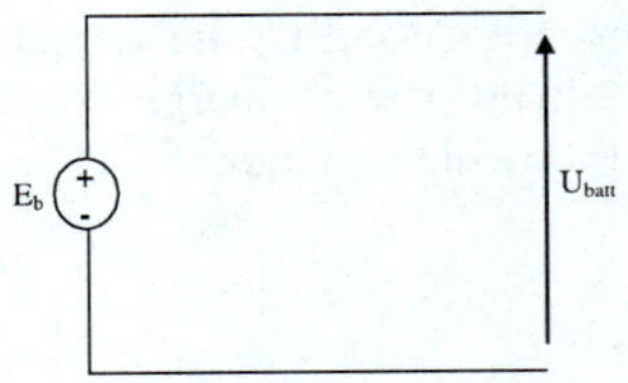

Fig. 5.5 Ideal model of battery [127]

5.1.2.4 Unnewehr Universal Model

Unnewehr and Nasar [136] suggest simplifying the Shepherd equations by replacing (5.25) with

$$R_{\text{batt}} = R_0 - K_R Q \tag{5.27}$$

where R_0 is the total internal resistance of a fully charged battery.

With all models which consider a linear variation of the voltage E_t versus the current, as are Shepherd and Unnewehr models, we can calculate the current for a given power P during discharge as:

$$I_{\text{batt}} = \frac{E_{\text{oc}} - \sqrt{E_{\text{oc}}^2 - 4 \cdot R_{\text{batt}} \cdot P}}{2 \cdot R_{\text{batt}}} \tag{5.28}$$

and during charge as:

$$I_{\text{batt}} = \frac{-E_{\text{oc}} + \sqrt{E_{\text{oc}}^2 + 4R_{\text{batt}}P}}{2R_{\text{batt}}}$$

During discharge, the maximum power $P_{\max}$ can be calculated as:

$$P_{\max} = \frac{E_{\text{oc}}^2}{4 \cdot R_{\text{batt}}} \tag{5.29}$$

5.1.3 Equivalent Circuit Battery Models

These models are modeling the batteries in the shape of electronic circuits. There are many models proposed by different scientists.

5.1.3.1 Ideal Model of Battery

In an ideal model, the battery is represented by a single voltage source and the other internal parameters are ignored [127] (Fig. 5.5).

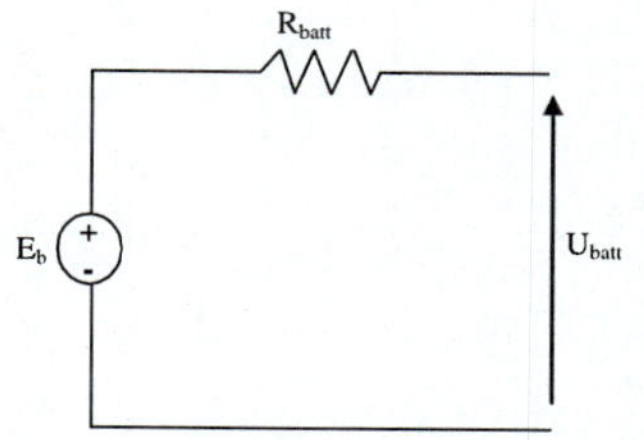

Fig. 5.6 Simple battery model

5.1.3.2 Simple Battery Model

The simple model consists of an ideal voltage source of value E_b (equal to open-circuit voltage) and an internal series resistance R_{batt}.

Where U_{batt} is the terminal voltage of battery.

This model is only suitable for applications where the state of charge is not important [129].

5.1.3.3 Improved Simple Model

Improved simple model varies the resistance of the battery. The resistance R_{batt} will be a function of the state of charge (SOC). We can write the resistance R_{batt} as [129];

$$R_{\text{batt}} = \frac{R_0}{\text{SOC}^k} \tag{5.30}$$

with

$$\text{SOC} = 1 - \frac{I_{\text{batt}} \cdot t}{C_{10}} \tag{5.31}$$

where SOC varies from 0 (for fully discharging up) to 1 (for fully charged), k is the capacity coefficient and takes the change in battery capacity under different discharge rates into account similar to the Peukert factor, R_0 is the initial resistance of the fully charged battery, t is the time of discharging and C_{10} is the 10-h capacity.

5.1.3.4 Modified Battery Model

Jean Paul Cun proposed an improved model based on the battery configuration given in Fig. 5.6. In this model the battery's state of charge is taken into account. He considers that the voltage E_b and the resistance of the battery R_{batt} are no longer constant but vary in accordance with its state of charge [127].

We have:

$$\begin{aligned} E_b &= E_0 - k_{E_b} \cdot \text{SOC} \\ R_{\text{batt}} &= R_0 - k_R \cdot \text{SOC} \end{aligned} \tag{5.32}$$

where k_{E_b} and k_R are coefficients that can be calculated experimentally.

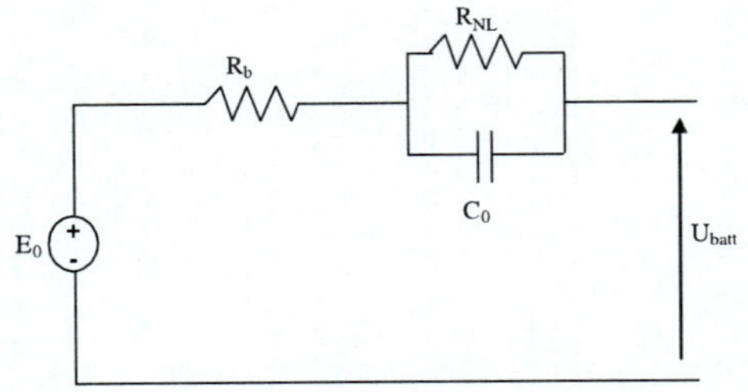

Fig. 5.7 Thevenin battery model [137]

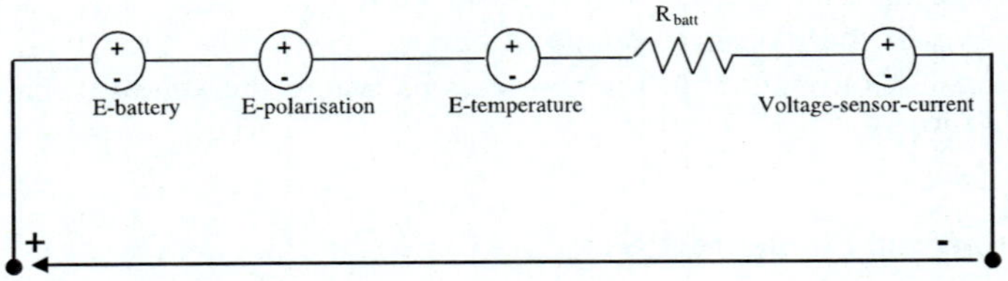

Fig. 5.8 Modified Thevenin model

5.1.3.5 Thevenin Battery Model

Thevenin battery model is a basic model, and consists of an ideal no-load voltage (E_0), an internal resistance (R_{batt}), a capacitance C_0 and an overvoltage resistance R_{NL} [137] (Fig. 5.7).

C_0 is the capacitance of the parallel plates and R_{NL} is the nonlinear resistance contributed by the contact resistance of plate to electrolyte.

The disadvantage of this model is that all the parameters are assumed to be constant but in reality these parameters change with temperature and battery conditions.

A new approach to evaluate batteries is a modified Thevenin model [133]. The electrical equivalent of the proposed model is shown in Fig. 5.8, where E-battery is a simple DC voltage source designating the voltage in the battery cells, E-polarization represents the polarization effects due to the availability of active materials in the battery, E-temperature represents the effect of temperature on the battery terminal voltage and R_{batt} is the battery internal impedance, the value of which depends primarily on the relation between cell voltage and state of charge (SOC) of the battery.

Voltage–sensor–current is basically a voltage source with a value of 0 V. It is used to record the value of battery current.

Thus, this model is comparatively more precise and can be extended for use with NiCd and Li-ion batteries [133]. Only a few modifications need to be carried

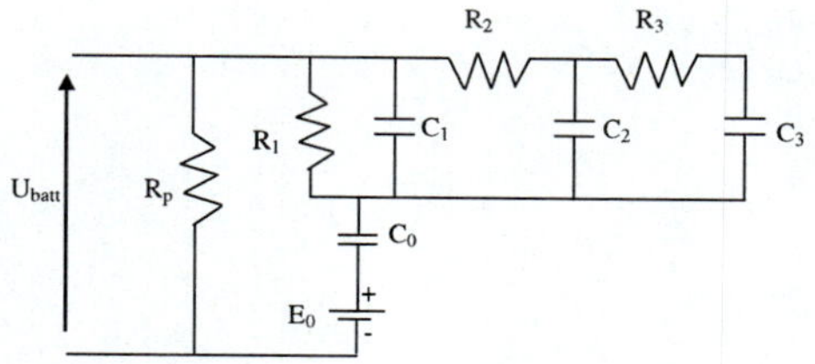

Fig. 5.9 Linear dynamic model [137]

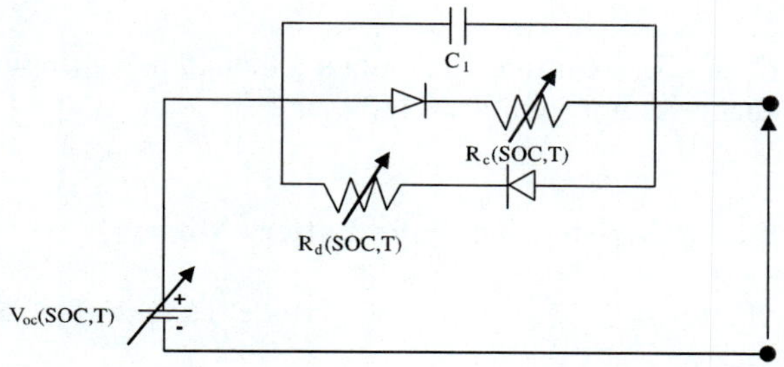

Fig. 5.10 Battery nonlinear model [137]

out in order to vary the parameters, such as state of charge, current density and temperature.

5.1.3.6 Linear Dynamic Model

An improved variant of the Thevenin model is a linear electrical battery model [138]. It models the behavior of the battery during overvoltage and self-discharge of the battery. This model is more accurate than the Thevenin model but it still does not consider the change of the value of parameters according to different operating conditions (Fig. 5.9).

5.1.3.7 Nonlinear Dynamic Model

The nonlinear dynamic model takes into account the variation of different parameters with state of charge of the battery, temperature and discharge rate (Fig. 5.10).

The circuit is composed of two sections:

- The battery open-circuit voltage which is represented by a controlled dc voltage source and its magnitude is changed by state of charge and temperature,
- Internal resistance which is modeled by R_c and R_d representing charging and discharging resistances, respectively. The value of these resistances is changed by the state of charge and temperature as well.

The model is nonlinear in the sense that the elements V_{oc}, R_d and R_c are not constants but are modeled as a function of state of charge and temperature. Only

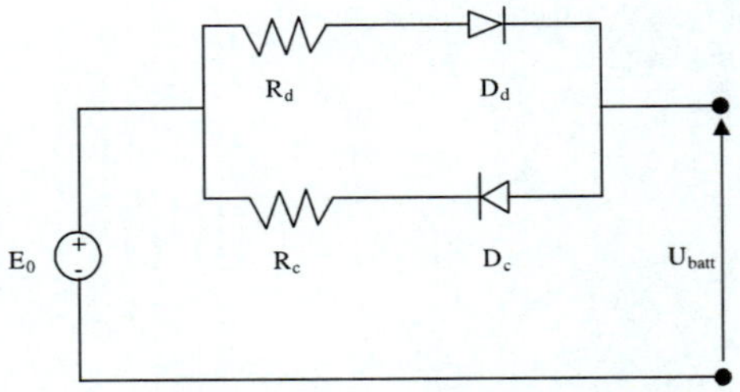

Fig. 5.11 Thevenin resistive model

C_1 has been considered constant although it is changing with state of charge but its change is not significant [139, 140].

5.1.3.8 Resistive Thevenin Battery Model

The resistive Thevenin battery model assumes the following [141]:

- The electrodes are made of porous materials.
- The electrolytic resistance is constant throughout discharge.
- The discharge occurs at constant current.
- Polarization is a linear function of the active material current density.

The circuit model for the resistive Thevenin battery is shown in Fig. 5.11, where R_c and R_d are internal resistances associated with the charging and discharging process of the battery, respectively, and D_c and D_d are diodes which are in series with R_c and R_d, respectively and have no physical significance in the battery. It implies that during charging or discharging only one of the resistances R_c or R_d will be used because when one diode is forward biased the other will be reverse biased.

5.1.3.9 Other Models

Modified Thevenin Equivalent Battery Model

A simple dynamic battery model is shown in Fig. 5.12 [141]. The dynamic equations of the circuit model for discharging and charging are given by:

$$\frac{\mathrm{d}V_p}{\mathrm{d}t} = -\frac{V_p}{R_d \cdot C_p} + \frac{U_{\text{batt}}}{R_d \cdot C_p} - \frac{I_{\text{batt}}}{C_p} \quad \text{if } V_p \le E_b \tag{5.33a}$$

and

$$\frac{\mathrm{d}V_p}{\mathrm{d}t} = -\frac{V_p}{R_d \cdot C_p} + \frac{U_{\text{batt}}}{R_C \cdot C_p} - \frac{I_{\text{batt}}}{C_p} \quad \text{if } V_p \ge E_b \tag{5.33b}$$

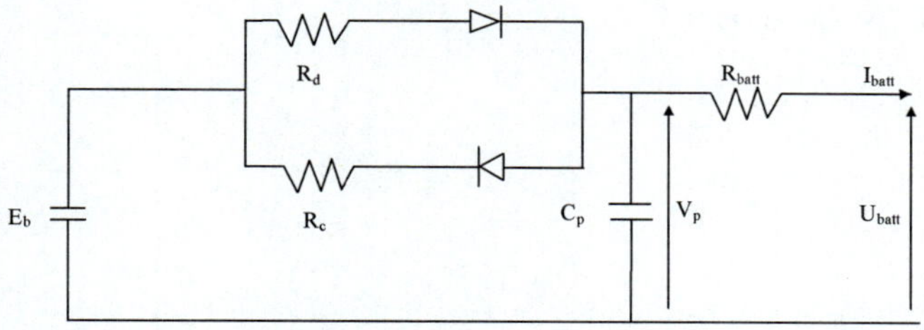

Fig. 5.12 Modified Thevenin equivalent battery model

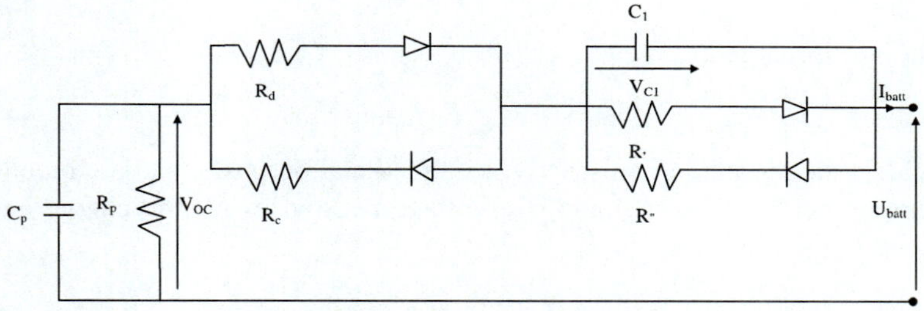

Fig. 5.13 Model from a series experimental test [137]

with

$$I_{\text{batt}} = \frac{V_p - U_{\text{batt}}}{R_{\text{batt}}}$$

Model From a Series Experimental Test

The following model is developed from a series of experimental tests performed through examination of the graphic plots of the experimental data, and manufacturer's specifications. This is the simplest model that meets most requirements for a good battery model. It contains most nonlinear characteristics as well as dependence on the state of charge [141] (Fig. 5.13).

$$C_1 \frac{\mathrm{d}V_{C1}}{\mathrm{d}t} = -I_{\text{batt}} - \frac{V_{C1}}{R''} \quad \text{if } V_{C1} \geq 0 \tag{5.34a}$$

and

$$C_1 \frac{\mathrm{d}V_{C1}}{\mathrm{d}t} = -I_{\text{batt}} - \frac{V_{C1}}{R'} \quad \text{if } V_{C1} \leq 0 \tag{5.34b}$$

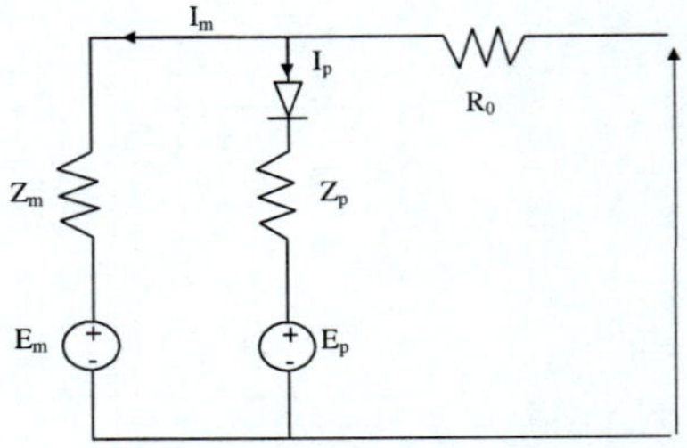

Fig. 5.14 Battery equivalent network

We obtain during charge:

$$U_{\text{batt}} = V_{\text{oc}} - R_c\, I_{\text{batt}} + V_{C1} \tag{5.35a}$$

and during discharge,

$$U_{\text{batt}} = V_{\text{oc}} - R_d\, I_{\text{batt}} + V_{C1} \tag{5.35b}$$

where ideal diodes are chosen strictly for directional purposes, and are required to differentiate between internal and overvoltage resistances for the charge and the discharge.

Dynamical Model

An empirical mathematical model is developed for lead acid traction battery [143]:

$$V_{\text{batt}} = V_{\text{oc}} - R_{\text{batt}} \cdot I_{\text{batt}} - \frac{K}{\text{SOC}} I_{\text{batt}} \tag{5.36}$$

where V_{oc} is the open-circuit voltage, I_{batt} is the discharge current, K a constant, typically 0.1 ohm and K/SOC represents the part of internal resistance which is SOC dependent.

Dynamic Model of Third-Order

The equivalent network of the following model of the battery is described by the Fig. 5.14 [142], where E_m in this the main branch is the electromotive force of the battery and Z_m is the main part of the internal impedance of the battery. This impedance Z_m can be approximated by two R–C blocks.

When a battery is being charged, a part of the electrical energy received is converted into other forms of energy such as water electrolysis. Because of this, a parasitic branch is added in the model. This part is used only when the battery is being charged.

$I_p(V_{\text{PN}})$ stands for the expression of simple resistor R_p because the nonlinear character of the parasitic branch is taken in account by the threshold voltage E_p.

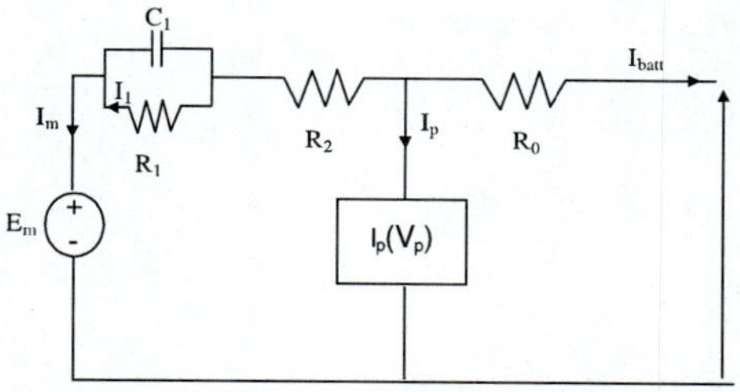

Fig. 5.15 Third-order battery model

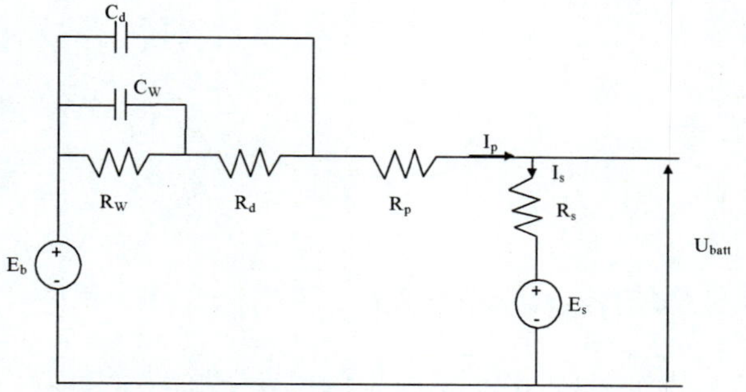

Fig. 5.16 Dynamic model of fourth order [143]

An equivalent network is represented in Fig. 5.15. This model is constituted by:

- An electrical equivalent with two *R–C* blocks and an algebraic parasitic branch.
- Algorithms for calculating the state of charge and internal (electrolyte) temperature.
- Equations for computation of the elements of the equivalent network as functions of state of charge and temperature.

C_1, R_1 and R_2 are the elements of Z_m.

Dynamic Model of Fourth Order

Giglioi [143] proposed a dynamic model which is an accurate and sophisticated model. It consists of two parts:

- Current I_p flowing through R_P (electrolyte reaction), R_d (ohmic effect) and its associated leakage capacitance C_d, and waste of energy R_w and its associated leakage capacitance C_W;
- Current I_s flowing through R_s (self-discharge) (Fig. 5.16).

This model is accurate but it has some drawbacks:

- It requires a longer time for computation due to the high order system.
- Modeling process is complicated because it involves a lot of empirical data.

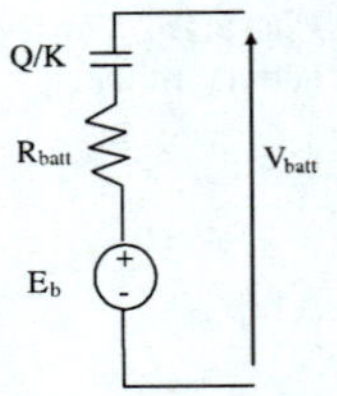

Fig. 5.17 Model traction

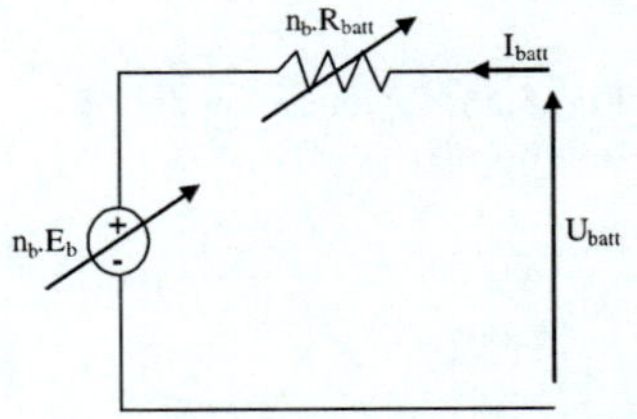

Fig. 5.18 Equivalent circuit of a battery of n_b elements

5.1.4 Traction Model

A mathematical model is used for a traction lead acid battery (Fig. 5.17) [127]. This model is used primarily in applications of electric and hybrid vehicles; it consists of the EMF E_b in series with a resistance R_{batt} and a capacitor (Q/K).

$$V_{\text{batt}} = E_b - R_{\text{batt}} \cdot I_{\text{batt}} - K \int \left(\frac{I_{\text{batt}}}{Q} \right) \mathrm{d}t \tag{5.37}$$

with V_{batt} terminal voltage of the battery, E_b zero-current voltage of the battery charged and K constant which depends on the battery. $\int \left(\frac{I_{\text{batt}}}{Q} \right) \mathrm{d}t$ indicates the state of battery discharge, Q being the battery capacity (Ah).

5.1.5 Application: CIEMAT Model

The advantage of the battery models developed by CIEMAT is their ability to cope with a wider range of lead acid batteries and requires few manufacturers' data technological parameters [129, 144, 145]. This model is based on the electrical scheme of Fig. 5.18, the battery is then described by two elements, a voltage source E_b and an internal resistance R_b, of which characteristics depend on a number of parameters (temperature and charge state).

For a number of n_b cells, the voltage equation is:

$$U_{\text{batt}} = n_b \cdot E_b \pm n_b \cdot R_{\text{batt}} \cdot I_{\text{batt}} \tag{5.38}$$

5.1.5.1 Capacitor Model

To highlight the physical phenomena that govern the operation of the storage system, the capacity should take into account the temperature. Therefore, we propose the model of capacity, giving the amount of energy that can be returned according to the mean discharge current, it is given by Eq. 5.39 The capacity model is established from the expression of the current I_{10}, which corresponds to the operating speed to C_{10}, where ΔT is the heating of the accumulator (assumed identical for all elements) over an ambient temperature which is equal to 25°C [146].

$$\frac{C_{\text{batt}}}{C_{10}} = \frac{1.67}{1 + 0.67\left(\frac{I}{I_{10}}\right)^{0.9}}(1 + 0.005\Delta T) \tag{5.39}$$

The capacity C_{batt} is used as a reference for determining the state of battery charge.

$$\text{SOC} = 1 - \frac{Q}{C_{\text{batt}}} \tag{5.40}$$

with

$$Q = I_{\text{batt}} \cdot t \tag{5.41}$$

t is the discharging time with a current I_{batt}.

5.1.5.2 Charge Voltage Equation

Expression of the battery voltage is function of the internal components of the battery depending on the electromotive force and the internal resistance.

$$V_{\text{batt-charge}} = n_b[2. + 0.16 \cdot \text{SOC}] + n_b.\frac{I_{\text{batt}}}{C_{10}}\left[\frac{6}{1 + I_{\text{batt}}^{1.3}} + \frac{0.27}{\text{SOC}^{1.5}} + 0.002\right] \cdot (1 - 0.007.\Delta T) \tag{5.42}$$

where n_b is the cells number, ΔT is the temperature variation

$$\Delta T = T - 25$$

and C_{10} is the rated capacity (I_{10})

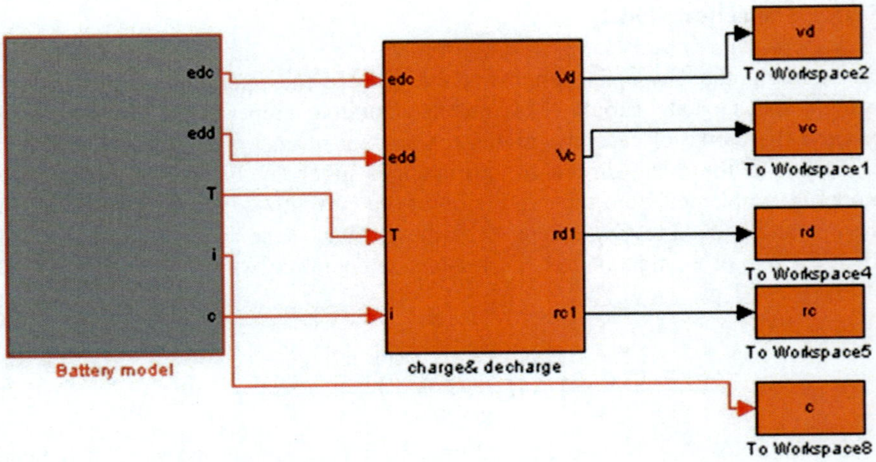

Fig. 5.19 Implementation of battery model in matlab/simulink

5.1.5.3 Discharge Voltage Equation

The equation of the voltage is similar to that obtained at charging.

$$V_{\text{batt-discharge}} = n_b[1.965. + 0.12.\text{SOC}] - n_b.\frac{I_{\text{batt}}}{C_{10}}\left[\frac{4}{1 + I_{\text{batt}}^{1.3}} + \frac{0.27}{\text{SOC}^{1.5}} + 0.002\right] \cdot (1 - 0.007.\Delta T)) \quad (5.43)$$

5.1.5.4 Efficiency

Two types of load efficiency are distinguished: the faradic efficiency (Coulombian) and energetic Efficiency (overall). The first concerns the ability of the battery to store electrical charge; it does not involve the Joule losses in the internal resistance. Energy efficiency takes into account the Coulombian efficiency and losses by Joule effect [129].

$$\eta = 1 - \exp\left[\frac{20.73}{\frac{I_{\text{batt}}}{I_{10}} + 0.55}(\text{SOC} - 1)\right] \quad (5.44)$$

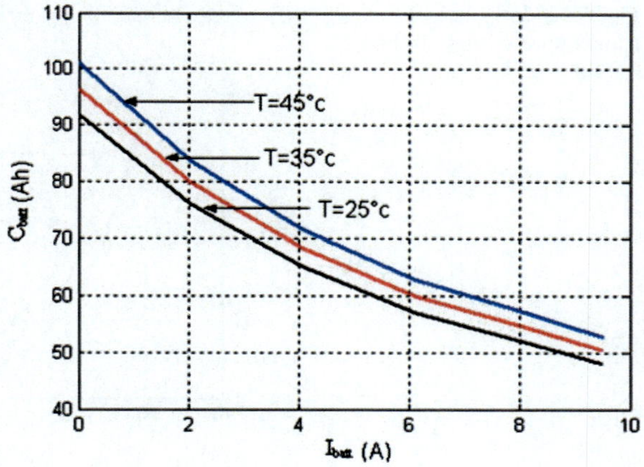

Fig. 5.20 Temperature influence on battery capacity

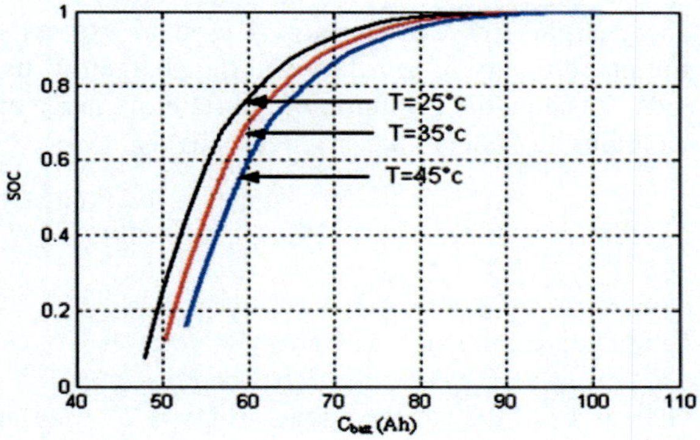

Fig. 5.21 Temperature influence on state of charge

5.1.5.5 Temperature Effects

We make application under Matlab/Simulink (Fig. 5.19).

Temperature Influence on Battery Capacity

The battery temperature affects the storage capacity. The capacity increases with the temperature of the element (Fig. 5.20). This is explained partly by the increased diffusion coefficient of sulfuric acid solutions, on the other hand by the decrease in resistivity of the electrolyte at commonly used concentrations.

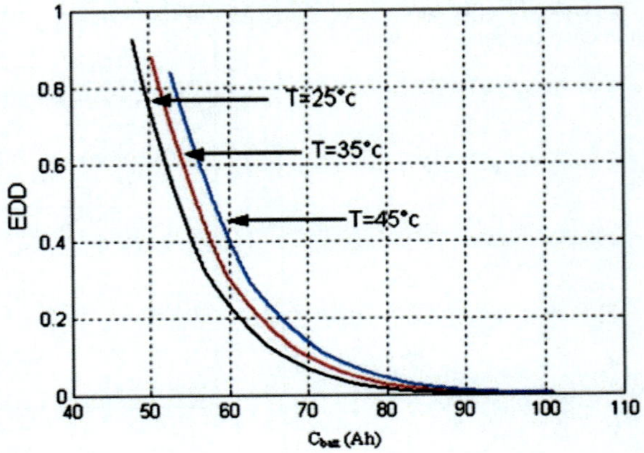

Fig. 5.22 Influence of temperature on the discharge state

Temperature Influence on State of Charge

The charging (or discharging) is used to express the relationship between the nominal capacity of a battery and the charge (or discharge) current. Figures 5.21 and 5.22 show the evolution of the state of charge and the discharge state (DOD) according to battery capacity, respectively.

$$\text{DOD} = 1 - \text{SOC} \tag{5.45}$$

Temperature Influence on Battery Voltage

The curves of the battery voltage (Figs. 5.23, 5.24) are influenced by temperature because voltage is a function of SOC, the charge and discharge time, temperature and current.

Influence of Charge and Discharge Time on the State of the Battery

The influence of charge time on the state of the battery is shown in Figs. 5.25 and 5.26

In the case of the charge, more charge time decreases, and more the battery charge state is close to 1. And in the case of discharge, more the discharge time is important more the discharge state reached its maximum value.

Influence of Charge Time and Discharge Time on the Battery Voltage

For both charge and discharge, the evolution of the voltage is directly proportional to the time of charge and discharge (Figs. 5.27, 5.28).

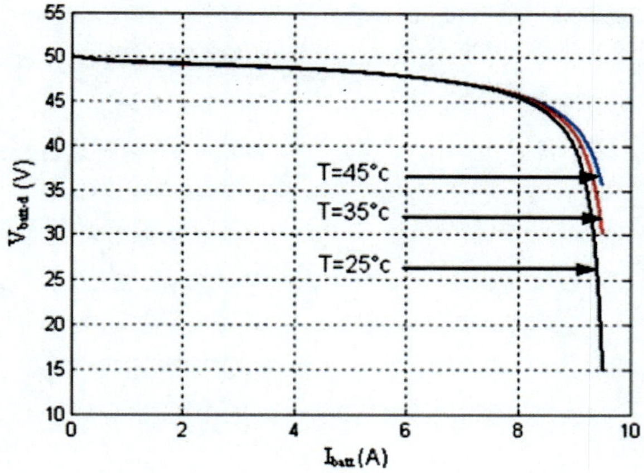

Fig. 5.23 Influence of temperature on the discharge voltage

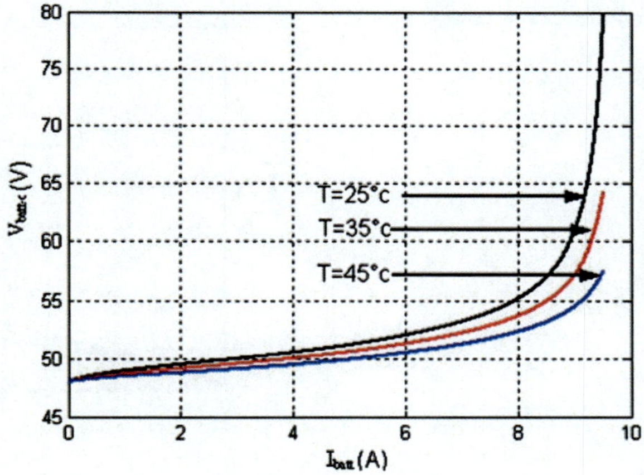

Fig. 5.24 Influence of temperature on the charge voltage

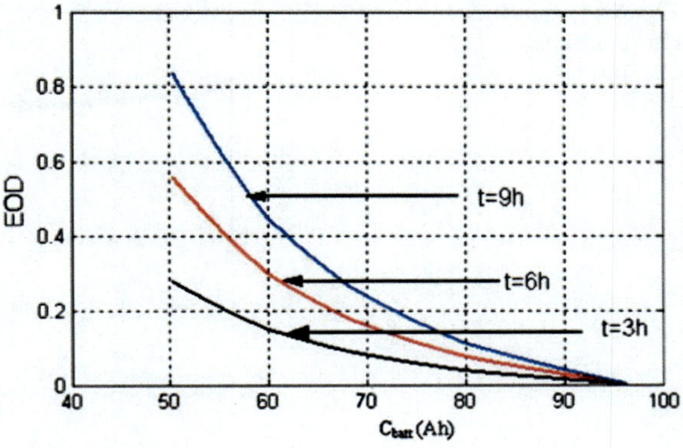

Fig. 5.25 Effect of time on the state of discharge (DOD)

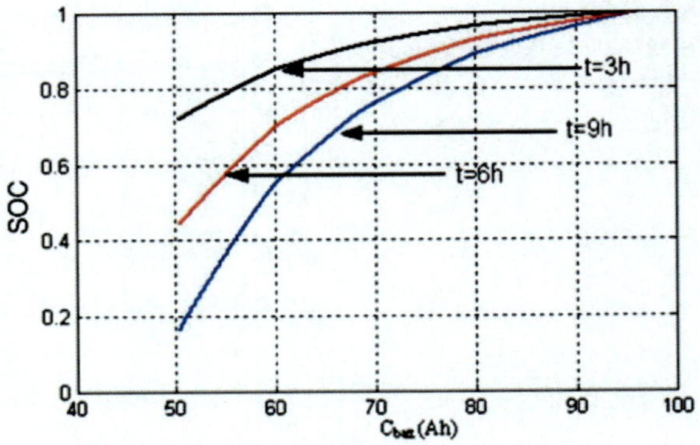

Fig. 5.26 Effect of time on the state of charge (SOC)

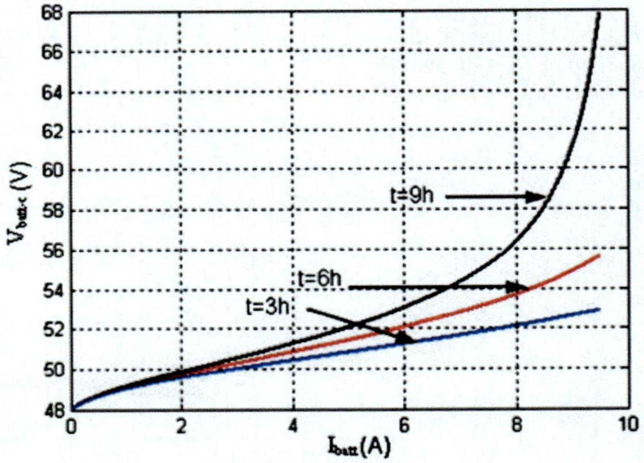

Fig. 5.27 Influence of time on the charge voltage [195]

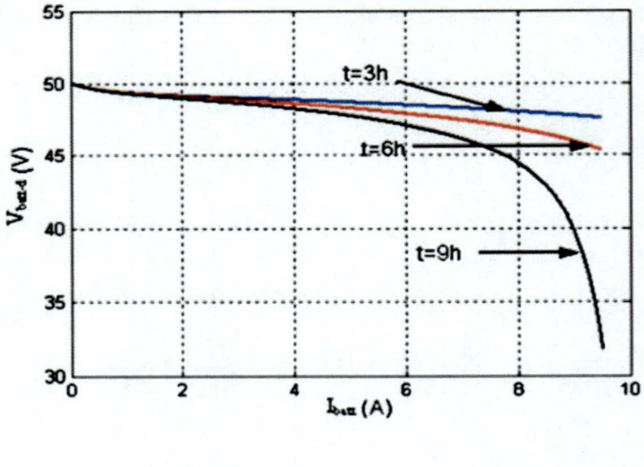

Fig. 5.28 Influence of time on the discharge voltage [195]

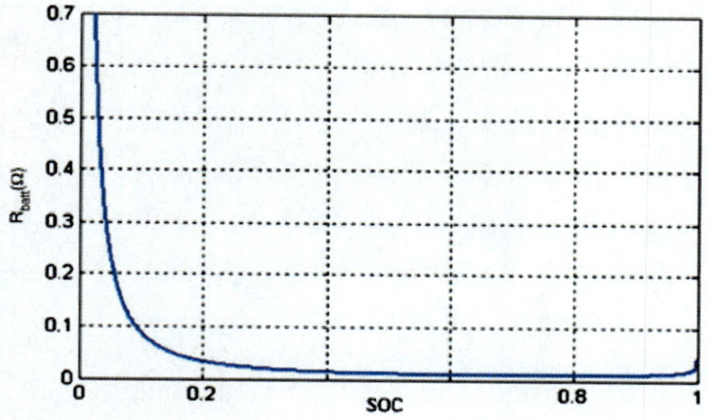

Fig. 5.29 Resistance of the battery (discharge) [195]

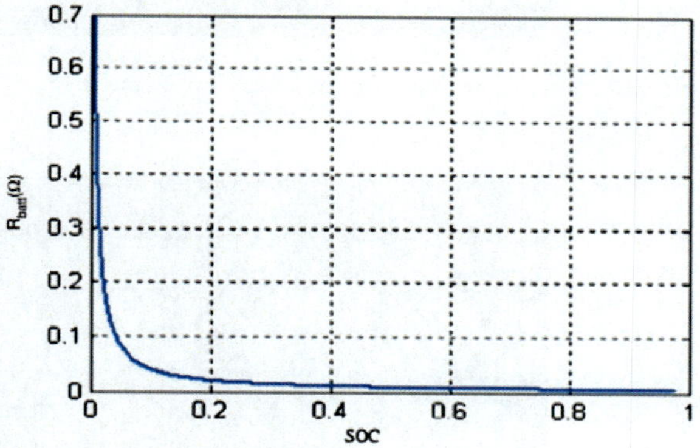

Fig. 5.30 Internal resistance of the battery (charge)

5.1.5.6 Internal Resistance

The internal resistance of the battery is directly related to the battery state of charge. When the battery is charged, the internal resistance is low and it becomes important for a state of charge near zero (Fig. 5.29).

The internal resistance of the battery (charge) depends on the state of charge. Its value becomes lower when the full charge is reached (Fig. 5.30).

The value of the internal resistance also increases rapidly with decreasing temperature, which is mainly due to the change in resistance of the electrolyte.

5.1.5.7 Practical Identification of Battery Impedance

In practice, the determination of batteries impedances is often made on stationary batteries. The basic principle is to impose on the battery an excitation in voltage or current [31]. Measurement at weak frequency gives more information on electrochemical operation because the internal kinetics of the battery has thus the time

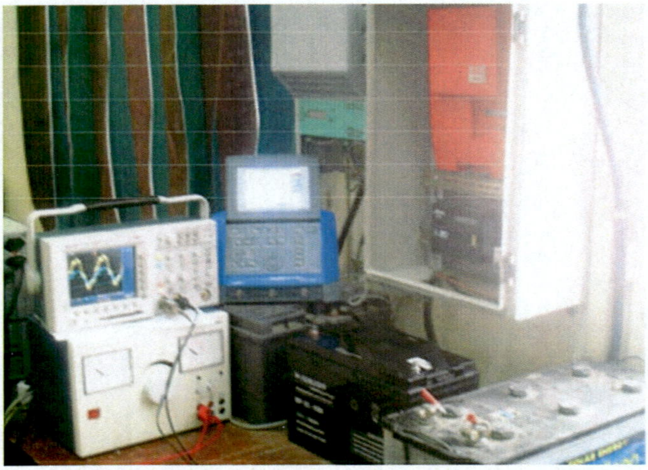

Fig. 5.31 Circuit measurement of battery impedances

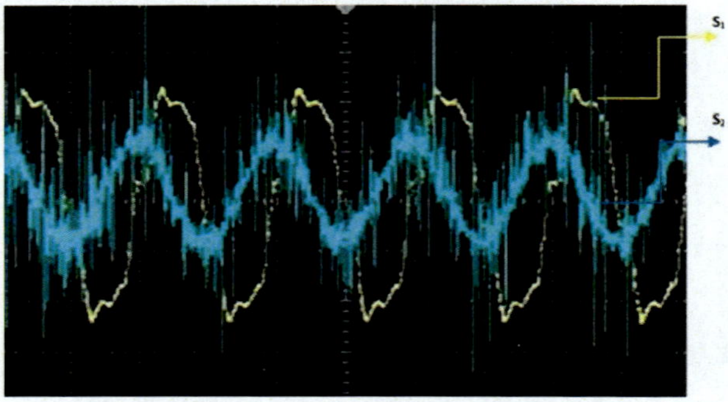

Fig. 5.32 Measured signals S_1 current and S_2 voltage

to react to the imposed disturbance. In order to determine the internal impedance of the battery (12 V, 100 Ah) for one state of charge of the battery, we superimpose an alternate sinusoidal signal of 50 Hz frequency to the continuous component of the battery. We made application with a system of storage which is composed of two lead acid batteries of 12 V, 92 Ah inter connected in series to have 24 V. We close the battery on a circuit including a rheostat for the current limitation (15.8Ω, 10 A), a shunt (250 V, 10 A) allowing to measure the circulating current in the circuit and a source of alternating voltage (0–36 V, 20 A). The measurement system is developed (Fig. 5.31).

We measure two voltages using an oscilloscope: the voltage at the shunt terminals which is a direct picture of the current circulating in the circuit and the voltage at the battery terminals. The ratio of these two voltages and their phase angle provides the absolute value of the internal impedance of the battery.

Measures are carried out by a digital oscilloscope, Tektronix TDS3032, 300 MHz with two channels, which makes the acquisition of signals possible, then to transfer them to a computer for analysis. The mass is common to both measures of voltage. The signal (V_1) permits to measure the battery voltage and the signal (V_2) the current. Figure 5.32 presents the signals recovered by Excel. The signal (S_1) is a direct image of the AC current of measurement (conversion factor: 50 V = 1A) while the second signal represents the resulting voltage on the battery's terminals.

The battery behaves as complex impedance Z_{batt} containing a resistance R_{batt} and a reactance X_{batt} to this disturbance.

$$Z_{\text{batt}} = R_{\text{batt}} - \text{j}X_{\text{batt}} \tag{5.46}$$

The module of the complex impedance is thus well defined by the ratio of the absolute values of the two signals. We deduce the dephasing by the temporal difference between the two signals with the passage by zero.

$$|Z_{\text{batt}}| = V_{\text{batt}}/I_{\text{batt}} \tag{5.47}$$

Knowing the module of Z_{batt} and its dephasing, we can thus deduce the real part R_{batt} and imaginary X_{batt} of the impedance for this state of charge, these values changes according to the latter.

Chapter 6
Photovoltaic Pumping Systems

Symbols

$a(h)$, $b(h)$, $c(h)$, $d(h)$	Coefficients corresponding to the working total head
a_i, b_i, c_i and d_i	Constants which depend on the type of solar pumping system
C_i	ith calculated value
h	Total head
hh p.u	Fundamental voltage value
i_a	Armature current
$(I_{sd}\ I_{sq})\ (d,\ q)$	Components of the stator current
i_F	Field current
J	Rotor and load moment of inertia
J_{motAC}	Inertia of the AC motor
Kh	Harmonics to be minimized
K_m	Constant related to the design of the machine
L_F	Field winding inductance
L_a	Armature winding inductance
M_i	ith measured value
n	Number of measurements
P_9	Electrical power input of the motor pump
p	Pole pair number of the AC machine
R_a	Armature resistance
$R_F + R_{adj}$	Total field resistance
R_m	Core loss resistance (Ω)
R_s	Stator resistance per phase (Ω)
R_r	Equivalent rotor resistance per phase (Ω)
T_L	Load torque
v	Terminal voltage
V_a, V_b and V_c	Phase voltages
V_s	rms Motor voltage (V)

D. Rekioua and E. Matagne, *Optimization of Photovoltaic Power Systems*,
Green Energy and Technology, DOI: 10.1007/978-1-4471-2403-0_6,

(V_{sd}, V_{sq}) (d, q)	Components of the stator voltage
X_m	Magnetizing reactance (Ω)
X_s	Stator leakage reactance (Ω)
X_r	Equivalent rotor leakage reactance (Ω)
σ	Leakage coefficient
ϕ	Flux per pole
(ϕ_{sd}, ϕ_{sq}) (d, q)	Components of the stator flux
ω	Rotational speed of the rotor
ω_{rAC}	AC motor velocity angular
ω_s	Angular frequency of the supply (rd/s)

Photovoltaic pumping has become one of the most promising fields in photovoltaic applications. To achieve the most reliable and economical operation, more attention is paid to their design and their optimal use. Depending upon the intended application, the pumping system can be selected from surface, submersible or floating pump types [147]. Submersible pumps remain underwater, surface pumps are mounted at water level at the vicinity of the well or, in the case of a floating pump, on top of the water. Pumps can be classified according to their operating modes. Mainly there are centrifugal and positive displacement pumps [147]. In the centrifugal pump, the rotation of an impeller forces water into the pipe. The water velocity and pressure depend on the available mechanical power at the rotating impeller and the total head. The displacement pump uses a piston or a screw to control the water flow. Compared to the centrifugal pump, the positive displacement pump presents a better efficiency under low power conditions.

The water pumps may be driven by DC or AC motors. The earlier PV pumping systems were principally based on DC motors [148]. The DC motors present the drawback of maintenance expenses due to frequent brush replacements [149]. Brushless permanent magnet DC motors have been introduced in some systems Photovoltaic array [150]. Recent developments in induction motor technology made this option attractive among the AC motors-based pumping setups. The induction motor is more robust, requires much less maintenance and is available at lower costs than DC motors [149, 150].

6.1 PV Pumping Systems Based on DC Machines

6.1.1 Description

Photovoltaic water pumping systems generally consist of a PV array, a motor pump subsystem, a controller and a tank. The motor pump subsystem includes a

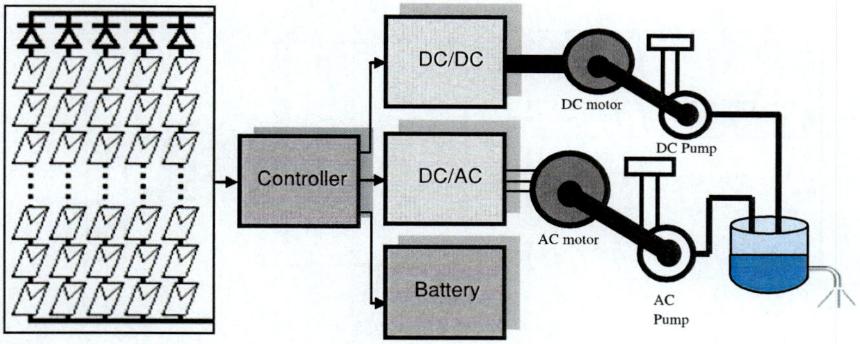

Fig. 6.1 General configuration of a photovoltaic pumping system [62]

Fig. 6.2 Photovoltaic system loaded by DC motor

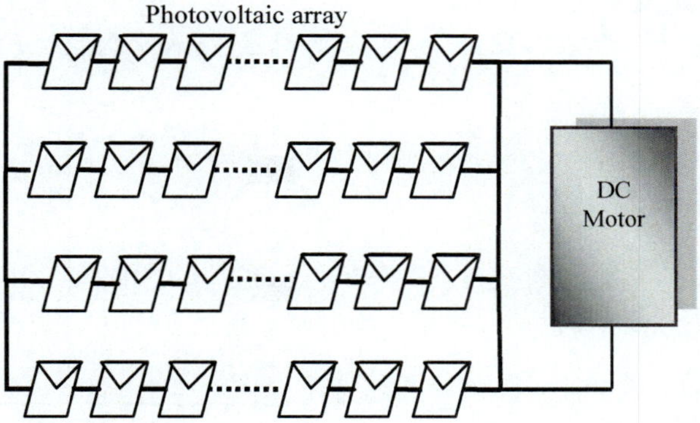

DC motor and a centrifugal pump. There are two configurations of PV pumping systems based on DC motor. In the first system (Fig. 6.2), the motor could be connected directly to the PV generator. For allowing operation of the pumping system at maximum power, the motor controller has to be able to perform the tracking of the MPP.

The second system comprises PV array feeding a DC motor via a DC–DC converter which may perform the MPPT (Fig. 6.3).

6.1.2 System Modeling

6.1.2.1 DC Machine Model

DC motors are electrical machines which consume DC electrical power and produce mechanical power. In shunt machines, the field circuit is connected in parallel with the armature circuit while DC series machines have their field circuit

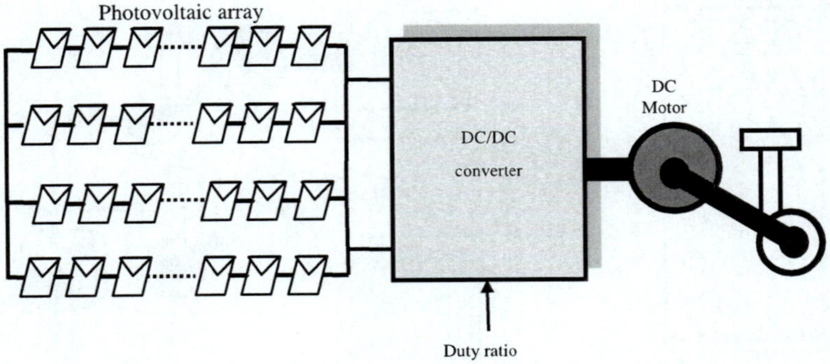

Fig. 6.3 Photovoltaic system loaded by DC motor via DC–DC buck-boost converter

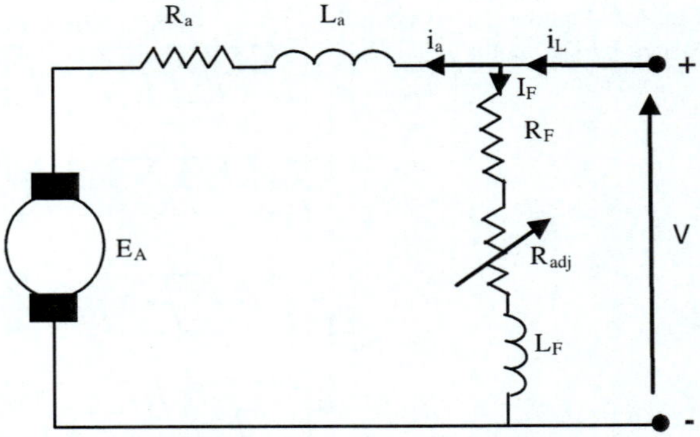

Fig. 6.4 Equivalent circuit of DC Shunt motor

in series with the armature where both field and armature currents are identical. Permanent magnet machines, on the other hand, have only one circuit (armature winding) and the flux generated by the magnets is constant.

DC Shunt Motor

The equivalent circuit of DC Shunt motor is shown in Fig. 6.4. Adjustable resistor R_{adj} is connected in series with the field circuit for speed control.

The dynamical model of DC shunt motor can be written as

$$L_F \frac{di_F}{dt} = V - (R_F + R_{\text{adj}})i_F \tag{6.1}$$

$$L_a \frac{di_a}{dt} = V - R_a i_a - K_m \phi \omega \tag{6.2}$$

$$J \frac{d\omega}{dt} = K_m \phi i_a - T_L \tag{6.3}$$

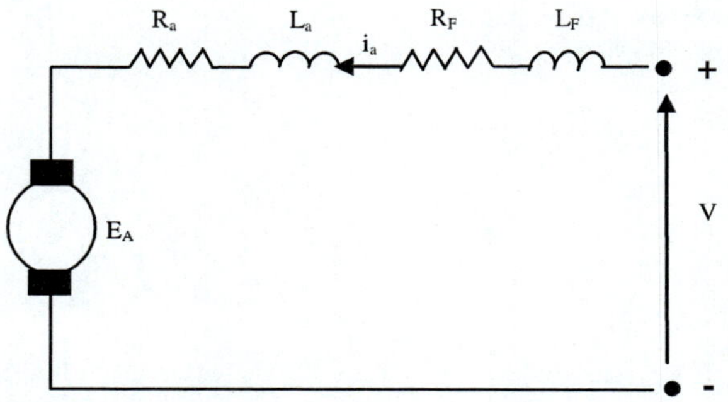

Fig. 6.5 Equivalent circuit of DC series motor

with L_F as the field winding inductance, i_F field current, V terminal voltage and $R_F + R_{\text{adj}}$ total field resistance, L_a armature winding inductance, i_a armature current, Ra armature resistance, K_m constant related to the design of the machine, ϕ flux per pole, ω rotational speed of the motor, J rotor and load moment of inertia and T_L load torque.

DC Series Motor

The equivalent circuit of DC series motor is shown in Fig. 6.5. The field circuit is connected in series with the armature and then the armature and field currents are identical.

The nonlinear dynamical mathematical model is given by:

$$(L_F + L_a)\frac{di_a}{dt} = V_T - (R_a + R_F)i_a - K_m\phi\omega \tag{6.4}$$

$$J\frac{d\omega}{dt} = K_m\phi i_a - T_L \tag{6.5}$$

Permanent Magnet DC Motor (PMSM)

Due to the absence of the field current and field winding, permanent magnet machines exhibit high efficiency in operation, simple and robust structure in construction and high power to weight ratio. The attractiveness of permanent magnet machines is further enhanced by the availability of high-energy rare earth permanent magnet materials like SmCo and NdFeB. However, the speed control of permanent magnet DC motor via changing the field current is not possible. Then, MPPT is possible only using a DC–DC converter as in Fig. 6.3.

The dynamical model can be written as:

$$L_a\frac{di_a}{dt} = V - R_a i_a - K_m.\phi\omega \tag{6.6}$$

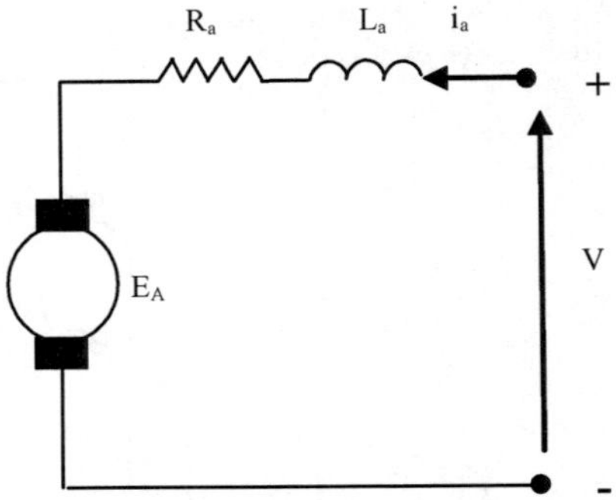

Fig. 6.6 Equivalent circuit of PMSM

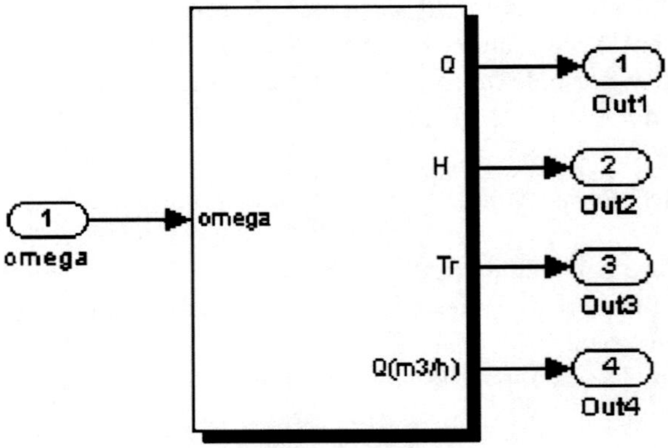

Fig. 6.7 Pump model

$$J\frac{d\omega}{dt} = K.\phi.i_a - T_L \tag{6.7}$$

6.1.2.2 Centrifugal Pump Model

Description

In the centrifugal pump, the rotation of an impeller forces water into the pipe. The water velocity and pressure depend on the available mechanical power at the rotating impeller and the total head. A centrifugal pump commonly requires a single quadrant drive.

Pump Model

There are still many obstacles inhibiting a larger implementation of PV pumping systems [149]. Among these problems is the lack of accurate tools for the prediction of the system performances. To achieve improvements in PV pumping design, it is necessary to study and model photovoltaic water pumping systems, particularly motor pump subsystems. Many research papers have dealt with modeling of PV pumping systems . Reference [151] has investigated a method to determine a PV pumping system performance by representing the water flow rate and efficiency of the system as function of supply frequency and pumping head. Reference [152] has used sets of differential equations to predict the behavior of a PV water pumping system depending on the level of solar global irradiance incident on the PV array. The characteristics have been represented by current, voltage, head (I, V, h) and water flow, current, head (Q, I, h) relationships [153]. The current, voltage (I_{pv}, V_{pv}) and water flow, voltage (Q, V) relations have been presented [154]. Ref. [155] has predicted the relation between water flow and solar radiation. The electrical power input has been expressed as a function of water flow at different heads [156]. The precedent models do not give the water flow output directly as a function of the electrical power input to the motor pump subsystem. The actual model relates directly the pumped water flow output Q to the motor pump subsystem electric power input P.

We use the model which expresses the water flow output (Q) directly as a function of the electrical power input (P) to the motor pump, for different total heads. The experimental data has been collected for several pumps by using the test bench. The collected data consists of measuring the water flow Q for different values of the electrical power input P and total head h. On the basis of these experiments, a model is elaborated by the use of the least squares method to the set of measurements data. A polynomial fit of the third order expresses the relationship between the flow rate and power input, as described by the following Eqs.[151, 152, 156]:

$$P_a\,(h, Q) = a\,(h)\,Q^{\,3} + \;b\,(h)\;Q^{\,2} + \;c\,(h)\;Q + \;d\,(h) \tag{6.8}$$

where P_a is the electrical power input of the motor pump, h is the total head and a(h), b(h), c(h), d(h) are the coefficients corresponding to the working total head.

$$a(h) = a_0 + a_1 h^1 + a_2 h^2 + a_3 h^3 \tag{6.9}$$

$$b(h) = b_0 + b_1 h^1 + b_2 h^2 + b_3 h^3 \tag{6.10}$$

$$c(h) = c_0 + c_1 h^1 + c_2 h^2 + c_3 h^3 \tag{6.11}$$

$$d(h) = d_0 + d_1 h^1 + d_2 h^2 + d_3 h^3 \tag{6.12}$$

with a_i, b_i, c_i and d_i constants which depend on the type of solar pumping system.

The calculation of the instantaneous flow in terms of power is calculated using Newton–Raphson method. Thus at the kth iteration, the flow Q is given by the following equation:

For Pa > 0:

$$Q_k = Q_{k-1} - \frac{F(Q_{k-1})}{F'(Q_{k-1})} \tag{6.13}$$

with

$$F(Q_{k-1}) = a\,Q^3_{k-1} + b\,Q^2_{k-1} + c\,Q_{k-1} + d - P_a \tag{6.14}$$

where $F'(Q_{k-1})$ is the derivative of the function $F(Q_{k-1})$.

Validation of the presented motor pump subsystem model is performed at different heads and input powers by comparing the measured water flow rates with the simulation results. The comparison uses the root mean square error (RMSE) and mean bias error (MBE) indicators; these coefficients are defined by Eqs. 15 and 16 as

$$RMSE = \left[\frac{\sum_{i=1}^{n}(C_i - M_i)^2}{\sum_{i=1}^{n} M_i^2}\right]^{0.5} \times 100 \tag{6.15}$$

$$MBE = \left[\frac{\sum_{i=1}^{n}\left[\frac{|M_i - C_i|}{M_i}\right]}{n}\right] \times 100 \tag{6.16}$$

where n is the number of measurements, C_i the ith calculated value and M_i the ith measured value.

6.1.3 Application

The experiments are carried out using the pump testing system at the Centre of Development of Renewable Energies located in Algiers-Algeria [60]. The main components of the test facility consist of 30 mono-crystalline PV modules with a total peak power of 2.2 kWp, several pumps, a water tank, an air/water regulated pressure vessel, an air compressor, a piping system and measuring devices. A motor operated valve is used to set the water head between 0 and 120 m with flow rates ranging between 0 and 30 m^3 per hour.

We present the experiments conducted for two centrifugal pumps C_1 (400 W) and C_2 (1000 W). The water flow Q and electric power input P values are logged at different total heads, for the tested pumps C_1 and C_2. Examples of Q–P curves are shown in Figs. 6.8, and 6.9 and the results concerning the RMSE and MBE indicators are given in Figs. 6.10,and 6.11. Good agreement between experimental and calculated data is seen in Figs. 6.8, 6.9, 6.10, 6.11. The low values of the RMSE and MBE validate the present model.

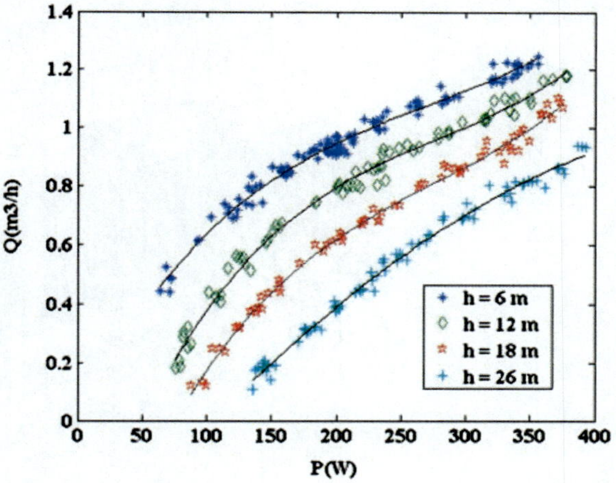

Fig. 6.8 Q–P characteristics for C_1 at heads: 6, 12, 18, 26 m [62]

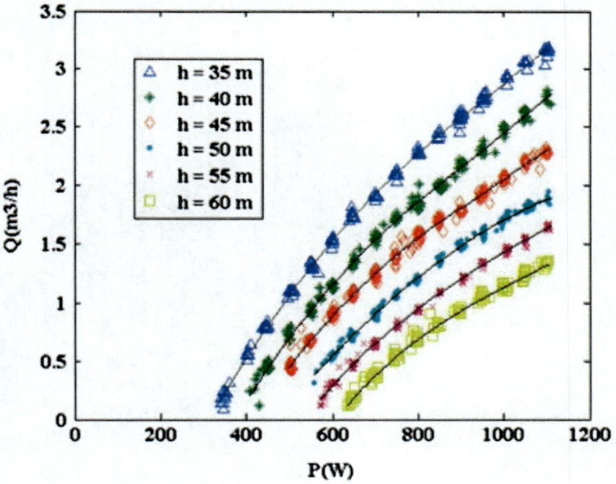

Fig. 6.9 Q–P characteristics for C_2 at heads: 35, 40, 45, 50, 55, 60 m [62]

6.2 PV Pumping Systems Based on AC Motor

6.2.1 Description

Photovoltaic water pumping systems based on induction machine consist of a PV array, a motor pump subsystem, a controller and a tank (Fig. 6.12). The motor pump subsystem includes a motor, a pump and a power converter.

The controller is an electronic interface between the photovoltaic array and the motor pump subsystem. In the case of AC motor use, an inverter may be included in the controller [158]. Usually a tank is commonly employed for water storage. The principle is to store water rather than electricity in batteries, thereby reducing the cost and complexity of the system.

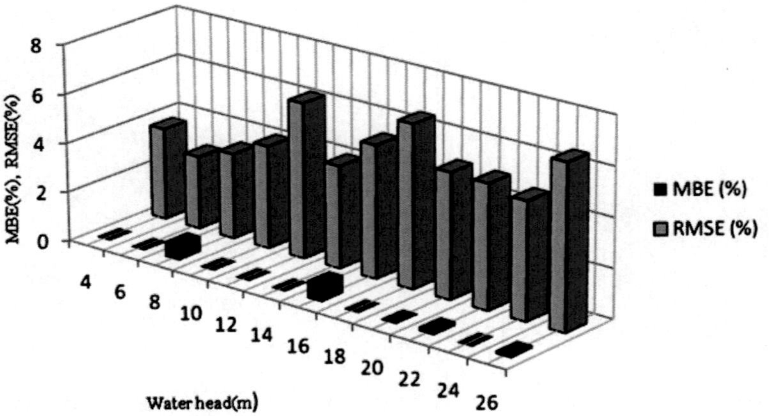

Fig. 6.10 RMSE and MBE results for the centrifugal pumps C_1 [62]

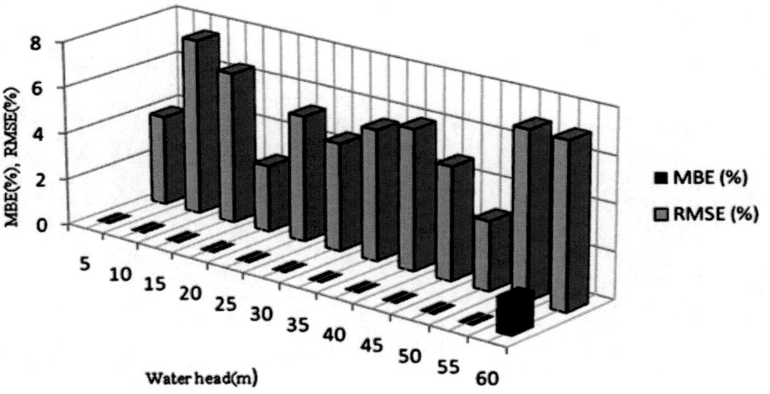

Fig. 6.11 RMSE and MBE results for the centrifugal pumps C_2 [62]

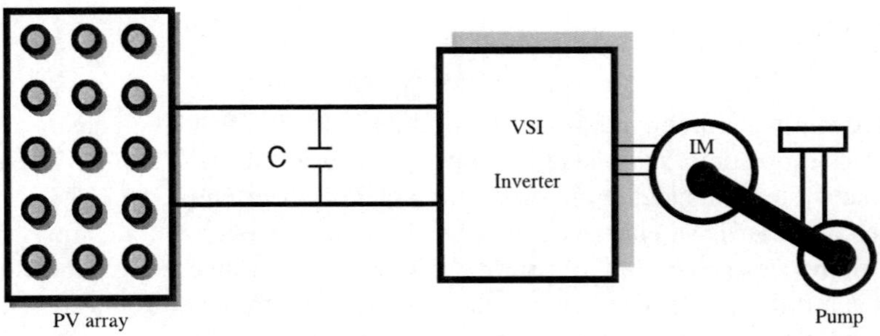

Fig. 6.12 Block diagram of PV pumping based on induction machine

6.2.2 System Modeling

6.2.2.1 The d–q Induction Motor Model

We use an induction motor which is modeled using voltage and flux equations referred in a general normalized stator frame [157]:

– Stator voltage equations:

$$\begin{cases} V_{ds} = R_S I_{ds} + \dfrac{d\Phi_{ds}}{dt} \\ V_{qs} = R_S I_{qs} + \dfrac{d\Phi_{qs}}{dt} \end{cases} \tag{6.17}$$

where (I_{ds} I_{qs}), (V_{ds}, V_{qs}) and (Φ_{ds}, Φ_{qs}) are the (d, q) components of the stator current, voltage and flux, R_s is the stator resistance.

– Rotor voltage equations:

$$\begin{cases} 0 = V_{dr} = R_r I_{dr} + \dfrac{d\Phi_{dr}}{dt} + p\,\omega_{rAC}.\Phi_{qr} \\ 0 = V_{qr} = R_r I_{qr} + \dfrac{d\Phi_{qr}}{dt} - p\,\omega_{rAC}\Phi_{dr} \end{cases} \tag{6.18}$$

where I_{dr}, I_{qr} are (d, q) rotor current, Φ_{dr}, Φ_{qr} are (d, q) rotor flux, R_r is the rotor resistance.

We obtain the following mathematical model:

$$\begin{bmatrix} \dfrac{di_{ds}}{dt} \\ \dfrac{di_{qs}}{dt} \\ \dfrac{di_{dr}}{dt} \\ \dfrac{di_{qr}}{dt} \end{bmatrix} = \frac{1}{\sigma} \begin{bmatrix} -\dfrac{R_s}{L_s} & \dfrac{p\cdot\omega_{rAC}\cdot L_m^2}{L_s\cdot L_r} & \dfrac{L_m.R_r}{L_s r} & \dfrac{p\cdot\omega_{rAC}\cdot L_m}{L_s} \\ -\dfrac{p\cdot\omega_{rAC}\cdot L_m^2}{L_s r} & -\dfrac{R_s}{L_s} & \dfrac{p\cdot\omega_{rAC}\cdot L_m}{L_s} & \dfrac{L_m.R_r}{L_s\cdot L_r} \\ \dfrac{L_m.R_s}{L_s\cdot L_r} & -\dfrac{p\cdot\omega_{rAC}\cdot L_m}{L_r} & -\dfrac{R_r}{L_r} & -p.\omega_{rAC} \\ \dfrac{p\cdot\omega_{rAC}\cdot L_m}{L_r} & \dfrac{L_m.R_s}{L_s\cdot L_r} & p.\omega_{rAC} & -\dfrac{R_r}{L_r} \end{bmatrix} \cdot \begin{bmatrix} i_{ds} \\ i_{qs} \\ i_{dr} \\ i_{qr} \end{bmatrix} + \frac{1}{\sigma} \begin{bmatrix} \frac{1}{L_s} & 0 \\ 0 & \frac{1}{L_s} \\ -\frac{L_m}{L_s.L_r} & 0 \\ 0 & -\frac{L_m}{L_s.L_r} \end{bmatrix} \cdot \begin{bmatrix} v_{ds} \\ v_{qs} \end{bmatrix} \tag{6.19}$$

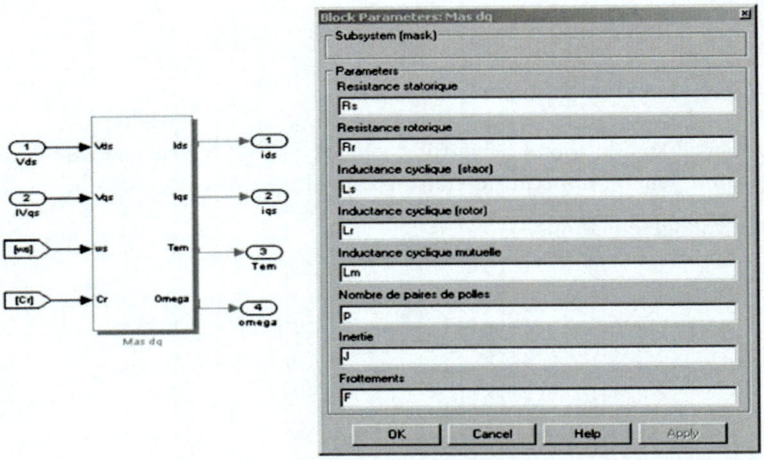

Fig. 6.13 Motor block diagram

with L_m as the magnetizing inductance of the motor and L_s and L_r are the stator and rotor inductances and

$$\sigma = 1 - \frac{L_m^2}{L_s\, L_r}$$

– Mechanical equation:

$$T_{\text{emAC}} - T_{\text{Load}} = J_{\text{motAC}} \cdot \frac{d\omega_{\text{r AC}}}{dt} \tag{6.20}$$

with ω_{rAC} as the AC motor angular velocity and J_{motAC} the inertia of the AC motor.

The electromagnetic torque can be written as:

$$T_{\text{emAC}} = p.(\phi_{\text{ds}}.i_{\text{qs}} - \phi_{\text{qs}}.i_{\text{ds}}) \tag{6.21}$$

where p is the pole pair number of the AC machine

$$\Phi_{\text{ds}} = L_s.i_{\text{ds}} + L_m.i_{\text{dr}}$$
$$\Phi_{\text{qs}} = L_s.i_{\text{qs}} + L_m.i_{\text{qr}}$$

The modelisation is made under Matlab/Simulink (Fig. 6.13).

6.2.2.2 The *d-q* Synchronous Motor Model

Using the normalized (*d-q*) rotating reference frame, with i_d andi_q as the (*d*, *q*) stator currents, the machine model based on Park's *d-q* components is described by the following equations [159]:

$$\begin{aligned} V_{\mathrm{ds}} &= R_S I_{\mathrm{ds}} - L_q \omega_{\mathrm{rAC}} . I_{\mathrm{qs}} + L_d \tfrac{dI_{\mathrm{ds}}}{dt} \\ V_{\mathrm{qs}} &= R_S I_{\mathrm{qs}} + L_d \omega_{\mathrm{rAC}} . I_{\mathrm{ds}} + L_q \tfrac{dI_{\mathrm{qs}}}{dt} + \omega_{\mathrm{rAC}} \phi_f \end{aligned} \quad (6.22)$$

with R_s as the armature winding, L_d and L_q are *d* and *q* axial inductances, Φ_f the interlinkage magnetic flux, V_d and V_q the *d* and *q* axial voltages, i_{ds} and i_{qs} the *d* and *q* axial currents and ω_{rAC} the angular velocity of rotor.

The mechanical equation is given as:

$$J \frac{d\Omega_{\mathrm{rAC}}}{dt} = T_{\mathrm{emAC}} - T_L \quad (6.23)$$

where

$$T_{\mathrm{emAC}} = p(\phi_f I_{\mathrm{qs}} + (L_d - L_q) I_{\mathrm{ds}} I_{\mathrm{qs}}) \quad (6.24)$$

with J as the rotor inertia moment, T_L the load torque and p pole the pair number.

6.2.2.3 Displacement Pump

Description

The displacement pump uses a piston or a screw to control the water flow. As compared to the centrifugal pump, the positive displacement pump presents a better efficiency under low power conditions.

Pump Model

We use the same model as above (see Section "Pump model").

6.2.2.4 Application

The experiments are carried out using the pump testing system at the Centre of Development of Renewable Energies located in Algiers-Algeria [62]. We present the experiments conducted for the displacement pump D_1. The water flow Q and electric power input P values are logged at different total heads, for the tested pump D_1.

Fig. 6.14 Example of displacement pump

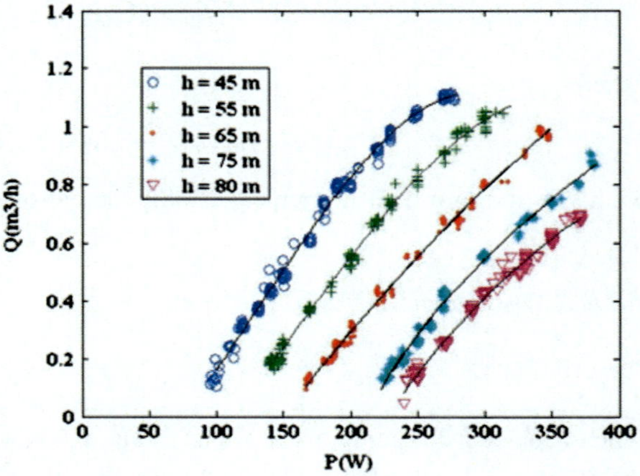

Fig. 6.15 Q–P curves for pump D1 at head: 45, 55, 65, 75, 80 m [62]

Examples of Q–P curves are shown in Fig. 6.15 and the results concerning the RMSE and MBE indicators are given in Fig. 6.16. Good agreement between experimental and calculated data is seen in Fig. 6.15. The low values of the RMSE and MBE validate the present model (Fig. 6.16).

6.2.2.5 Voltage Source Inverter Model

Natural PWM Strategy

A natural PWM strategy is used to drive the full bridge DC–AC inverter with a modulation index m and the ratio between the frequencies of the carrier and modulating waveforms r.

The capacitor across the inverter input terminals as shown in Fig. 6.12, serves to smooth the output voltage of the DC source and to reduce the source impedance.

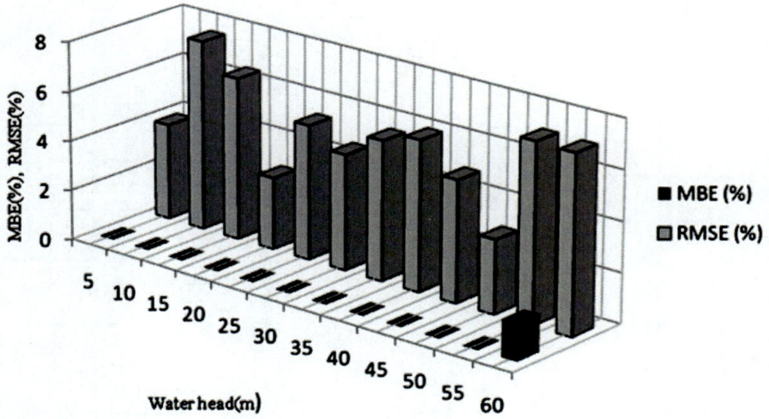

Fig. 6.16 RMSE and MBE results for the centrifugal pumps D_1 [62]

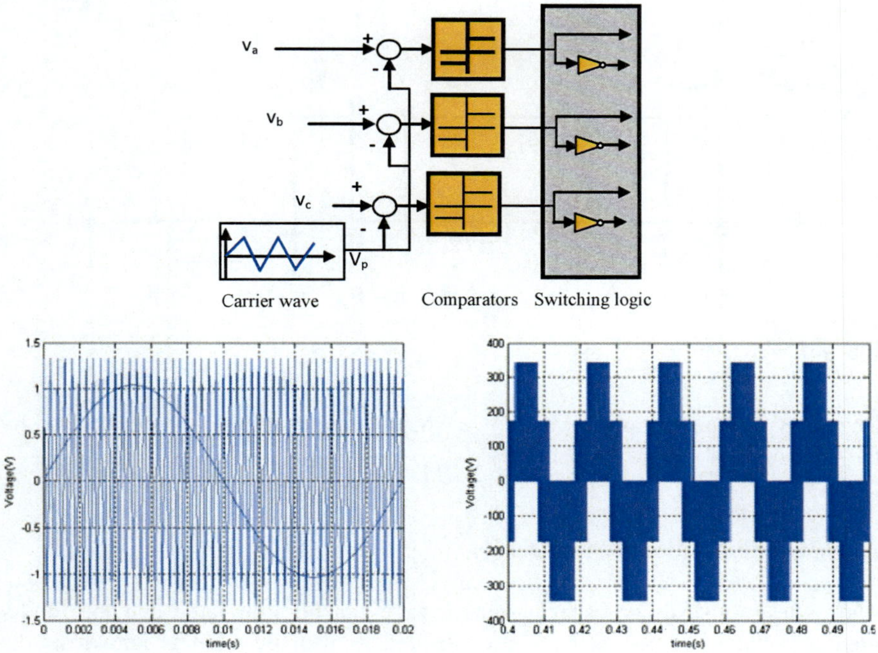

Fig. 6.17 Sinusoidal PWM strategy and voltage waveforms

The three-phase inverter consists of three legs, one for each phase. It is assumed that the switches and diodes are ideal devices. The control signals are generated by modulating three low frequency sinusoidal (v_a, v_b, v_c) with a common high

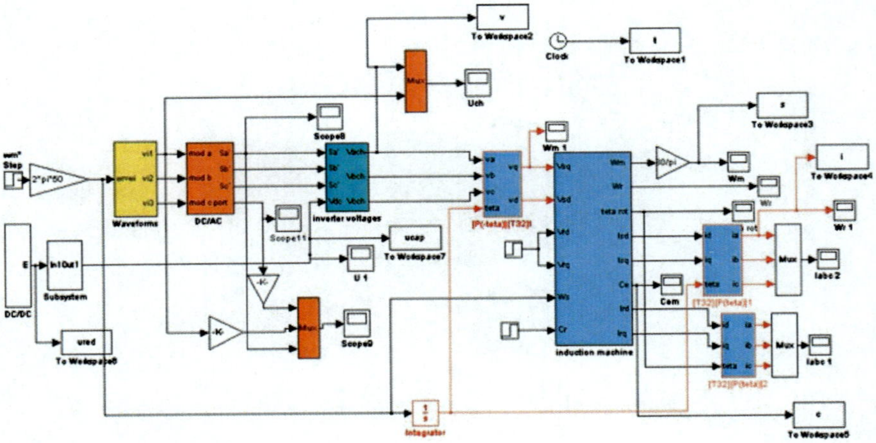

Fig. 6.18 Subsystem of an IM fed by an MLI inverter

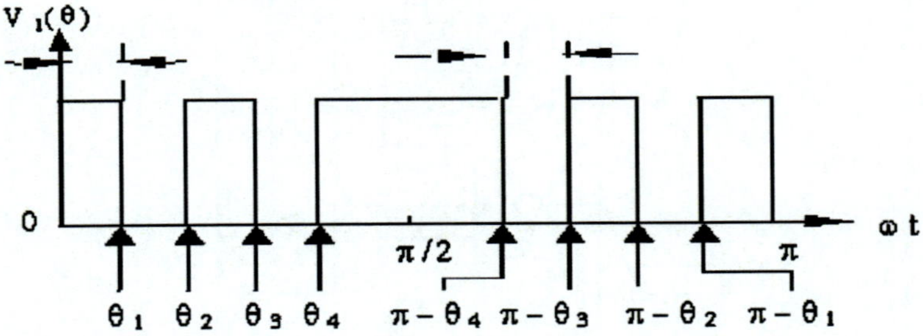

Fig. 6.19 Optimized P.W.M signal

frequency triangular carrier wave (V_p). The switching instants are determined by the crossover of the two waveforms.

Harmonic Elimination PWM Strategy

The undesirable harmonics of a square wave can be eliminated and the fundamental voltage component can be controlled as well by what is known as the harmonic elimination method. In this method, notches are created in the square wave at predetermined angles (Fig. 6.19). It can be shown that the notch angles can be controlled to eliminate certain harmonic components and control the fundamental voltage [158].

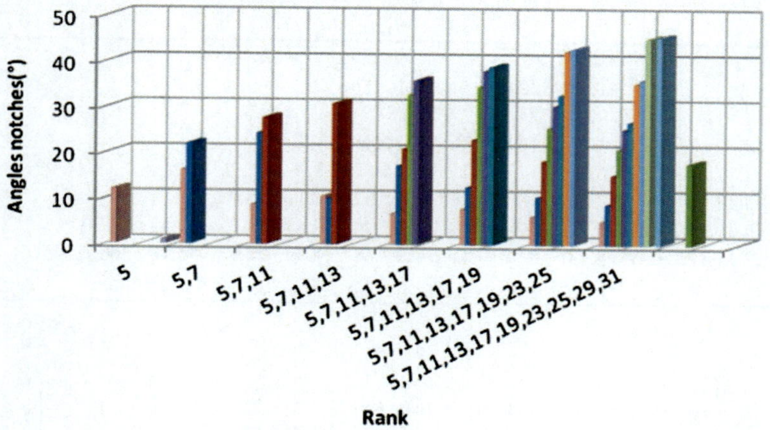

Fig. 6.20 Angles notches for different harmonic ranks

The Fourier series development of a control signal $C_1(\omega t)$ is:

$$C1(\omega t) = a_0 + \sum_{n=1}^{\infty} a_n.\cos(n.\theta) + \sum_{n=1}^{\infty} b_n.\sin(n.\theta) \tag{6.25}$$

where

$$\begin{bmatrix} a_0 = 0 \\ a_n = \frac{2}{\pi}.\int_{\frac{-\pi}{2}}^{\frac{\pi}{2}} C_1(\theta).\cos(n\theta)d(\theta) \\ b_n = \frac{2}{\pi}.\int_{\frac{-\pi}{2}}^{\frac{\pi}{2}} C_1(\theta).\sin(n\theta)d(\theta) \\ \theta = \omega.t \end{bmatrix} \tag{6.26}$$

In a quarter-cycle symmetrical waveform, only the sine odd harmonics exist. Therefore, the coefficients are given by:

$$\begin{bmatrix} a_n = 0 \\ b_n = \frac{4}{\pi}.\int_{0}^{\frac{\pi}{2}} \left[C_1(\theta) - \frac{1}{2}\right].\sin(n\theta).d(\theta) \end{bmatrix} \tag{6.27}$$

Then, the following system of transcendental equations is obtained:

Table 6.1 Angles Notches for 5, 7, 11 ... 31th harmonics[157]

Rank	θ_1	θ_2	θ_3	θ_4	θ_5	θ_6	θ_7	θ_8	θ_9	θ_{10}
5	12.00									
5,7	16.25	22.05								
5,7,11	8.74	24.39	27.75							
5,7,11,13	10.54	10.09	30.90	32.86						
5,7,11,13,17	6.79	17.30	21.03	34.67	35.99					
5,7,11,13,17,19	7.80	12.67	23.08	25.62	38.11	38.99				
5,7,11,13,17,19,23,25	6.19	10.45	18.40	21.04	30.49	32.85	42.43	42.90		
5,7,11,13,17,19,23,25, 29,31	5.13	8.90	15.29	17.88	25.39	26.98	35.40	36.22	45.33	45.61

$$\begin{bmatrix} \frac{1}{2} - \cos(\theta_1) + \cos(\theta_2) - \cos(\theta_3) + \ldots\ldots\ldots + \cos(\theta_n) = hh \\ \frac{1}{2} - \cos 5(\theta_1) + \cos 5(\theta_2) - \cos 5(\theta_3) + \ldots\ldots + \cos 5(\theta_n) = 0 \\ \frac{1}{2} - \cos 7(\theta_1) + \cos 7(\theta_2) - \cos 7(\theta_3) + \ldots\ldots + \cos 7(\theta_n) = 0 \\ \frac{1}{2} - \cos 11(\theta_1) + \cos 11(\theta_2) - \cos 11(\theta_3) + \ldots + \cos 11(\theta_n) = 0 \\ \frac{1}{2} - \cos 13(\theta_1) + \cos 13(\theta_2) - \cos 13(\theta 3) + \ldots\ldots + \cos 13(\theta_n) = 0 \\ \ldots\ldots\ldots\ldots\ldots\ldots\ldots\ldots\ldots\ldots\ldots\ldots\ldots\ldots\ldots\ldots\ldots\ldots \\ \frac{1}{2} - \cos j(\theta_1) + \cos j(\theta_2) - \cos j(\theta_3) + \ldots\ldots + \cos j(\theta_n) = 0 \end{bmatrix} \tag{6.28}$$

where hh is p.u. fundamental voltage value.

This system of nonlinear transcendental equations is solved employing an algorithm written in language Matlab, θ_1, θ_2.... θ_n are then determined (Table 6.1).

– Application:

An application is made on AC motor supplied from a voltage source inverter. The machine is controlled using a position sensor which produces the inverter control signals. This sensor is fixed on to the rotor shaft and allows for supply frequency setting to the value of the rotor frequency. Simulation and experimental results are presented in Fig. 6.21 [159].

Pondered Harmonic Minimization Strategy

We can use another alternative where performance indexes are considered. This approach is to define a performance index related to the undesirable effects of the harmonics and to select the switching angles so that the fundamental voltage is controlled and the performance index is minimized. This is called a distortion

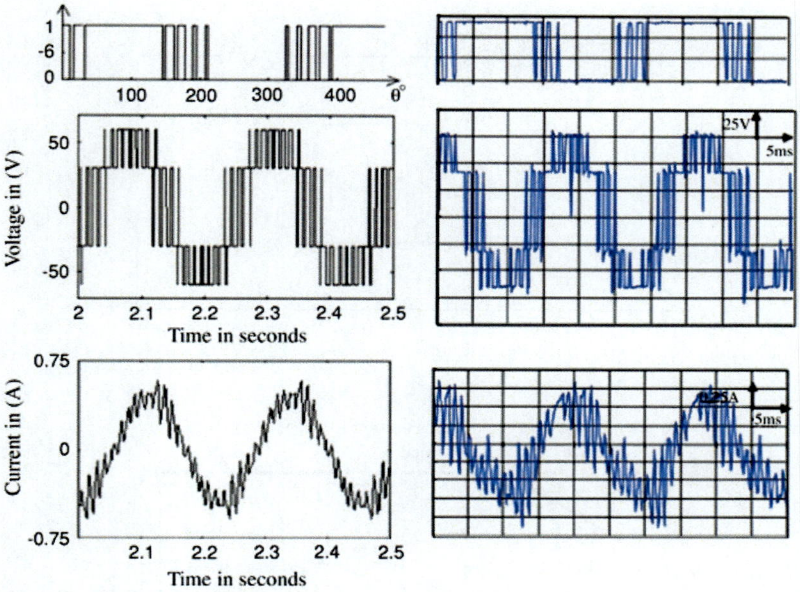

Fig. 6.21 Simulation and experimental results in the case of 5, 7, 11 and 13th harmonics elimination

minimization PWM technique. In this method, the total harmonic voltage distortion factor (THD) is minimized. The minimization of this quantity and the control of the fundamental are achieved by an appropriate positioning of the switching angles. Considering that higher harmonics are easier to filter, we take into account harmonics until *n*th with a weigh decreasing with frequency. The THD performance factor is then defined as:

$$\mathrm{THD} = \sqrt{\sum_{kh \neq 1}^{n} \left(\frac{V_{\mathrm{kh}}}{kh}\right)^2} \tag{6.29}$$

6.2.3 Scalar Control of the PV System

The steady-state performance of an induction motor is modeled using the conventional equivalent circuit of Fig. 6.22, where V_s is the rms phase motor voltage (V), R_s the stator resistance per phase (Ω), R_r the equivalent rotor resistance per phase (Ω), R_m the core loss resistance (Ω), $X_s = \omega_s\ (L_s - L_m)$ the stator leakage reactance (Ω), $X_r = \omega_{\mathrm{s}}\ (\ L_r - L_m)$ the equivalent rotor leakage reactance (Ω),

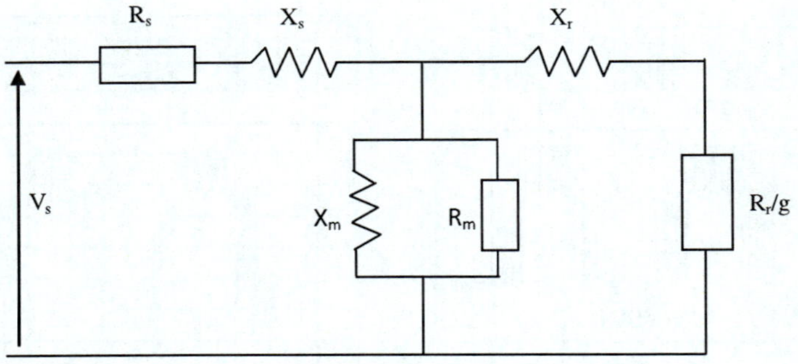

Fig. 6.22 Induction motor equivalent circuit

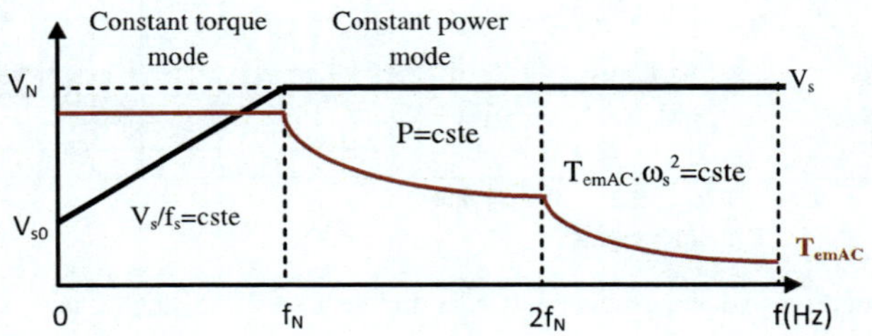

Fig. 6.23 Scalar control

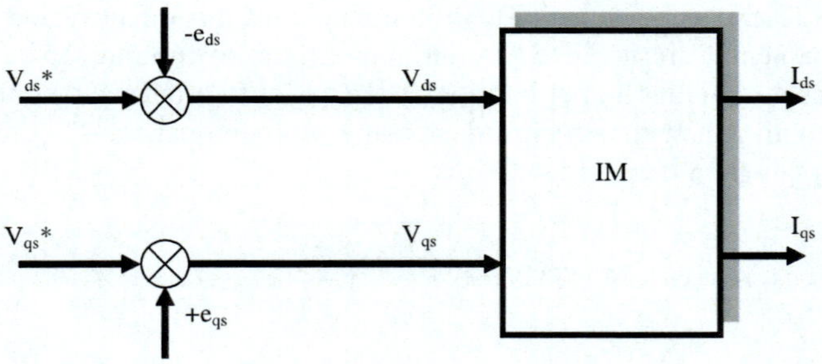

Fig. 6.24 Structure of the decoupling regulator

$X_m = \omega_s L_m$ the magnetizing reactance (Ω), ω_s the angular frequency of the supply (rd/s), ω_{rAC} the motor speed (rd/s) and

$$g = \frac{\omega_s - p.\omega_{rAC}}{\omega_s} \tag{6.30}$$

with p as the pole pair number.

Assuming thatR_m is infinite, the stator current I_s and the rotor current I_r in the elements X_r and R_r/g are related by the phasorial expression

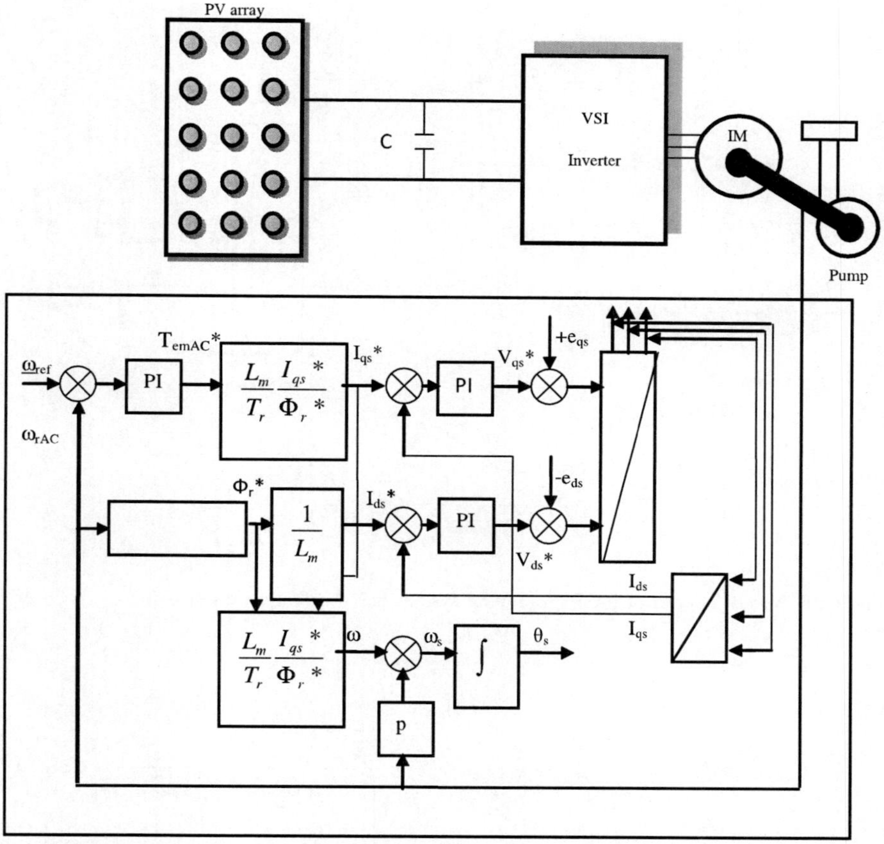

Fig. 6.25 Block diagram of indirect vector control

$$\bar{I}_s = \bar{I}_r + \frac{(R_r/g + jX_s)\,\bar{I}_r}{jX_m} = \frac{R_r/g + j\,L_r\,\omega_s}{j\,L_m\,\omega_s}\,\bar{I}_r$$

It is then easy to compute I_r as a function of V_s. One obtains

$$\bar{I}_r = \frac{j\,\omega_s\,L_m\,V_s}{R_sR_r/g - \omega_s^2(L_rL_s - L_m^2) + j\omega_s(R_rL_s/g + R_sL_r)}$$

The expression of electromagnetic torque can be written as:

$$T_{\mathrm{em\,AC}} = 3\,\frac{R_r}{g}\,I_r^2\,/\,(\omega_s/p)$$

or

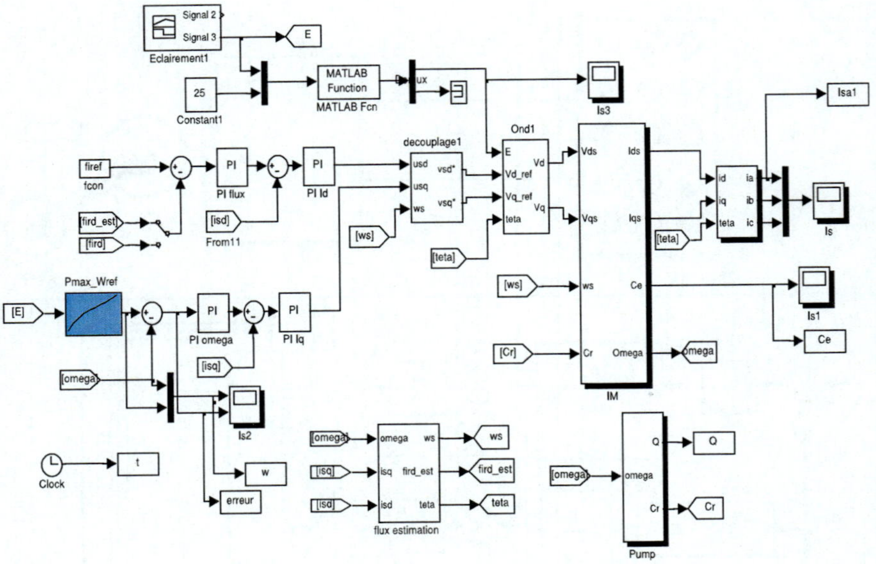

Fig. 6.26 Vector control of the PV pumping system based on induction machine

$$T_{\text{emAC}} = 3p.V_s^2 \frac{\frac{R_r}{g}\omega_s L_m^2}{\left[\frac{R_r R_s}{g} - \omega_s^2.(L_s L_r - L_m^2)\right]^2 + \omega_s^2\left[\frac{R_r L_s}{g} + R_s L_r\right]^2} \tag{6.31}$$

Power converted in mechanical power, stator current and stator power will be:

$$P_{\text{conv}} = 3\frac{R_r}{g} I_r^2 (1-g) \tag{6.32}$$

$$I_s = \left|1 + \frac{R_r/g + j(L_r - L_m)\,\omega_s}{j\,L_m\,\omega_s}\right| I_r = \left|\frac{R_r/g + j\,L_r\,\omega_s}{j\,L_m\,\omega_s}\right| I_r \tag{6.33}$$

$$P_s = P_{\text{conv}} + P_{jr} + P_{js} \tag{6.34}$$

From the previous equations, we obtain the total power absorbed by the stator:

$$P_s = 3\left[\frac{R_r}{g} + R_s\frac{(R_r/g)^2 + L_r^2\omega_s^2}{L_m^2\omega_s^2}\right] I_r^2$$

or, according to speed:

$$P_s = 3k^2 \left[\frac{(\omega_s - p\,\omega_{\mathrm{rAC}})\,\omega_s}{R_r \left(\frac{L_s}{L_r}\right)^2} + R_s \frac{R_r^2 + L_r^2\,(\omega_s - p\,\omega_{\mathrm{rAC}})^2}{L_s^2 \omega_s^2} \right] V_s^2 \tag{6.35}$$

with

$$k^2 = \frac{L_s^2 R_r^2}{[R_r R_s - g\,\omega_s^2\,(L_r L_s - L_m^2)]^2 + \omega_s^2\,[R_r L_s + g\,R_s L_r]^2}$$

Generally, in variable speed drives, motor air-gap flux is maintained constant at all frequencies so that the motor can deliver a constant torque. This will occur if the V_s/f_s ratio (or V_s/ω_s) is kept constant at its nominal value. To compensate the voltage drop due to stator resistance effect at low frequencies, a boost voltage V_{s0} is added to phase voltages. For aerodynamic loads, the stator voltage as function of frequency is given by

For $0 \leq f \leq f_N$ $\quad V_s = V_{s0} + x.f_s$

For $f \geq f_N$ $\quad V_s = V_N$

where $x = (V_N - V_{s0})/f_N$ is the slope of the linear part.

6.2.4 Vector Control of the PV System Based on Induction Machine

We oriented the rotoric flux ϕ_r along the direct axis

$$\begin{cases} \Phi_{\mathrm{dr}} = \Phi_r \\ \Phi_{\mathrm{qr}} = 0 \end{cases} \tag{6.36}$$

We obtain

$$\begin{cases} V_{\mathrm{ds}} = R_s I_{\mathrm{ds}} + \sigma L_s \dfrac{dI_{\mathrm{ds}}}{dt} + \dfrac{L_m}{L_r}\dfrac{d\Phi_r}{dt} - \omega_s.\sigma L_s.I_{\mathrm{qs}} \\ V_{\mathrm{qs}} = R_s I_{\mathrm{qs}} + \sigma L_s \dfrac{dI_{\mathrm{qs}}}{dt} + \omega_s \dfrac{L_m}{L_r}\Phi_r + \omega_s.\sigma L_s.I_{\mathrm{ds}} \\ \omega_{\mathrm{rAC}} = \dfrac{L_m}{T_r}\dfrac{I_{\mathrm{qs}}}{\Phi_r} \\ T_{\mathrm{emAC}} = p\dfrac{L_m}{L_r}(\Phi_r.I_{\mathrm{qs}}) \end{cases} \tag{6.37}$$

The new control is as follows:

$$\begin{cases} V_{ds}* = (R_s + s.\sigma L_s)I_{ds} = V_{ds} + \omega_s.\sigma L_s.I_{qs} = V_{ds} + e_{ds} \\ V_{qs}* = (R_s + s.\sigma L_s)I_{qs} = V_{qs} - \left(\omega_s \frac{L_m}{L_r} \Phi_r + \omega_s.\sigma L_s.I_{ds}\right) = V_{qs} - e_{qs} \end{cases} \quad (6.38)$$

with σ as the leakage coefficient:

$$\sigma = 1 - \frac{L_m^2}{L_s L_r}$$

The structure of the indirect vector control is given by

The pump torque and speed are set in relation in the following equation [160]:

$$T_{emAC} = KP.\omega_{rAC}^2 \quad (6.40)$$

where KP is the pump constant.

Thus, the mechanical output power of induction motor is given by:

$$P_{mec} = KP.\omega_{rAC}^3 \quad (6.41)$$

Neglecting inverter and motor losses, from energy conservation (P_{in} = -$P_{out} = P_{pv}$) the angular speed is:

$$\omega_{rAC} = \sqrt[3]{\frac{P_{pv}}{KP}} \quad (6.42)$$

The mechanical torque of induction motor pump can be written as:

$$T_{emAC} = \sqrt[3]{KP.P_{pv}^2} \quad (6.43)$$

6.2.5 DTC Control of the PV System

6.2.5.1 DTC Principles

Direct torque control (DTC) of induction machines (IM) is a powerful control method for motor drives. Featuring a direct control of the stator flux and torque instead of the conventional current control technique, it provides a systematic solution to improve the operating characteristics of the motor and the voltage inverter source [147, 148]. In principle, the DTC method is based mainly on instantaneous space vector theory. By optimal selection of the space voltage vectors in each sampling period,the DTC achieves effective control of the stator flux and torque [148]. Consequently, the number of space voltage vectors and switching frequency directly influence the performance of DTC control system.

Table 6.2 Switching table for the conventional DTC

E_T	E_Φ	N					
		1	2	3	4	5	6
$E_T = 1$	$E_\Phi = 1$	$V_2(110)$	$V_3(010)$	$V_4(011)$	$V_5(001)$	$V_6(101)$	$V_1(100)$
	$E_\Phi = 0$	$V_6(101)$	$V_1(100)$	$V_2(110)$	$V_3(010)$	$V_4(011)$	$V_5(001)$
$E_T = 0$	$E_\Phi = 1$	$V_3(010)$	$V_4(011)$	$V_5(001)$	$V_6(101)$	$V_1(100)$	$V_2(110)$
	$E_\Phi = 0$	$V_5(001)$	$V_6(101)$	$V_1(100)$	$V_2(110)$	$V_3(010)$	$V_4(011)$

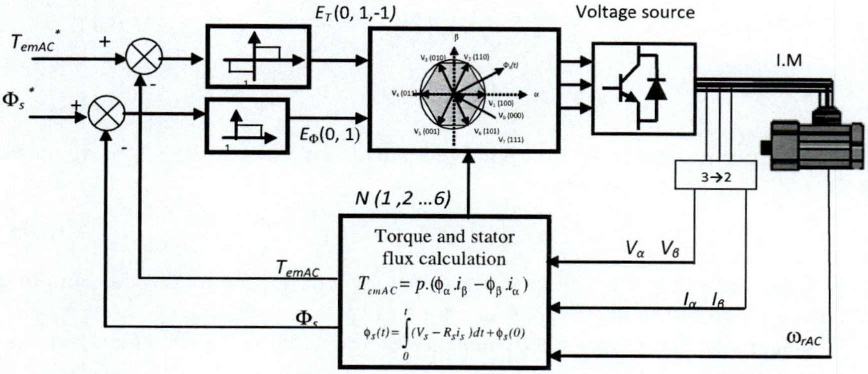

Fig. 6.27 Block diagram of the conventional direct torque control of induction motor drives

For a prefixed switching strategy, the drive operation, in terms of torque, switching frequency and torque response, is quite different at low and high speeds.

6.2.5.2 DTC Structure

A configuration of the DTC scheme is represented in Fig. 6.27. In this system the instantaneous values of flux and torque can be calculated from stator variables and mechanical speed or using only stator variables. Stator flux and torque can be controlled directly and independently by properly selecting the inverter switching configurations. With a three-phase voltage source inverter, six non-zero voltage vectors and two zero-voltage vectors can be applied to the machine terminals.

The stator flux can be estimated using measured current and voltage vectors [161]:

$$\phi_s(t) = \int_0^t (V_s - R_s i_s)dt \tag{6.44}$$

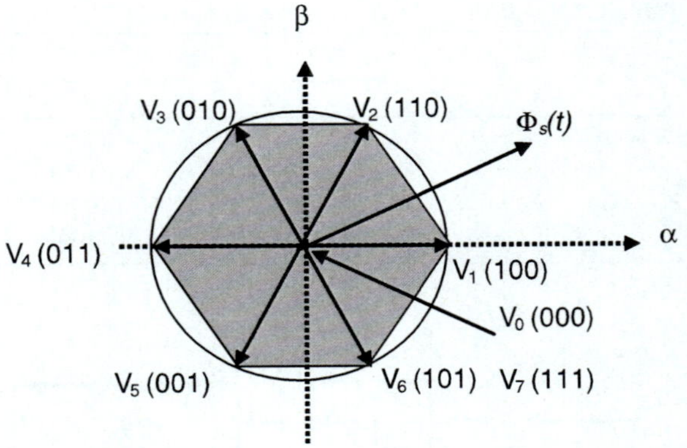

Fig. 6.28 Movement of the inverter voltage in the space-vector plane

Since stator resistance R_s i_s is relatively small, the voltage drop $R_s.i_s$ might be neglected $(V_s \gg R_s i_s)$, we obtain:

$$\phi_s(t) = Vs.T + \phi_s(0) \tag{6.45}$$

$\phi_s(0)$ is the stator flux initial value at the switching time and T the sampling period in which the voltage vector is applied to stator windings.

It is clear that stator flux directly depends on the space voltage vector V_s and the system sampling period T.

The stator voltage vector V_s is selected using Table 6.2., where signs of torque and flux errors E_T and E_Φ are determined with a zero hysteresis band (Fig. 6.27).

$$E_T = T_{\text{eref}} - T_{\text{emAC}} \tag{6.46}$$

$$E_\varphi = \phi_{\text{sref}} - \phi_s \tag{6.47}$$

where

$$\phi_s = \sqrt{(\phi_{s\alpha})^2 + (\phi_{s\beta})^2} \tag{6.48}$$

Table 6.2. shows the associated inverter switching states of the conventional direct torque control strategy.The definition of flux sector and inverter voltage vectors are shown in Fig. 28, where the stator flux vector is rotating with a speed of ω_{rAC}. For each possible switching configuration, the output voltages can be represented in terms of space vector, according to the following equation:

$$V_S = V_{S\alpha} + jV_{S\beta} = \sqrt{\frac{2}{3}}\left[V_a + V_b \exp\left(j\frac{2\pi}{3}\right) + V_c \exp\left(j\frac{4\pi}{3}\right)\right] \tag{6.49}$$

where V_a, V_b and V_c are voltage phases.

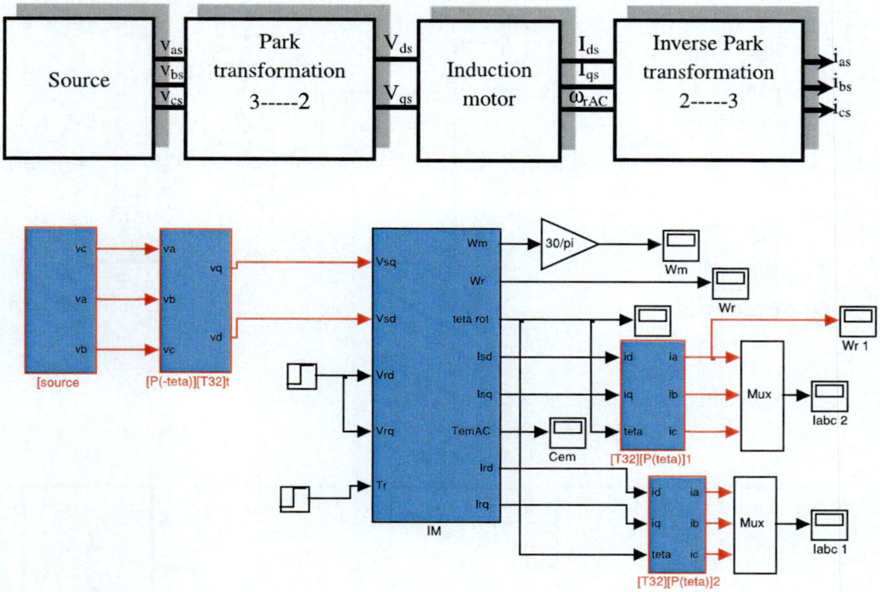

Fig. 6.29 Simulation diagram of induction motor

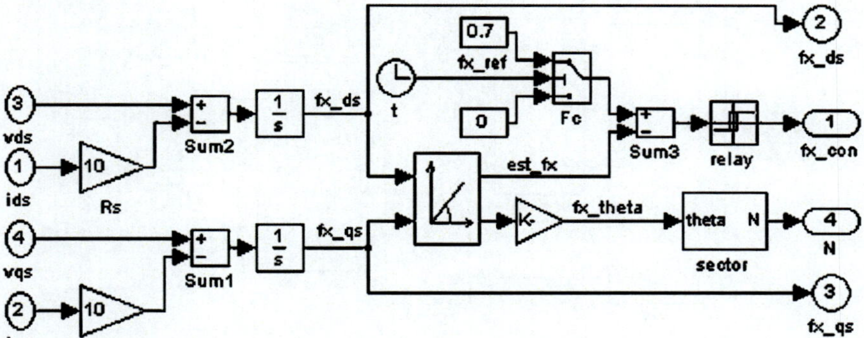

Fig. 6.30 Simulation diagram of stator flux control

6.2.5.3 Application to Induction Motor

The DTC control applied to the induction motor is simulated through Matlab Simulink. The simulation model of the induction motor is given in Fig. 6.29.

The simulation diagram of the stator flow control stator is illustrated in Fig. 6.30. The control signal flow (fx_con) is generated by a comparator with

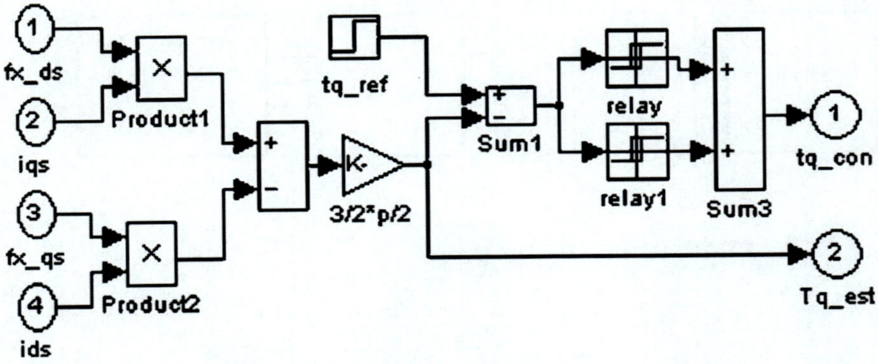

Fig. 6.31 Simulation diagram of the electromagnetic torque control

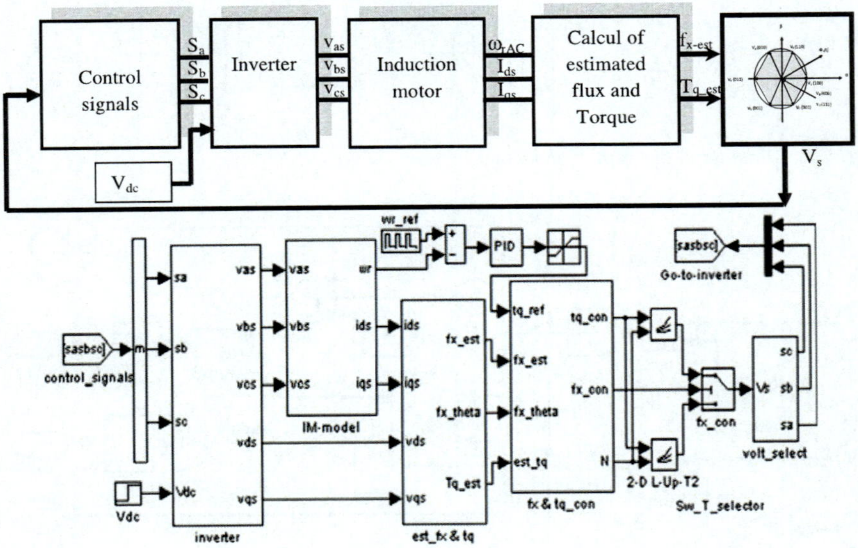

Fig. 6.32 Simulation diagram of the overall structure of the DTC control

hysteresis, after estimating the vector flux and compares the flow module to its reference value.

Fig. 6.32. shows the block diagram of the overall system.

Some simulation results are presented.

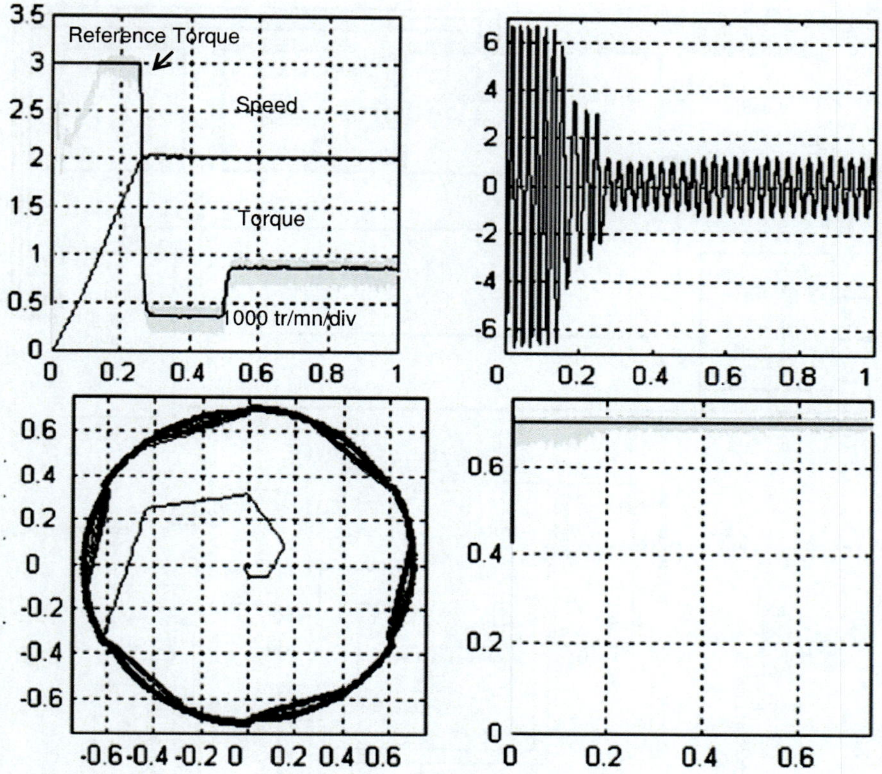

Fig. 6.33 Evolution of the torque, speed, stator current and stator flux with a speed reference

6.2.5.4 DTC of IM Fed by a PV Generator

Description of the Global System

The block diagram is given in Fig. 6.34.

Application Under Matlab/Simulink

We make an application (Fig. 6.35) under Matlab/simulink. Some simulation results are presented in (Fig. 6.36).

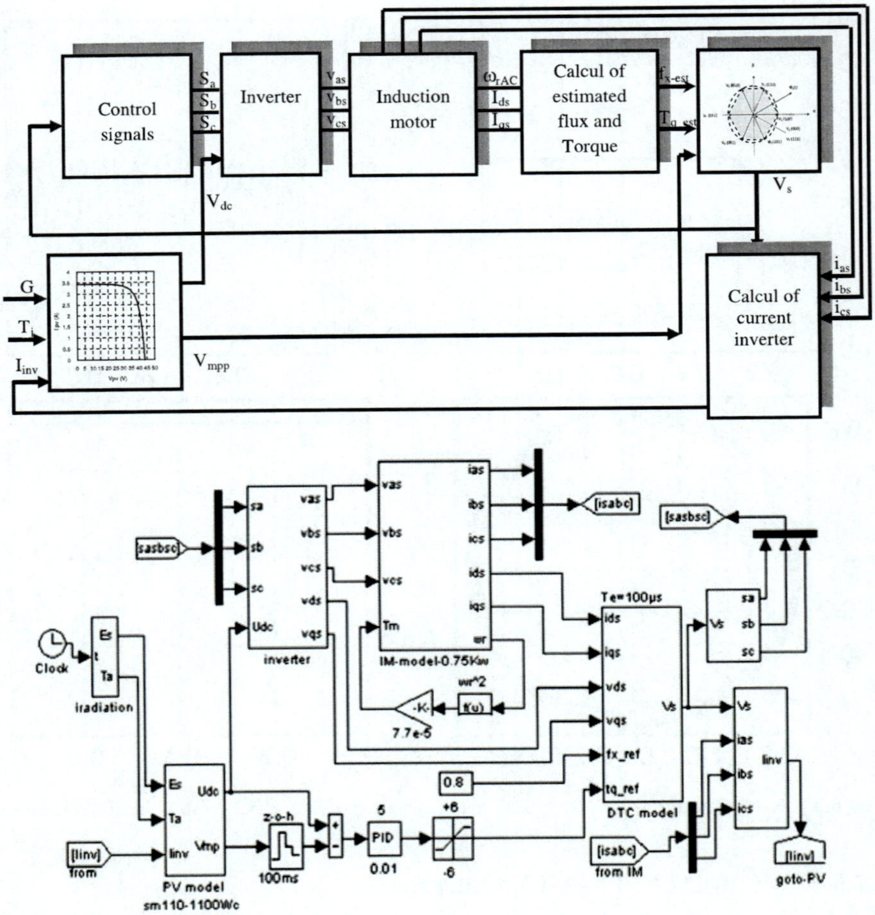

Fig. 6.34 Simulation of direct torque control of the induction generator fed by PV generator

6.3 Maximum Power Point Tracking for Solar Water Pump

6.3.1 With DC Machine

Figure 6.37 shows the maximum power tracking system with DC/DC converter. The system consists of a PV array, a DC to DC boost converter with MPPT and the DC motor coupled to a centrifugal pump.

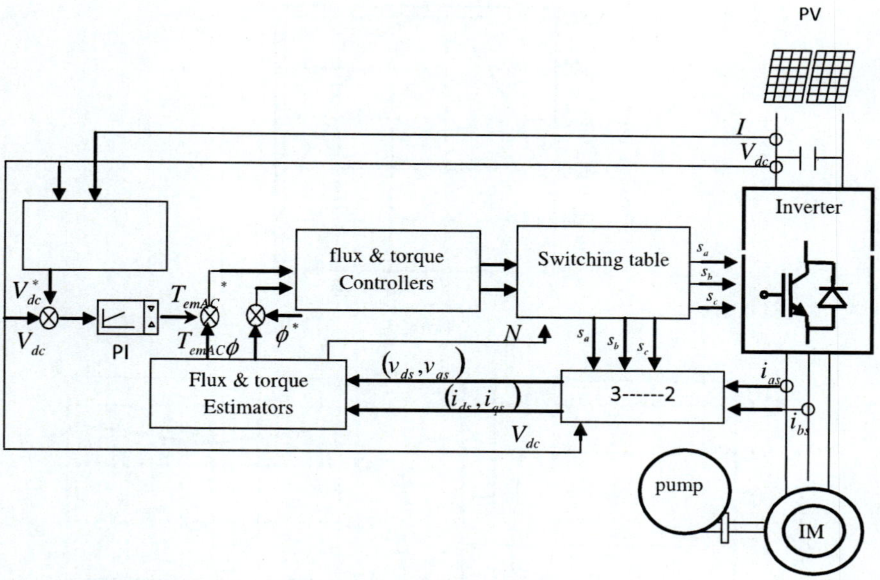

Fig. 6.35 Studied system

6.3.2 With AC Machine

Figure 6.38 illustrates the maximum power tracking system for three-phase AC motors.

6.4 Economic Study

This section presents an analysis of the economic feasibility of the PV pumping system in comparison with systems using diesel generators. The economic study consists of determining first the capital cost for the PV system and diesel alternative and then calculating the life cycle costs. The life cycle cost uses a combination of costs to measure the cost effectiveness of a specific pumping system. The study below is only a rough estimation because it uses relations with very simplified formulae and keeps between mean values relations obtained for instantaneous values. It is however valuable for fast comparison between different designs.

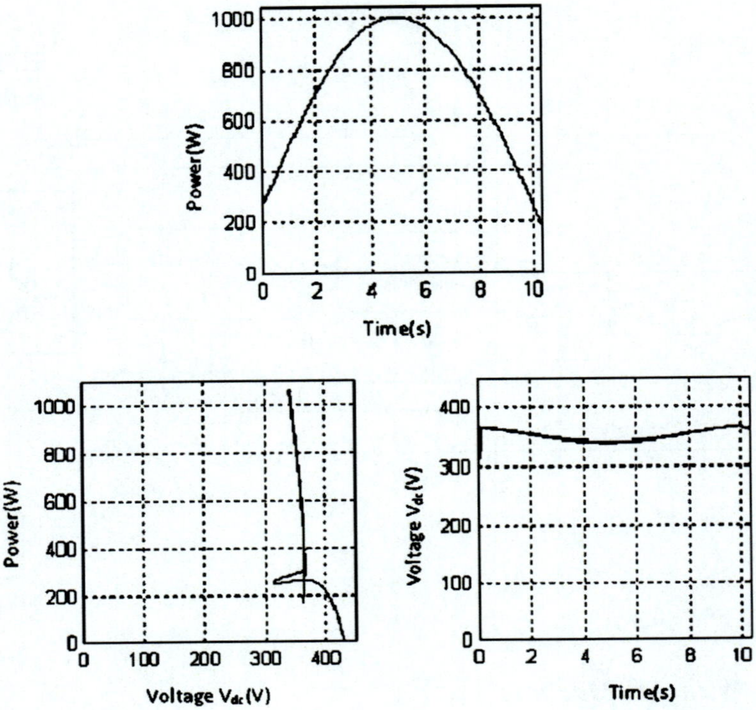

Fig. 6.36 Simulation results obtained from the PV generator according to the variation in irradiance

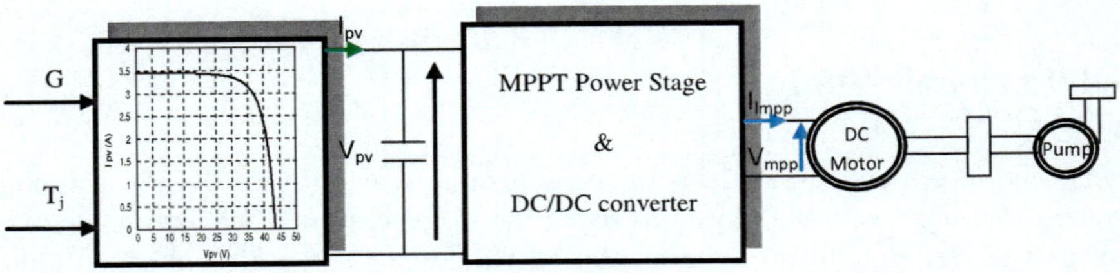

Fig. 6.37 Maximum power point tracking with DC motor

6.4.1 Estimation of the Water Pumping Energy Demand

The required electric energy E_a that is to be delivered by either the PV or the diesel system is determined by:

$$E_a = P_a t_s \tag{6.50}$$

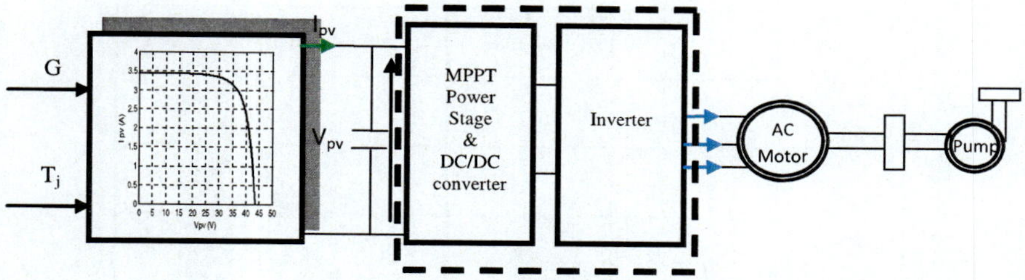

Fig. 6.38 Maximum Power Point Tracking with AC motor

where P_a is the average electrical input power and t_s is the average working time.

The necessary energy to be delivered by a PV generator taking into account the mismatch factor is calculated as follows:

$$E_{pv} = F\,E_a \tag{6.51}$$

where F is the mismatch factor.

The peak power of the PV generator is given by:

$$P_{\text{pv}} = E_{\text{pv}}\,\frac{G_{\text{ref}}}{G} \tag{6.52}$$

where G_{ref} is the peak solar irradiance intensity (1 kW/m^2) and G is the annual average of solar irradiance on a horizontal surface.

The water flow Q is calculated by the corresponding daily pumped water volume V is given by:

$$V = Q\,t_s \tag{6.53}$$

6.4.2 Life Cycle Cost (LCC) Calculations

The LCC analysis uses a combination of the initial capital cost, operation, maintenance costs and replacement costs with economic assumptions. The life cycle cost calculates the present worth of all expenses expected to occur over the life cycle of the system by using the following equation [153]:

$$\text{LCC} = C + M + R \tag{6.54}$$

where C is the capital cost ($), M is the sum of all operation and maintenance costs ($), R is the sum of all equipment replacements costs ($).

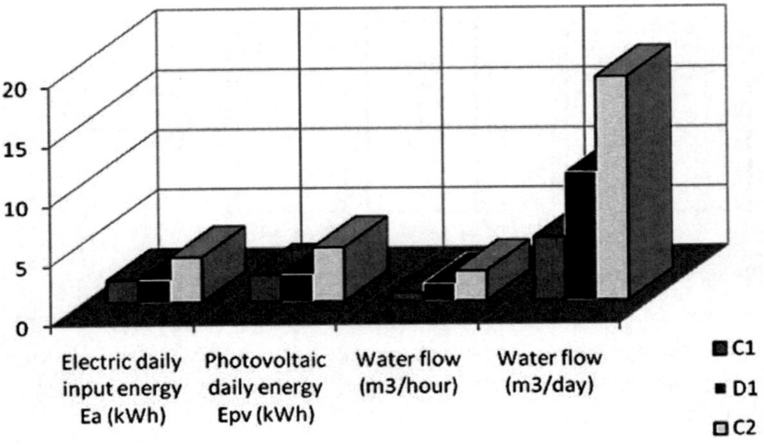

Fig. 6.39 Working parameters of three-pumps with a total head of 20 m [62]

6.4.2.1 LCC Analysis of PV Pumping System

The initial capital cost of the PV pumping system can be calculated by adding up the costs of the PV components (modules, the motor pump subsystem) and auxiliary costs. The auxiliary costs include the costs of the system engineering and planning, panel structure, wiring and miscellaneous items. The engineering and planning cost is assumed as 9% of the PV components cost. The panel supporting structure cost is taken to be about 5% of the PV equipments cost. The wiring and other items cost is considered to be 0.2% of the PV subsystems cost. The total initial investment of the PV pumping system is calculated as follows:

$$C = C_{pv} + C_{sub} + C_{aux} \tag{6.55}$$

where C_{pv} is the initial capital cost of the photovoltaic modules, C_{sub} is the initial capital cost of the motor pump subsystem and C_{aux} is the initial auxiliary capital cost.

Normally, a life cycle analysis deals with totals of annual costs during the life cycle period, but it is more convenient to deal with the life cycle cost on an annual basis. The annualized life cycle costs are calculated using the following formulas:

$$\mathrm{CRF} = \frac{d}{1-(1+d)^{-T}} \tag{6.56}$$

$$\mathrm{PWF} = \frac{1+d}{1+i} \tag{6.57}$$

$$C_y = C\ \mathrm{PWF} \tag{6.58}$$

$$C_k = R_k\,PWF \tag{6.59}$$

$$R_y = \sum_k C_k \tag{6.60}$$

$$A_y = C_y + M_y + R_y \tag{6.61}$$

where CRF is the capital recovery factor, d is the discount rate, T is the lifetime period, PWF is the present worth factor, i is the interest rate, C_y is the annualized capital cost, C_k is the present worth of replacement at year k, R_k is the cost of replacement of a system component at year k, R_y is the present worth of all replacements incurred during the lifetime T, M_y is the yearly operating and A_y is annualized life cycle cost.

6.4.2.2 LCC Analysis of Diesel Pumping System

Usually, diesel generators are oversized as compared to the power needs of water pumping systems [153]. In Algeria, the available diesel generators are rated at 5 kW or larger resulting in high investment cost, operating cost, maintenance and replacement costs. These costs are added up to the diesel fuel cost to calculate the total life cycle cost. To produce the electric energy E_a determined, it is necessary to use a fuel volume V_d. This volume as calculated by Eq. 6.2, depends on efficiency η_d of the diesel generator and the heating value k_d of the fuel. The average efficiency η_d of various generators varies between 0.299 and 0.353 as reported by [163].

$$V_d = \frac{E_a}{\eta_d\,k_d} \tag{6.62}$$

where η_d is the diesel generator energy conversion efficiency and k_d is the heating value of diesel.

The annual cost for diesel fuel consumption is found by:

$$A_f = N_{\text{days}}\,F_d\,V_d \tag{6.63}$$

where F_d is the actual diesel price;
N_{days} Number of days (365 days).

6.4.2.3 Pumped Water Cost

The cost of m^3 of the pumped water C_w by PV and diesel system pumping systems is calculated, using the cost annuity method based on the LCC analysis by Eq. 6.64:

$$C_w = \frac{A_y}{N_{\text{days}} V} \tag{6.64}$$

where A_y is the annualized cost of the considered PV or diesel pumping system and V is the daily pumped water.

6.4.3 Environmental Aspects of PV Power Systems

6.4.3.1 Introduction

Generally, PV systems do not emit substances that may damage human health or the environment. PV systems do not produce any noise, toxic gas emissions or greenhouse gases. In fact, PV systems can lead to significant emission reductions.

6.4.3.2 Water Pumping Planning and Evaluation of Carbon Dioxide Reduction

The reduction of CO_2 emissions is estimated in the case of photovoltaic pumping facilities installed instead of diesel pumping systems.

Water Planning

For water planning, the available electrical power input is assumed to be known and the working conditions are specified (type of pumps, pumping heads, average working time). Using Eq. 6.65 with the average values, the average water quantity is calculated by

$$Q_a = a\,(h_a)\,P_a^3 + \; b\,(h_a)\;P_a^2 + \; c\,(h_a)\;P_a \; + \; d\;(h_a) \tag{6.65}$$

where Q_a is the average water flow (m^3 per hour), P_a is the average electrical power input to the motor pump (W) and $a(h_a)$, $b(h_a)$, $c(h_a)$, $d(h_a)$ are coefficients corresponding to the working head h_a.

The number N of pumps needed to deliver a fixed water volume Q_w is simply expressed by:

$$N = \frac{Q_w}{N_{\text{days}}\; Q_a t_a} \tag{6.66}$$

where N is the number of pumps, Q_w is the average water production objective (m^3 per year) and t_a is the average working time (hours).

Evaluation of CO_2 Savings

The greenhouse gas emissions from Arab countries was 986 million tons in 1999 [162]. The main component of these gases is CO_2 [162]. Electric energy in remote Algerian locations is usually produced by means of diesel generators [163]. The carbon dioxide emissions are expected to increase due to population growth and economical development. To limit CO_2 emissions, it is necessary to encourage the development of water pumping systems powered by photovoltaic energy. The method used in this work compares between the PV water pumping systems with diesel-based generators water installations in Algerian remote locations. The calculation of the savings is carried out as follows:

First, the total electrical energy needed by N pumps in a PV system is given by:

$$E_w = N\,P_a t_a \tag{6.67}$$

where E_W is the total electrical energy (kWh), N is the number of pumps, P_a is the average power input of a pump (W) and t_a is the average working time of a pump (hours).

The second step is to find the quantity of diesel fuel that is necessary to provide the same electric energy E_w by using Eq. 6.67. In the third step, the exhausts volume V_c of carbon dioxide is calculated by the following relation:

$$V_c = k_c\,V_d \tag{6.68}$$

where V_d is the consumption of the diesel fuel and k_c is the carbon dioxide weight equivalent for the diesel fuel [164].

By combining Eqs. 6.67 and 6.68, the volume of carbon dioxide reductions V_c is expressed as a function of the average pump power, average working time and number of pumps, as follows:

$$V_c = \frac{k_c}{\eta_d\,k_d} N P_a t_a \tag{6.69}$$

where P_a is the average power of a pump (W), t_a is the average working time of a pump (hours), η_d is the diesel generator energy conversion efficiency and k_d is the heating value of diesel.

Application

In Algeria the total peak power of the installed pumping systems is not precisely known. Nevertheless, according to the Ministry of Energy and Mines, the reported peak power of the installed water pumping sets is around 54 kW [165]. This corresponds approximately to only 54 water pumps systems, showing that much further developments may be achieved. Installing much more photovoltaic powered water sets is encouraged by the recent governmental policy in the domain of

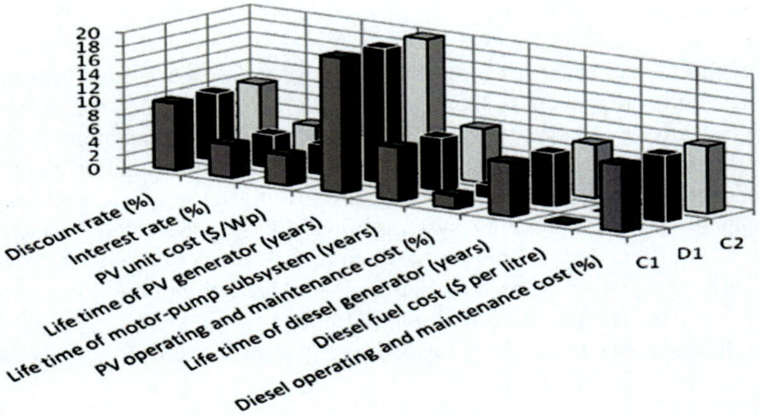

Fig. 6.40 Economic assumptions for three different pumps (C1, D1, C2)

renewable energies. The Algerian government stated that 10% of the electric energy in 2020 is to be provided by renewable energies sources. Increasing the number of water pumping systems is an interesting way in order to reach the above-mentioned 20% objective. In this context, raising the peak power of installed PV water systems from 54 kW to 1 MW, corresponding to 1000 pumps of C_2 type, is possible and constitutes a reasonable goal. Not only will satisfy this choice the water demands in remote locations and improve the life of isolated populations, but it also alleviates the negative impact of diesel sets ups on the climate. For the case of 1000 pumps of type C_2, working at the average conditions of a power input equal to 950 W and during 5 h a day, the carbon dioxide exhausts reductions are evaluated by Eq.6.23 and attain the amount of 4.2 ton/year. This shows that larger amounts of carbon gas emissions might be avoided by an extended application of PV systems in the water pumping field. There is a vast potential for the diffusion of the PV pumping technology in Algeria, not only for drinking water applications but for irrigation as well. The potential number of PV pumping systems is estimated at about 13,300 to provide 40 l/day per person to satisfy the drinking water demand of the 2.66 million inhabitants living in the remote regions of Algeria [60].

The case study considers PV and diesel pumping facilities using pumps C_1(400 W), C_2(1000 W) and D_1(400 W), assuming for the three systems that the total head is 20 m and an average daily working time t_s equal to 7.5 h.

C1 and C2 are centrifugal pumps and D1 a positive displacement pump. The average electrical input power P_a is supposed to be equal to 250 W for pumps C_1 and D_1, and equal to 500 W for pump C_2. Figure 6.40 presents the working parameters and the size of the PV array for the three pumps C_1, C_2 and D_1.

For C_1, D_1 and C_2, the corresponding volumes of daily pumped water at the specified conditions are respectively equal to about 5, 19 and 11 m^3. Assuming a

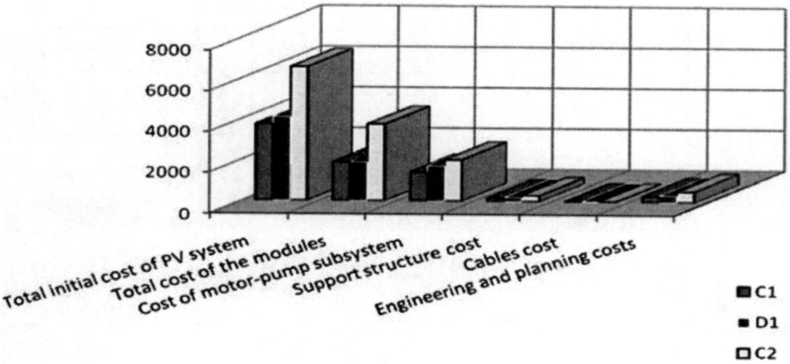

Fig. 6.41 Total initial investment costs of the PV systems ($)

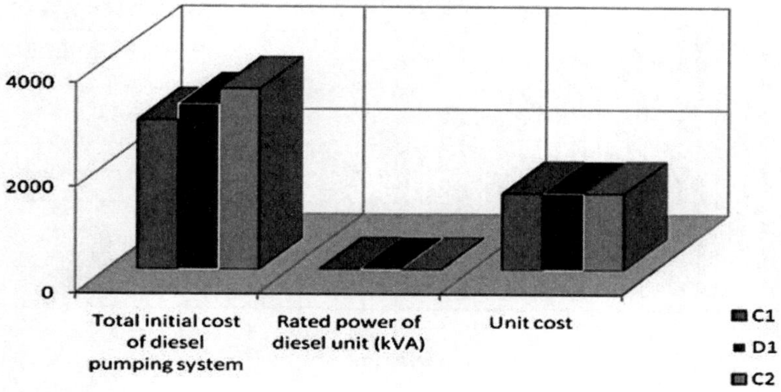

Fig. 6.42 Total initial investment costs of the Diesel systems ($)

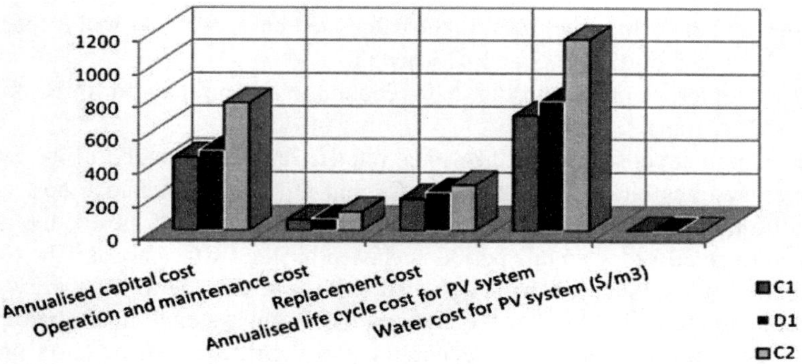

Fig. 6.43 Annuity and water cost calculations for the PV systems ($)

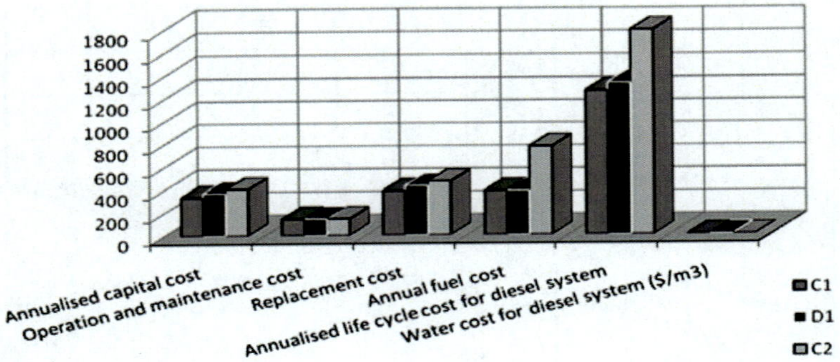

Fig. 6.44 Annuity and water cost calculations for diesel systems ($)

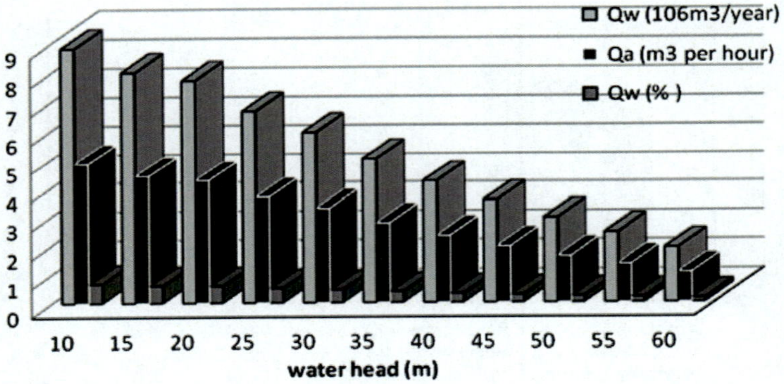

Fig. 6.45 Water produced by 1000 pumps of type C_2 with an average working time of 5 h, an input power of 950 W at different heads

daily demand of 40 liters per person, the calculated pumped water will respectively satisfy the needs of 125, 275 and 475 people.

For example, we present in Fig. 6.40. economic assumptions for three different pumps (C_1, C_2 and D_1).

The equipment costs provided by local retailers are represented in Fig. 6.41. for the three systems based on pumps C_1, C_2 and D_1.The maintenance cost is considered to be 2% of the total investment costs. The lifetime of the PV modules is assumed to be 20 years, so this period is chosen for the LCC analysis. The lifetime of the motor pump subsystems is assumed to be 8 years, due to the hard working conditions in the major parts of the country. These subsystems should be replaced to match the lifetime of the PV generator and their replacement costs must be considered.

The total initial costs of diesel systems are presented in Fig. 6.42.

For the diesel pumping system, the initial capital cost includes the diesel generator and the motor pump subsystem. The annual fuel cost A_f and the maintenance cost is assumed to be equal to 10% of the diesel generator cost. Replacements of the diesel generator and motor pump subsystem are assumed to occur after equal lifetimes of 8 years.

As expected, the calculated cost of initial investment for PV systems is higher than the initial cost of systems based on diesel generators. But on an annual basis, PV systems are more economical than diesel systems. The cost of water pumped by a photovoltaic system is lower than that pumped by a facility based on a diesel generator. We present in Fig. 6.45 the results of calculations corresponding to 1000 C_2 pumps, working 5 h per day with a power input of 950 W for each one. At each total head, the required energy to cover between 0.14 and 0.66% of the Algerian drinking water consumption (in 2000), is approximately 4.8 MWh [60].

In Fig. 6.45, Q_w is also expressed as a ratio of the PV pumped water to the withdrawals (1.33×10^9 m^3) in 2000 [62].

Chapter 7
Hybrid Photovoltaic Systems

Symbols

a_1, a_2 and a_3	Constants
A_{pv}	Total area of the photovoltaic generator
b_1, b_2, b_3 and b_4	Constants
c	Load consumption
C^*O_2	Oxygen concentration in the cathode area (mol/cm^3)
C_p	Heat capacity of gas (kJ/°K)
$Cp(\lambda, \beta)$	Power coefficient
DOD	Depth of discharge
E	Tolerance predefined
E_{Nernst}	Cell thermodynamic potential associated to reversible voltage (V)
f	Fraction of load supplied by the photovoltaic energy
F	Faraday constant (C/mol)
f_{comp}	Compressor molar flow (mol/s)
$(F_{H_2})_{purg}$	Purge flow system if the anode compartment is not closed
F_{mass}	Compressor mass flow (g/s)
F_{molar}	Molar flow of the compressor (mol/s)
F_{ref}	Molar flow reference in the valve (mol/s)
F_{steam}	Corresponds to the amount of steam supplied by the humidification system (mol/s)
F_{valve}	Molar flow through the valve (mol/s)
G_{in}	Solar irradiance on an inclined plane
G_x	Overall gear ratio
I_{batt}	Battery current
i_{ds} and i_{qs}	d and q axial current
I_{load}	Load current
I_{pac}	Current in the battery (A)
K_a and K_b	Constants to be determined

D. Rekioua and E. Matagne, *Optimization of Photovoltaic Power Systems*,
Green Energy and Technology, DOI: 10.1007/978-1-4471-2403-0_7,

K_{opt}	Coefficient which depends on the ratio of tip speed and optimal power coefficient maximum
L_d and L_q	d and q axial inductance
M	Molar mass of the air
M_{H_2O}	Molar molecular weight of water (g/mol)
n_{cath}	Moles number in the cathode compartment (mol)
n_{H_2}	Number of hydrogen moles in the compartment (mol)
$(n_{H_2O})_{steam}$	Number of water steam moles in the compartment (mol)
N_{cell}	Number of cells in the stack
N_j	Number of days of autonomy
P_{atm}	Atmospheric pressure (output pressure)
P_{ch}	Power required by the load
P_{cath}	Pressure in the cathode compartment (Pa)
$P_{H_2}(t)$	Hydrogen partial pressure
P_{pv}	Power supplied by the photovoltaic generator
$P_{required}$	Total required power
P_{wind}	Wind power
R	Molar gas constant (J/K/mol)
R_s	Stator winding resistance
R_c	Contact resistance equivalent of the electrodes (Ω)
S_{cell}	Cell active area (m^2)
S_{open}	Maximum opening of the valve (m^2)
St_{O2}	Ratio of the stoichiometry
T_e	Temperature input (°K)
T_{pac}	Cell temperature (K)
T_j	Temperature of photovoltaic cells
T_{pac}	Absolute operating temperature of the stack (°K)
T_s	Output temperature (K)
U_{batt}	Battery voltage
U_{act}	Activation losses (V)
U_{conc}	Loss of concentration (V)
U_{ohm}	Ohmic losses (V)
v	Volume of fuel consumed
V_{anod}	Anode volume of compartment (m^3)
V_{cath}	Volume of the cathode compartment (m^3)
V_{ds} and V_{qs}	*d* and *q* axial voltage
V_{pac}	Unit cell voltage (V)
V_{wind}	Wind velocity
$(X_{H_2})_{anod}$	Molar fraction of hydrogen in the compartment
$(X_{O2})_{air}$	Molar fraction of oxygen in the air
$(X_{O2})_{valve}$	Molar fraction of oxygen in the valve
X_{steam}	Mole fraction of water vapor in the humidification system
β	Pitch angle control
γ	Polytropic exponent

ΔP	Power excess value
η_{gen}	Photovoltaic generator efficiency
η_{is}	Isentropic efficiency
η_{r}	Reference efficiency of the photovoltaic generator
λ	Tip speed ratio
ρ	Air density
ρ_{e}	Gas density (kg/m^3)
ω_{flow}	Control bandwidth (rad/s)
$\Omega_{r\ AC}$	Angular velocity of rotor
Ω_{t}	Turbine angular velocity

Hybrid power systems (HPS) combine two or more sources of renewable energy as one or more conventional energy sources [167–169]. The renewable energy sources such as photovoltaic and wind do not deliver a constant power, but due to their complementarities their combination provides a more continuous electrical output. Hybrid power systems are generally independent of large interconnected networks and are often used in remote areas [170, 171]. The purpose of a hybrid power system is to produce as much energy from renewable energy sources to ensure the load demand. In addition to sources of energy, a hybrid system may also incorporate a DC or AC distribution system, a storage system, converters, filters and an option to load management or supervision system. All these components can be connected in different architectures. The renewable energy sources can be connected to the DC bus depending on the size of the system. The power delivered by HPS can vary from a few watts for domestic applications up to a few megawatts for systems used in the electrification of small villages. Thus, the hybrid systems used for applications with very low power (under 5 kW) generally feed DC loads. Larger systems, with power greater than 100 kW, connected to an AC bus, are designed to be connected to large interconnected networks [172]. Hybrid systems are characterized by several different sources, several different loads, several storage elements and several forms of energy (electrical, thermal).

7.1 Advantages and Disadvantages of a Hybrid System

7.1.1 Advantages of Hybrid System

The most important advantages of hybrid power systems are:

- Not dependent on one source of energy.
- Simple to use.

- Efficiency, low life cycle cost of the components.
- Lower needs for storage.

7.1.2 Disadvantages of a Hybrid System

We can resume some disadvantages of hybrid power systems as:

- More complex than single-source systems.
- High capital cost compared to diesel generators.

7.2 Configuration of Hybrid Systems

Photovoltaic and wind generators in a hybrid system can be connected in three configurations, DC bus architecture, AC bus architecture and DC–AC bus architecture [172, 173].

7.2.1 Architecture of DC Bus

In the hybrid system presented in Fig. 7.1, the power supplied by each source is centralized on a DC bus. Thus, generators which provide AC power have to be connected first to a rectifier in order to obtain DC power. The inverter should supply the alternating loads from the DC bus and must follow the set point for the amplitude and frequency [174]. The batteries are sized to supply peak loads. The advantage of this topology is the simplicity of operation and the load demand is satisfied without interruption even when the generators are unable to provide the load demand.

7.2.2 Architecture of AC Bus

In AC bus topology, all components of the HPS are related to AC loads, as shown in Fig. 7.2. This configuration provides superior performance compared to the previous configuration, since each converter can supply the load independently and simultaneously with other converters [173]. This provides flexibility for the energy sources which fed the load demand. In the case of low load demand, all generator and storage systems are in stand-by except for example the photovoltaic generator to cover the load demand. However, during heavy load demands or during peak

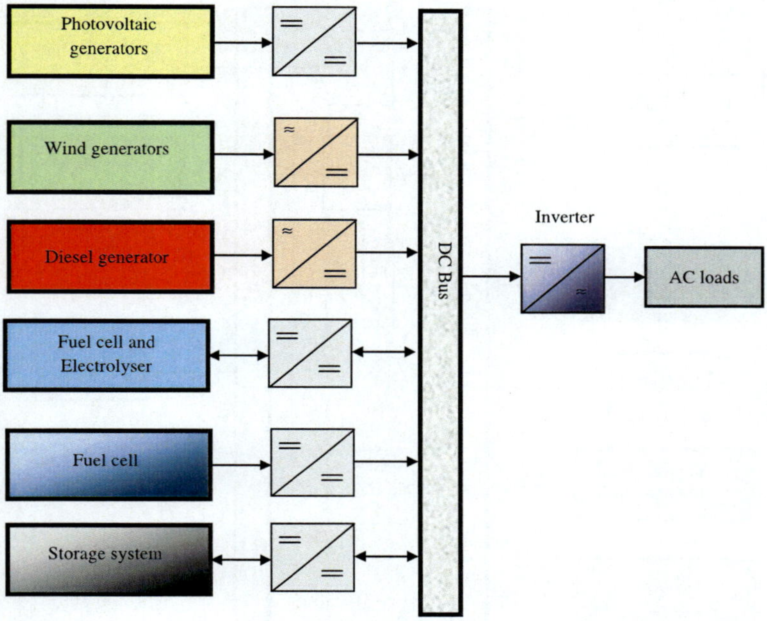

Fig. 7.1 Configuration of the hybrid system with DC bus

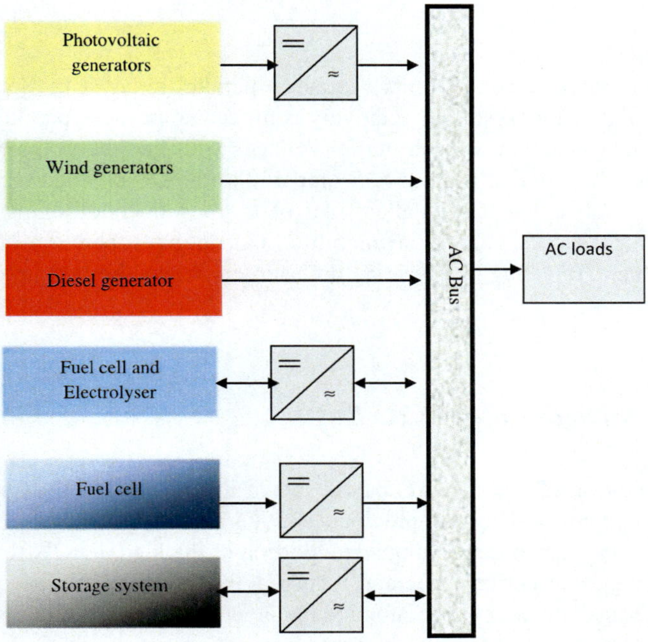

Fig. 7.2 Configuration of the hybrid system with AC topology

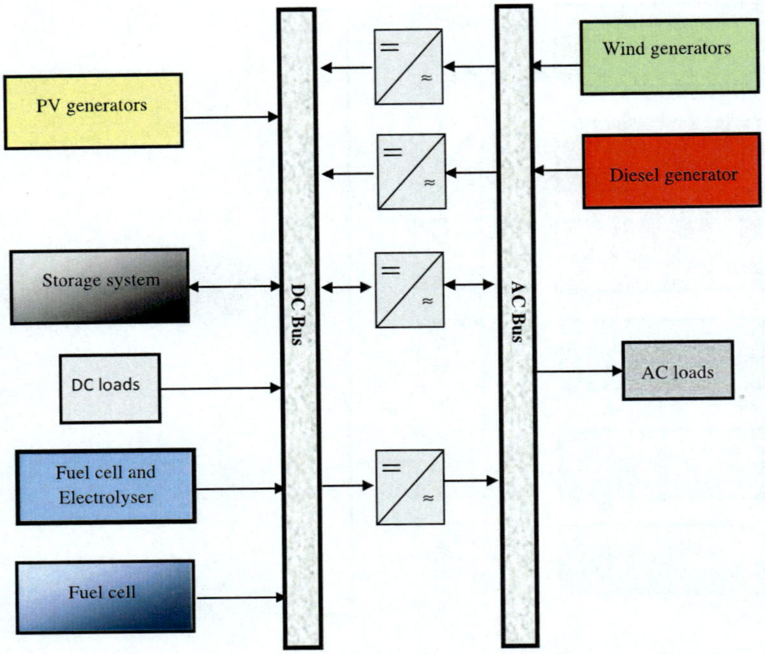

Fig. 7.3 Configuration of the hybrid system with DC and AC buses

hours, generators and storage units operate in parallel to cover the load demand. The realization of this system is relatively complicated because parallel operation requires synchronization of each output voltages with the charge voltages [172]. This topology has several advantages compared to the DC coupled topology such as higher overall efficiency, smaller sizes of the power conditioning unit while keeping a high level of energy availability, and optimal operation of the diesel generator due to reducing its operating time and consequently its maintenance cost [173].

7.2.3 Architecture of DC/AC Bus

The configuration of DC and AC bus is shown in Fig. 7.3 [170]. It has superior performance compared to the previous configurations. In this case, renewable energy and diesel generators can power a portion of the load directly to AC, which can increase system performance and reduce power rating of the diesel generator and the inverter. The diesel generator and the inverter can operate independently or in parallel by synchronizing their output voltages. Converters located between two buses (rectifier and inverter) can be replaced by a bidirectional converter which,

Table 7.1 Classification of hybrid systems by power range [194]

Power of hybrid system (kW)	Applications
Low power <5	Autonomous system: telecommunication stations, pumping water, other isolated applications.
Average power 10–250	Micro isolated systems: feeding a remote village, rural…
High power >500	Large isolated systems (Islands)

in normal operation, performs the conversion DC/AC (inverter operation). When there is a surplus of energy from the diesel generator, it can also charge batteries (operating as a rectifier). The bidirectional inverter can supply the peak load when the diesel generator is overloaded.

The advantages of this configuration are:

- The diesel generator and the inverter can operate independently or in parallel. When the load level is low, one or the other can generate the necessary energy. However, both sources can operate in parallel during peak load.
- The possibility of reducing the nominal power of the diesel generator and the inverter without affecting the system ability to supply peak loads.

The disadvantages of this configuration are:

- The implementation of this system is relatively complicated because of the parallel operation (the inverter should be able to operate autonomously and to operate with synchronization of its output voltages with output voltages of diesel generator).

7.2.4 Classifications of Hybrid Energy Systems

The power delivered by hybrid systems can vary from a few watts for domestic applications up to a few megawatts for systems used in the electrification of small islands [167]. For hybrid systems with power below 100 kW, the configuration with AC and DC bus, with battery storage, is the most used. The storage system uses a high number of batteries to be able to cover the average load for several days. This type of hybrid system uses small renewable energy sources connected to the DC bus. Another possibility is to convert the direct-current power into an alternating-current one by using inverters. Hybrid systems used for applications with very low power (below 5 kW) supply generally DC loads (Table 7.1).

7.3 The Different Combinations of Hybrid Systems

7.3.1 Hybrid Photovoltaic/Diesel Generator Systems

Systems based on a combination of photovoltaic generators and Diesel generators may also include energy storage such as battery. Photovoltaic panels and generators are highly complementary to each other. Photovoltaic systems do not cause fuel consumption and costs are maintained generally low. Diesel generators cause costs of operating [176], but can produce energy demand. The combination of energy sources allows the continuous cover of energy demand (electric power generation using photovoltaic generators when the weather is sunny and Diesel generators when weather is dark). However, energy production is not entirely renewable and for applications in remote sites, transportation and fuel costs and the environmental costs are high [176].

We present in Fig. 7.4 the block diagram of the hybrid system model.

During the day, the inverter converts DC power from the solar PV into AC power for the load. The extra power produced is stored in battery system. During the night, the inverter converts DC power from the battery into AC power for the load. The battery will supply the load to its maximum discharge level. When the battery reaches its maximum discharge level, the diesel generator serves the load and charges the battery. The controller unit manages the load demand and the energy supplied.

7.3.1.1 Photovoltaic Generator Model

See Chap. 2.

7.3.1.2 Diesel Generator Model

Diesel generators can boost up the electricity supply during sudden increase in energy demand or when the batteries, capacity decreases and thus for not facing supply interruption. The generator is characterized by its efficiency defined as the ratio of the electrical power supplied to the equivalent energy of the fuel used to produce it. The fuel consumption of a generator can be characterized by quantities corresponding to different points of motor operation. The diesel generator can be modeled by a polynomial of first or second order [177, 178] such as Eq. 7.1.

$$v = a \cdot p^2 + b \cdot p + c \tag{7.1}$$

where p is the electric power, a, b and c are constants and v is the volume of fuel consumed.

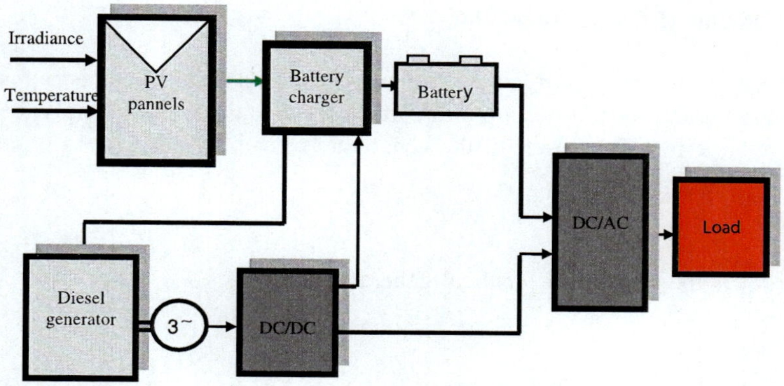

Fig. 7.4 Block diagram of the hybrid photovoltaic/diesel generator system

The curve cuts the y-axis in $v = c$, which corresponds to no-load operation. The slope of the line (b) represents the increase in consumption with the load, but near the nominal operating point, the increase in consumption is not linear. This is due to the influence of Diesel motor efficiency. Making measurements, we have made it possible to fit the values of the parameters a, b and c.

7.3.1.3 Control Strategy of Diesel Generator

There are several strategies for controlling the power delivered by the generator. We present two strategies:

(a) Strategy 1

The generator starts if $V_{\text{batt}} \leq V_{\text{min.}}$ and feeds only the battery.
The generator stops if $V_{\text{bat}} = V_{\text{nom.}}$

(b) Strategy 2

It is to supply a portion of the load by the generator if:

$$\Delta P = P_{\text{ch}} - P_{\text{pv}} \geq \varepsilon \tag{7.2}$$

where P_{ch} is the power required by the load, P_{pv} is the power supplied by the photovoltaic generator, ε is the tolerance predefined and ΔP is the power excess value.

7.3.1.4 Sizing of Diesel Generator

The Diesel generator is used in hybrid energy systems as backup power in the case where the weather conditions are unfavorable for a while. It must supply the load during this period. The power of the generator is given by [180]

$$P_{G} = \frac{P_{ch}}{N_{G}} \tag{7.3}$$

where N_G is the number of identical generators.

7.3.2 *Hybrid Wind/Photovoltaic/Diesel Generator System*

Hybrid PV/wind/Diesel generator systems are well suited for decentralized production of electricity, and can contribute to solving the problem of connecting to the electrical power networks (cases of isolated sites) [167, 168]. The initial data in the implementation of such a system of production from renewable sources of energy like any other energy system is the demand, which will be determined by comparing the load to be supplied. This request must be estimated as accurately as possible both from a point of view of energy balance and of power temporal distribution, even if its random nature often makes this a difficult task. Adding a generator to a system of renewable energy production may on the one hand increase the reliability of power system loads and on the other hand reduce significantly the cost of electricity, producing a significant decrease of the size of the storage system [170, 181, 182].

Reference [169] proposed that there are multiple types of electrical circuit architectures which could be used depending on people needs and site capabilities. In the first architecture (Fig. 7.5), the generators and the battery are all installed in one place and are connected to a main AC bus bar before being connected to the grid. The power delivered by all the energy conversion systems and the battery is fed to the grid through a single point. In this case, the power produced by the PV system and the battery is converted into AC before being connected to the main AC bus. This system is called a centralized AC bus architecture.

The energy conversion systems can also be connected to the grid in another manner (Fig. 7.6). This system is called a decentralized AC bus architecture. The power sources in this case do not need to be connected to one main bus bar. The power generated by each source is conditioned separately to be identical to that required by the grid.

The third architecture uses a main centralized DC bus bar (Fig. 7.7). The energy conversion systems producing AC power (wind energy converter and the diesel generator) deliver their power to rectifiers in order to convert it into DC and then it is delivered to the main DC bus bar. A main inverter feeds the AC grid from this main DC bus.

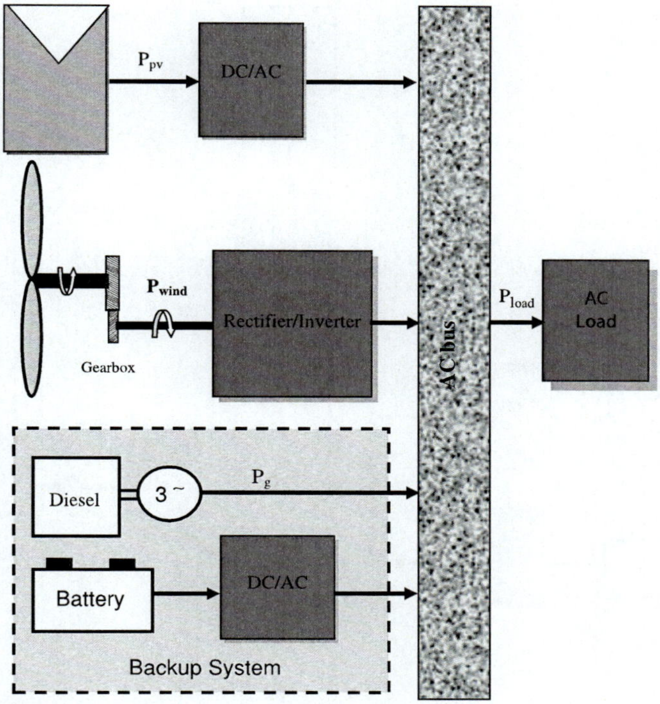

Fig. 7.5 Centralized AC bus architecture

7.3.2.1 Wind Generator System

Wind Turbine

The total kinetic power of the wind through a wind disc of radius R is given by [183]:

$$P_{\text{wind}} = \frac{1}{2}\,\rho\,\pi\,R^2\,V_{\text{wind}}^3 \tag{7.4}$$

where ρ represents the air density and V_{wind} the wind velocity (Fig. 7.8).

A wind turbine can only convert just a certain percentage of the captured wind power. This percentage is represented by $C_P(\lambda, \beta)$ where β is the pitch angle of specific wind turbine blades and λ is the ratio between the turbine angular velocity Ω_t and the wind speed v_{wind}. This ratio is called the tip speed ratio.

$$\lambda = \frac{R \cdot \Omega_t}{v_{\text{wind}}} \tag{7.5}$$

with

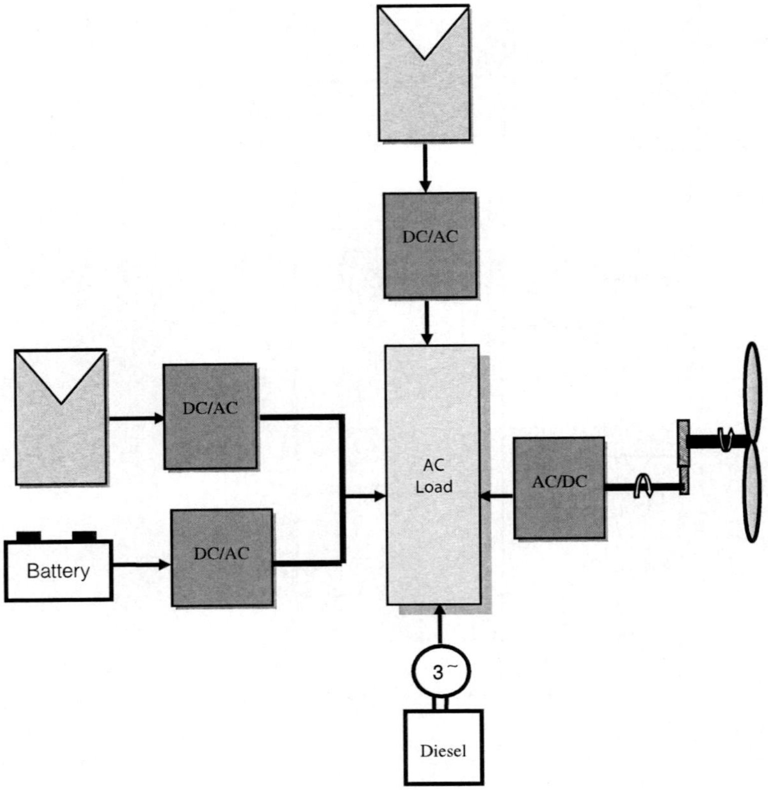

Fig. 7.6 Decentralized AC bus architecture

$$\Omega_t = \frac{\Omega_{mec}}{G_x} \tag{7.6}$$

where G_x is the overall gear ratio and Ω_{mec} the machine angular speed.

We obtain then the mechanical power which is converted by a wind turbine:

$$P_{wind} = \frac{1}{2} C_P(\lambda, \beta) \cdot \rho \cdot \pi \cdot R^2 \cdot v_{wind}^3 \tag{7.7}$$

The aerodynamic torque is defined as the ratio between the aerodynamic power and the angular velocity of the turbine

$$T_{aero} = \frac{P_{wind}}{\Omega_t} = C_P(\lambda, \beta) \cdot \frac{\rho}{2} \cdot S \cdot v_{wind}^3 \cdot \frac{1}{\Omega_t} \tag{7.8}$$

The electrical generator shaft torque T_g is given by

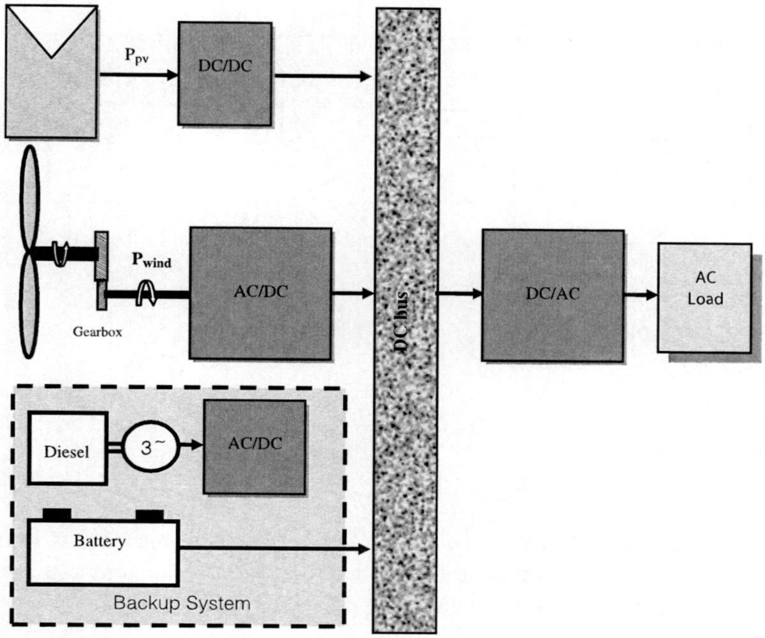

Fig. 7.7 Centralized DC bus architecture

Fig. 7.8 Wind turbine [185]

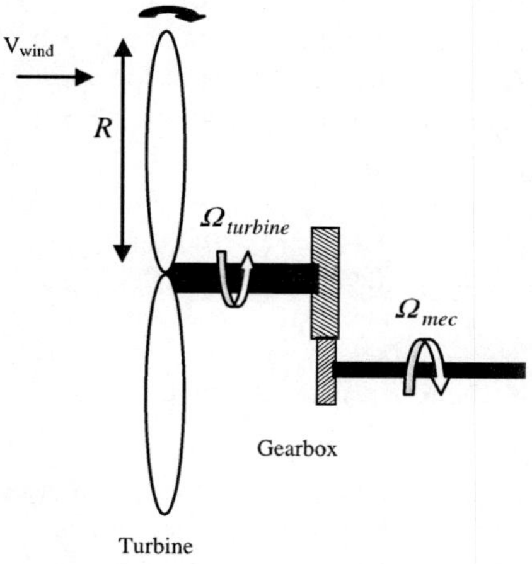

$$T_g = \frac{T_{aero}}{G_x} \tag{7.9}$$

The mechanical speed evolution is determined from the total torque T_{mec} applied to the electrical generator rotor

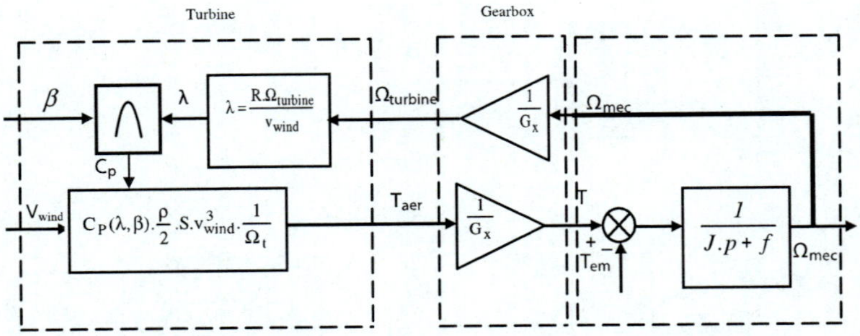

Fig. 7.9 Simplified block diagram

$$J\frac{d\Omega_{\text{mec}}}{dt} = T_{\text{mec}} \tag{7.10}$$

$$T_{\text{mec}} = T_g - T_{\text{emAC}} - T_{\text{visq}} \tag{7.11}$$

where J is the effective inertia of rotor, T_{emAV} is the electromechanical torque of the machine and T_{visq} the resistant torque due to frictions. This torque is modeled by a coefficient of viscous friction f:

$$T_{\text{visq}} = f\Omega_{\text{mec}} \tag{7.12}$$

The peak power for each wind speed occurs at the point where $C_P(\lambda, \beta)$ is maximized. To maximize the generated power, it is therefore desirable for the electrical generator to have a power characteristic that will follow the maximum $C_{p-\max}$ line. The action of the speed corrector must control the mechanical speed Ω_{mec} in order to get a speed reference $\Omega_{\text{mec}-\text{ref}}$. The simplified block diagram is given in Fig. 7.9.

Electrical Generators

- Induction generators (IG)

Induction machines are largely used in the field of wind energy conversion. For standalone operations, the squirrel induction machine is preferred because it is robust, needs little maintenance and does not need an auxiliary supply for magnetizing it. The simplest way to use it as an autonomous generator consists in connecting its stator windings to a capacitor bank in parallel to the load. The remaining magnetic flux, added to the magnetizing current through the capacitor bank yields the built up of the electromotive force and its increase to a useful value. This approach is very cheap and is well adapted to convert the wind energy into an electrical one for isolated areas or areas faraway from the distribution grid

[184]. However, the magnitude of the stator voltage and frequency are very sensitive to both speed and load values. The used model for the induction machine is expressed in $d-q$ frames (see Sect. 6.2.2.1)

- Doubly fed induction generator (DFIG)

The application of the doubly fed induction generator (DFIG) in the modern wind turbine becomes an imposing reality, by its offered performances. Generally, if a wind turbine contains the DFIG, the stator is connected directly to the grid; the rotor is connected to the grid by means of a static converter. Modern high-power wind turbines are equipped with adjustable speed generators; the constraints on the static converter will be decreased, which leads to the reduction in the size and cost of the converter. The energy storage system associated to a grid connected variable speed wind generation scheme using a DFIG is investigated [185]. By choosing a general d–q reference frame, stator voltages and fluxes can be rewritten as follows:

$$\begin{aligned}
V_{sd} &= R_s \cdot I_{sd} - \omega_s \cdot \Phi_{sq} + \frac{d\Phi_{sd}}{dt} \\
V_{sq} &= R_s I_{sq} + \omega_s \cdot \Phi_{sd} + \frac{d\Phi_{sq}}{dt} \\
V_{rd} &= R_r \cdot I_{rd} - \omega_r \cdot \Phi_{rq} + \frac{d\Phi r_{sd}}{dt} \\
V_{rq} &= R_r \cdot I_{rq} + \omega_r \cdot \Phi_{rd} + \frac{d\Phi_{rq}}{dt}
\end{aligned} \tag{7.13}$$

with

$$\begin{aligned}
\Phi_{sd} &= L_s \cdot I_{sd} + L_m I_{rd} \\
\Phi_{sq} &= L_s \cdot I_{sq} + L_m \cdot I_{rq} \\
\Phi_{rd} &= L_m \cdot I_{sd} + L_r \cdot I_{rd} \\
\Phi_{rq} &= L_m \cdot I_{sq} + L_r \cdot I_{rq}
\end{aligned} \tag{7.14}$$

- Permanent magnet synchronous generator (PMSG) (see Sect. 6.3.2.2).

Control of Wind Generator System with IG

We propose to present two strategies to control the DC voltage of an autonomous induction generator connected to a rectifier when the input speed varies. The control strategies are based on the choice of the reference frame orientation. The d–q frame must be chosen in order to maintain the d-axis always along the rotor flux. This choice implies an important simplification in the elaboration of the control. Normally, the saturation effect, as well as the cross magnetizing effect, must be taken into account in the model of the induction generator. This obviously complicates the model as it can be seen through electrical equations. Then, the machine control can seem to be difficult. However, when the flux is controlled to be constant, the model of the induction generator can be considered as a linear one. The inductance values, constant, depend on the given flux but there is no more cross magnetizing effect which has to be considered. Besides, the variations which can occur around the reference value of the flux can be considered as negligible. The operating of the machine can then be considered practically linear. The electrical equations of the induction machine are given in Eq. 7.13 [184].

As $\Phi_{rd} = \Phi_r$ and $\Phi_{rq} = 0$, this implies that the expressions of the flux Φ_r and its derivative take the forms:

$$\begin{aligned} \frac{d\Phi_r}{dt} &= -R_r \cdot I_{rd} \\ \Phi_r &= -\frac{R_r \cdot I_{rq}}{\omega_r} \end{aligned} \tag{7.15}$$

Moreover, the rotor currents can be expressed in the following way:

$$\begin{aligned} I_{rd} &= \frac{\Phi_r - L_m \cdot I_{sd}}{L_r} \\ I_{rq} &= -\frac{L_m \cdot I_{sq}}{L_r} \end{aligned} \tag{7.16}$$

Indeed, by introducing the leakage flux coefficient we obtain:

$$\begin{aligned} V_{sd} &= R_s \cdot I_{sd} + \sigma \cdot L_s \cdot \frac{dI_{sd}}{dt} - \omega_s \cdot \sigma \cdot L_s \cdot I_{sq} + L_m \cdot \frac{d\Phi_r}{dt} \\ V_{sq} &= R_s \cdot I_{sq} + \sigma \cdot L_s \cdot \frac{dI_{sq}}{dt} + \omega_s \cdot \sigma \cdot L_s \cdot I_{sd} + \omega_s \cdot L_m \end{aligned} \tag{7.17}$$

Finally, as the chosen frame implies $\Phi_{rq} = 0$, the expression of the electromagnetic torque becomes:

$$T_{emAC} = p \cdot \frac{L_m}{L_r} \cdot \Phi_r \cdot I_{sq} \tag{7.18}$$

We can write the rotor flux as a function of the current i_{sd} and the rotor time constant $T_r = L_r/R_r$

$$\Phi_r = \frac{L_m}{1 + T_r \cdot s} i_{sd} \tag{7.19}$$

where s represents the derivative operator

Thus, we can obtain:

$$\omega_r = \frac{L_m I_{sq}}{T_r \cdot \Phi_r} \tag{7.20}$$

By controlling the flux $\Phi_{rd\text{-}ref}$ at a wished value, it is possible to determine the magnetizing inductance L_m for the saturation level. This is given by the following equation [184]:

$$L_m \approx \frac{\Phi_{r-ref}}{i_{md}} = \sqrt{\frac{2}{3}} \frac{\Phi_{r-ref}}{|i_m|} \tag{7.21}$$

The intersection between the characteristic L_m ($|I_m|$) and the curve described by Eq. 7.19 makes it possible to determine the value of L_m to use in the control. Then, the rotor flux in the machine is adjustable by I_{sd}.The electromagnetic torque can be controlled by I_{sq}.

The strategies proposed are relatively close. The first is classical with a constant $\Phi_{rd\text{-}ref}$. The second is based on a rotor flux reference value obtained from the following relation:

$$\Phi_{r-re'f} = \frac{\omega}{\omega_{rat}} \Phi_{rd-rat} \tag{7.22}$$

where Φ_{rd-rat} and ω_{rat} are the rated values of the rotor flux and the rotation speed respectively.

Control of Hybrid Wind/Photovoltaic/Diesel Generator System

The monitoring equipment includes data loggers, wind speed and direction sensors, ambient and battery temperature sensors, and various AC and DC current/voltage/power sensors. The purposes for using monitoring systems are to [186]

- Determine the components and system efficiencies
- Verify the proper system functioning
- Provide system trouble shooting
- Detect and analyze significant load changes
- Calculate the actual cost of utilized energy.

We propose a control system of the hybrid PV/wind/Diesel system [187]. It is based on the overall energy balance equation.

$$P_{diesel} = P_{load} - P_{wind} - P_{pv} + \Delta P - P \tag{7.23}$$

where P_{diesel} is the power delivered from the Diesel generator(s), P_{Load} is the power required by the load, P_{Wind} is the power delivered from the wind turbine,

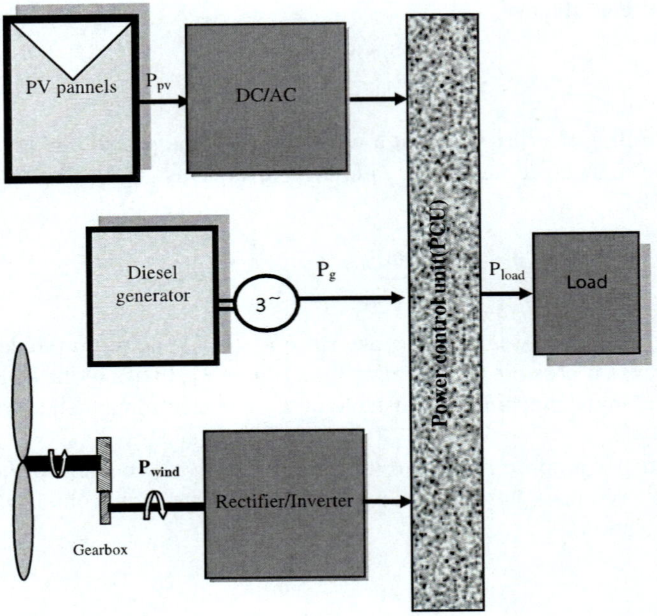

Fig. 7.10 Control of hybrid wind/photovoltaic/diesel generator system [9]

P_{pv} is the power delivered from the PV, ΔP is the power excess value and P is the unmet load (Fig. 7.10).

The power control unit (PCU) has a central location for making the various connections of subsystems (wind, photovoltaic, diesel generator).

The monitoring system role is to manage and control the operation of a hybrid power system, depending on weather (irradiance, wind speed) and the power required. The manager controls the opening and closing of three relays under the following conditions:

- The relay of the PV generator is open if:
  - the power output of the PV generator is zero;
  - the load power is zero and the batteries are charged.
- The relay wind generator is open if:

– the wind speed is less than the threshold wind speed of the turbine;
– the wind speed is greater than the stall wind speed of the turbine;
– the load power is zero and the batteries are charged.

- The relay of the Diesel generator is open if:
  - the generators (wind and PV) give a power higher than the load power;

– the load power is zero and the batteries are charged.

And the closure of this relay is when the battery state of charge reaches the minimum level.

From these conditions, we find that the monitoring system includes 06 inputs, 03 outputs and 06 tests.

- The input variables are:
  - The irradiance (G)
  - The wind Speed(V_{wind})
  - The power of PV generator(P_{pv})
  - The wind power generator(P_{wind})
  - The load power (P_{load})
  - The battery voltage(V_{batt})
- The output variables are:
  - T_{pv} the relay control signal of PV generator
  - T_{wind} the relay control signal of wind generator
  - T_{diesel} the relay control signal of Diesel generator
- The different tests are:
  - Test on the PV power $P_{\text{pv}} = 0$ or $G = 0$ ($\Longleftrightarrow$A)
  - Test on the wind speed ($\Longleftrightarrow$B)
  - Test on the load power $P_{\text{load}} = 0$($\Longleftrightarrow$C)
  - Test on PV and wind power $P_{\text{pv}} + P_{\text{wind}} \geq P_{\text{load}}$ ($\Longleftrightarrow$D)
  - Test on voltage battery $V_{\text{batt}} \leq V_{\min}$ ($\Longleftrightarrow$E)
  - Test on voltage battery $V_{\text{batt}} \geq V_{\max}$ ($\Longleftrightarrow$F).

From the number of tests, we determined the number of possible combinations that we calculated using the following equation:

$$X = 2^n \tag{7.24}$$

where X is the number of possible combinations and n the number of inputs.

Table 7.2 summarizes the 64 combinations. According to the table we see that the number of possible combinations is reduced to 36. The logical equations are determined and give the control signals of the relays from each source (Figs. 7.11, 7.12, 7.13):

$$\begin{aligned} T_{\text{pv}} &= (E \cdot \overline{F} + \overline{E} \cdot \overline{F})(\overline{A} \cdot D + \overline{A} \cdot \overline{B} \cdot \overline{C} \cdot \overline{D} + \overline{A} \cdot B \cdot \overline{C} \cdot \overline{D} \cdot \overline{E} \cdot F) \\ T_{\text{wind}} &= (E \cdot \overline{F} + \overline{E} \cdot \overline{F})(\overline{B} \cdot D + \overline{A} \cdot \overline{B} \cdot \overline{C} \cdot \overline{D} + A \cdot \overline{B} \cdot \overline{C} \cdot \overline{D} \cdot \overline{E} \cdot F) \\ T_{\text{diesel}} &= (E \cdot \overline{F \cdot D})(A \cdot B + A \cdot \overline{B} \cdot \overline{C} + \overline{A} \cdot B \cdot \overline{C}) \end{aligned} \tag{7.25}$$

Table 7.2 The different combinations

A	B	C	D	E	F	T_{pv}	T_{wind}	T_{diesel}
1	1	1	1	1	1	*	*	*
1	1	1	1	1	0	0	0	1
1	1	1	1	0	1	0	0	0
1	1	1	1	0	0	0	0	0
1	1	1	0	1	1	*	*	*
1	1	1	0	1	0	0	0	1
1	1	1	0	0	1	0	0	0
1	1	1	0	0	0	0	0	0
1	1	0	1	1	1	*	*	*
1	1	0	1	1	0	*	*	*
1	1	0	1	0	1	*	*	*
1	1	0	1	0	0	*	*	*
1	1	0	0	1	1	*	*	*
1	1	0	0	1	0	0	0	1
1	1	0	0	0	1	0	0	0
1	1	0	0	0	0	0	0	0
1	0	1	1	1	1	*	*	*
1	0	1	1	1	0	0	1	0
1	0	1	1	0	1	0	0	0
1	0	1	1	0	0	0	1	0
1	0	1	0	1	1	*	*	*
1	0	1	0	1	0	*	*	*
1	0	1	0	0	1	*	*	*
1	0	1	0	0	0	*	*	*
1	0	0	1	1	1	*	*	*
1	0	0	1	1	0	0	1	0
1	0	0	1	0	1	0	0	0
1	0	0	1	0	0	0	1	0
1	0	0	0	1	1	*	*	*
1	0	0	0	1	0	0	1	1
1	0	0	0	0	1	0	1	0
1	0	0	0	0	0	0	1	0
0	1	1	1	1	1	*	*	*
0	1	1	1	1	0	1	0	0
0	1	1	1	0	1	0	0	0
0	1	1	1	0	0	1	0	0
0	1	1	0	1	1	*	*	*
0	1	1	0	1	0	*	*	*
0	1	1	0	0	1	*	*	*
0	1	1	0	0	0	*	*	*
0	1	0	1	1	1	*	*	*
0	1	0	1	1	0	1	0	0
0	1	0	1	0	1	0	0	0
0	1	0	1	0	0	1	0	0
0	1	0	0	1	1	*	*	*

(continued)

Table 7.2 (continued)

A	B	C	D	E	F	T_{pv}	T_{wind}	T_{diesel}
0	1	0	0	1	0	1	0	1
0	1	0	0	0	1	1	0	0
0	1	0	0	0	0	1	0	0
0	0	1	1	1	1	*	*	*
0	0	1	1	1	0	1	1	0
0	0	1	1	0	1	0	0	0
0	0	1	1	0	0	1	1	0
0	0	1	0	1	1	*	*	*
0	0	1	0	1	0	*	*	*
0	0	1	0	0	1	*	*	*
0	0	1	0	0	0	*	*	*
0	0	0	1	1	1	*	*	*
0	0	0	1	1	0	1	1	0
0	0	0	1	0	1	0	0	0
0	0	0	1	0	0	1	1	0
0	0	0	0	1	1	*	*	*
0	0	0	0	1	0	1	1	1
0	0	0	0	0	1	0	0	0
0	0	0	0	0	0	1	1	0

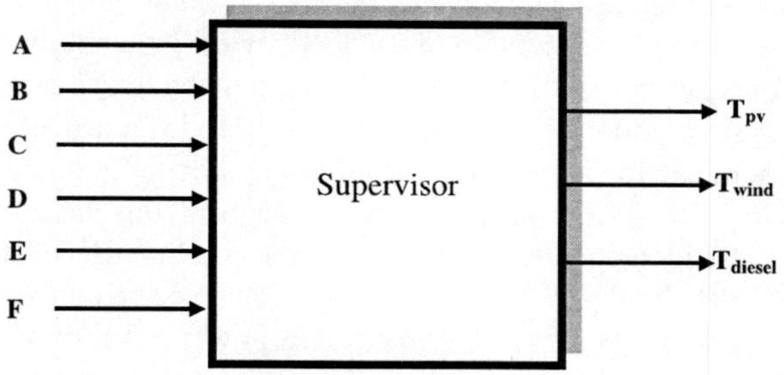

Fig. 7.11 Supervisor of hybrid wind/photovoltaic/diesel generator system

Application of Hybrid Wind/Photovoltaic/Diesel Generator System

Wind/Photovoltaic/Diesel systems can be used for desalination of sea water [188]. Different distinct approaches to Wind/Photovoltaic/Diesel exist, each with its own architecture. Generally, these systems are used to supply village power.

7.3.3 Hybrid Wind/Photovoltaic System

The optimization of wind and photovoltaic energy with electrochemical storage (batteries) depends on many economic models of each system separately (wind and

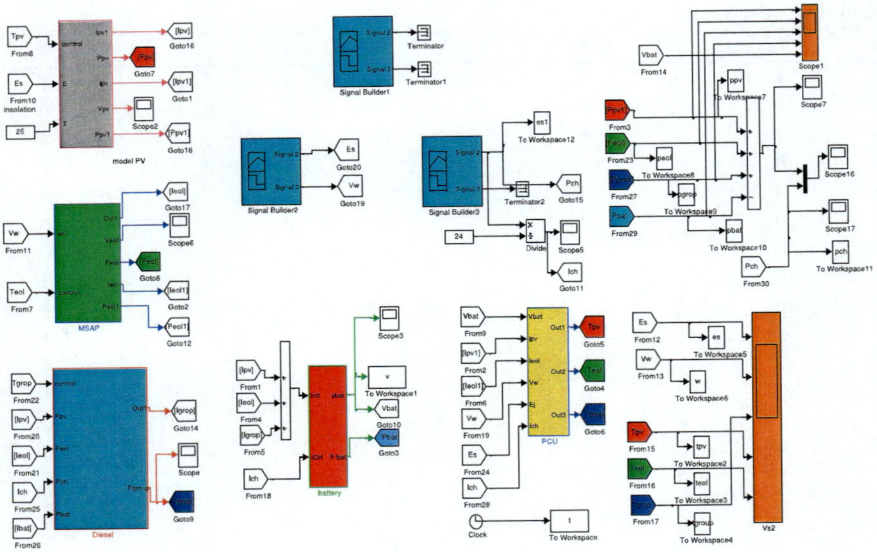

Fig. 7.12 Block diagram of the hybrid photovoltaic/wind/diesel generator system [9]

photovoltaic). The advantage of a hybrid system depends on many important factors: the shape and type of load, wind, solar radiation, cost and availability of energy, the relative cost of the wind machine, solar array, electrochemical storage system and other efficiency factors [181]. Photovoltaic systems are currently economical for low power installations. For autonomous systems the cost of energy storage is the biggest constraint. Minimizing the cost of storage and reducing its capacity are the main reasons for the combination of wind and photovoltaic systems [189]. In Fig. 7.14, both energy sources are connected to a DC bus. A DC/DC converter can track the maximum power point of a photovoltaic subsystem. Similarly, a controlled rectifier is connected between the wind generator and the DC bus. Battery is included as part of back-up and storage system.

7.3.3.1 Sizing of Hybrid Wind/Photovoltaic System

The effectiveness of any electric system depends on its sizing and use. The sizing should be based on meteorological data, solar radiation and wind speed and the exact load profile of consumers over long periods.

- Determination of the load profile of consumers

The exact knowledge of the customers, load profile determines the size of the generators [9] (Fig. 7.15).

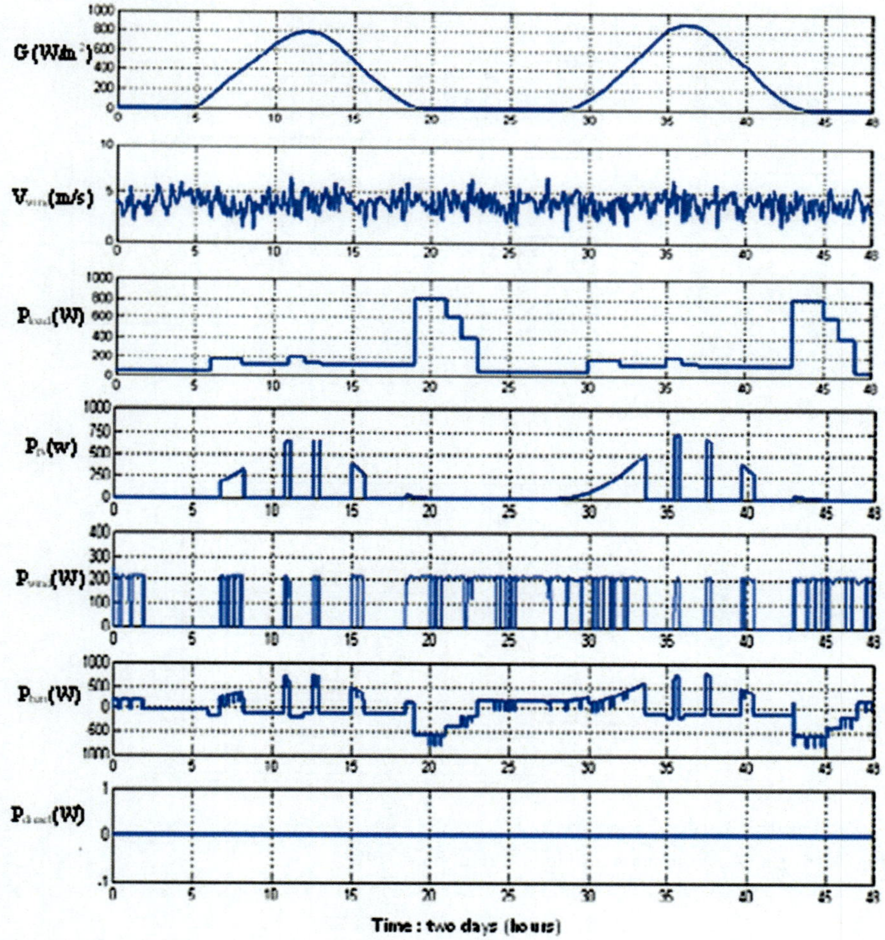

Fig. 7.13 Simulation results of hybrid wind/photovoltaic/diesel generator system

- Analysis of solar and wind energy potential

We make applications in Bejaia (Algeria) which is a coastal region. The curve in Fig. 7.16 is the superposition of two characteristics (wind speed and radiation), and shows their complementarity; we can say that the coupling of a photovoltaic system and wind is very interesting for electricity production throughout the year.

- Photovoltaic energy calculation:

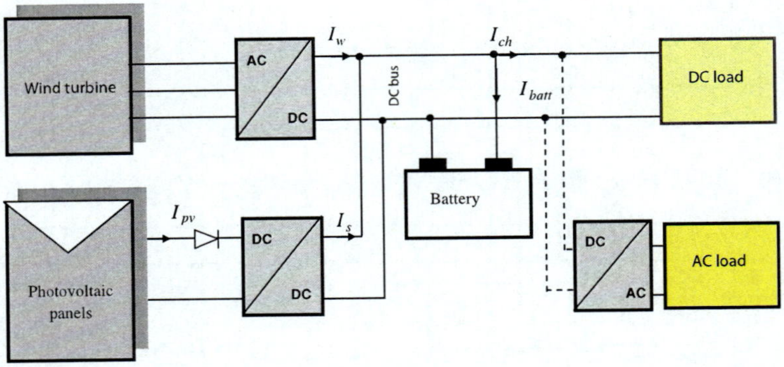

Fig. 7.14 Hybrid wind/photovoltaic system

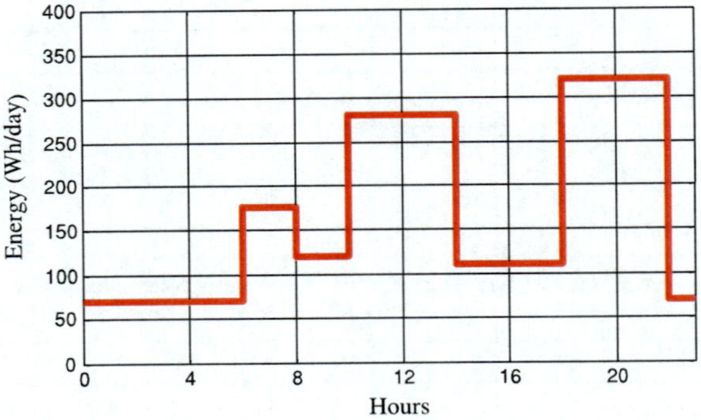

Fig. 7.15 Daily consumption profile of a rural house [9]

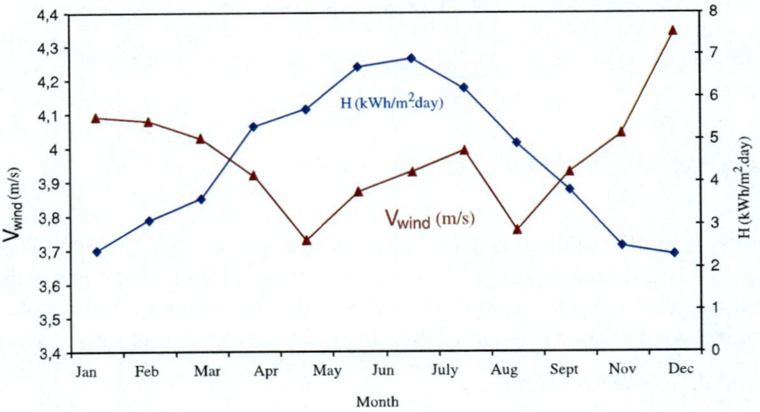

Fig. 7.16 Monthly average global radiation and wind speed monthly average of Bejaia (Algeria) site from 1998 to 2007

The energy produced by a photovoltaic generator per unit area is estimated using data from the global irradiance on an inclined plane, ambient temperature and the data sheet for the used photovoltaic panel.

The electrical energy produced per unit area by a photovoltaic generator is given by:

$$\Delta E_{pv} = \eta_{pv} . G\,\Delta t \tag{7.26}$$

where G is a solar radiation on tilted plane module and η_{pv} the efficiency of the photovoltaic generator:

$$\eta_{pv} = \eta_{r-pv} . \eta_{pc}\left[1 - \alpha_{sc}\left(T_{j} - T_{jref}\right)\right] \tag{7.27}$$

with η_{r-pv} the efficiency of the photovoltaic generator (power electronic converter included) and η_{pc} the power conditioning efficiency which is equal to one if a perfect maximum power tracker (MPPT) is used. α_{sc} is the temperature coefficient of short-current (A/°K) as found on the data sheet, T_j the cell temperature and T_{jref} the reference cell temperature. We put the emphasis on the fact that η_{r-pv} is not a constant, but depends on the climatic conditions (temperature, irradiance…).

- Wind energy calculation

The power contained in the form of kinetic energy per unit area in the wind is expressed by:

$$P_{wind} = \frac{1}{2} \cdot \rho \cdot v_{wind}^{3}\, \eta_{wind} \tag{7.28}$$

with η_{wind} the efficiency of the photovoltaic generator (power electronic converter and power conditioning efficiency included).

The energy produced by wind generator is expressed by:

$$\Delta E_{wind} = P_{wind} \cdot \Delta t \tag{7.29}$$

- Pre-sizing of photovoltaic and wind systems:

The monthly energy produced by the system per unit of area is denoted $E_{pv,m}$ (kWh/m^2) for photovoltaic energy and $E_{wind,m}$ (kWh/m^2) for wind energy and $E_{L,m}$ represents the energy required by load every month (where m = 1, 2,…,12 represents the month of the year). One has:

$$E_{pv,m} = \sum_{\text{month m}} \Delta E_{pv} \tag{7.30a}$$

$$E_{\text{wind, m}} = \sum_{\text{month m}} \Delta E_{\text{wind}} \tag{7.30b}$$

and

$$E_{\text{L, m}} = \sum_{\text{month m}} \Delta E_{\text{L}} \tag{7.31}$$

Pre-sizing is sometimes based on the worst month of the year. Then, the total area of the photovoltaic generator A_{pv} and the total area of the wind generator S_{wind} are chosen in such a way that

$$E_{\text{L, worth m}} = E_{\text{pv, worth m}} \cdot A_{\text{pv}} + E_{\text{wind, worst m}} \cdot S_{\text{wind}} \tag{7.32}$$

One can introduce the parameter f which is the fraction of load supplied by the photovoltaic energy, $(1 - f)$ being the fraction of load supplied by the wind energy. Then:

$f = 1$ indicates that the entire load is supplied by the photovoltaic source.

$f = 0$ indicates that the entire load is powered by the wind source.

Using f, one has

$$A_{\text{pv}} = \frac{f\, E_{\text{L, worst m}}}{E_{\text{pv, worst m}}} \tag{7.33a}$$

$$S_{\text{wind}} = \frac{(1-f)\, E_{\text{L, worst m}}}{E_{\text{wind, worst m}}} \tag{7.33b}$$

The pre-sizing is also often based on a monthly annual average [169, 190]. The calculation of the size of wind generator and photovoltaic (A_{pv} and S_{wind}) is established from the annual average values of each monthly contribution $\left(\overline{E_{\text{pv}}} \text{ and } \overline{E_{\text{wind}}}\right)$. The load is represented by the monthly annual average $\overline{E_{\text{L}}}$.

$$\begin{aligned} A_{\text{pv}} &= f \cdot \frac{\overline{E_{\text{L}}}}{\overline{E_{\text{pv}}}} \\ S_{\text{wind}} &= (1-f) \cdot \frac{\overline{E_{\text{L}}}}{\overline{E_{\text{wind}}}} \end{aligned} \tag{7.34}$$

The number of photovoltaic and wind generators to consider, is calculated according to the area of the system unit taking the integer value of the ratio by excess.

$$\begin{aligned} N_{\text{pv}} &= ENT\left[\frac{A_{\text{pv}}}{A_{\text{pv,u}}}\right] \\ N_{\text{wind}} &= ENT\left[\frac{S_{\text{wind}}}{S_{\text{wind,u}}}\right] \end{aligned} \tag{7.35}$$

- Batteries pre-sizing (see Sect. 1.3.5).

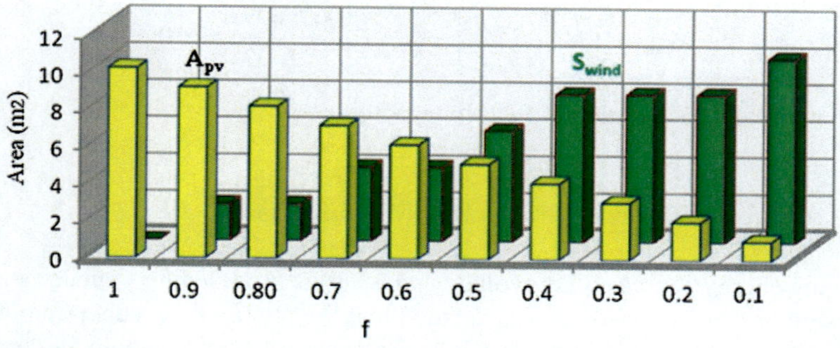

Fig. 7.17 Areas of photovoltaic panels and wind turbines

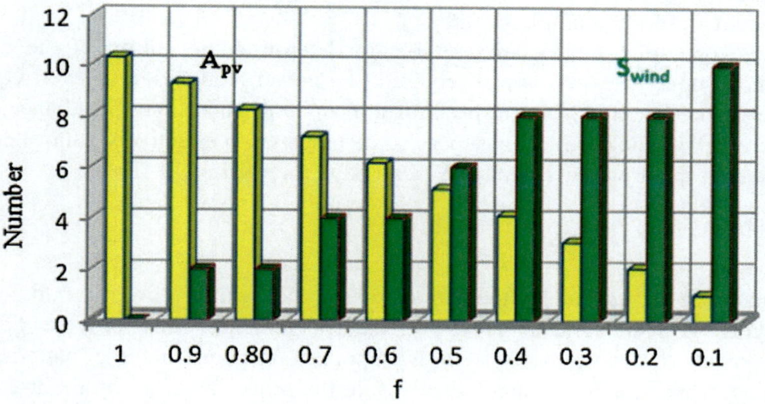

Fig. 7.18 Number of photovoltaic panels and wind turbines as a function of the fraction of the load

7.3.3.2 Application

$$E_{\mathrm{Lmax}} = 121{,}52kWh,\ U_{\mathrm{batt}} = 12V,$$
$$N_{\mathrm{m}} = 31days\ N_{\mathrm{j}} = 2\ \eta_{\mathrm{batt}} = 0.94$$

Figures 7.17 and 7.18 show respectively photovoltaic panels and wind turbines area and the obtained number to install according to the fraction of the load (f) see Eq. 1.22.

$$C_{\mathrm{batt}} = \frac{121,52\ .1000}{0,9\ .\ 12\ .\ 0,8\ .\ 31}\ .\ 2 = 907,407\mathrm{Ah}$$

The number of batteries is determined from the capacity ($C_{\mathrm{batt,\ u}} = 92$ Ah) taking the integer value of the ratio by excess.

$$N_{\text{batt}} = ENT\left[\frac{C_{\text{batt}}}{C_{\text{batt,u}}}\right] = ENT\left[\frac{907,407}{92}\right] = 10$$

We will use ten batteries of 92 Ah capacity.

7.3.3.3 Control of Hybrid Photovoltaic/Wind System

Managing energy sources (photovoltaic and wind) is provided by a supervisor. For the design of the supervisor, it was decided that the photovoltaic subsystem would be the main generator, while the wind generator subsystem would be complementary. This choice is motivated by the design already made based on the monthly averages annual site rating. However, the supervisor applications extend to considering the wind subsystem as the main generator and the photovoltaic subsystem would be complementary.

Three operating modes are possible to determine the ability of the hybrid system to supply the total power required (the power load and the power required to charge the batteries) on the basis of atmospheric conditions (irradiance, temperature and wind speed). This supervisor is essential to effectively control energy subsystems (photovoltaic, wind). We can have three cases [9, 191]:

- Case 1:

This mode corresponds to the periods where photovoltaic power is sufficient for supplying the load demand. However, the PV generator must provide the total power while the wind subsystem is supposed stopped and the batteries are charging. This situation is maintained while the power required by the load does not exceed the maximum PV power. Beyond this limit, the supervisor switches in Case 2 and activates the wind generator. In this case, the objective of the photovoltaic system is under power control according to this reference:

$$P_{\text{ref1\_S}} = P_{\text{required}} = V_{\text{batt}} \cdot (I_{\text{load}} + I_{\text{batt}}) \tag{7.36}$$

with I_{load} the load current, I_{batt} the battery current, and P_{required} the total required power.

- Case 2:

In this case, the photovoltaic system generates the maximum power (operating at maximum power point ($\text{MPPT}_{\text{w}} = 1$) and the wind system is controlled to produce a reference power. This one is the power required to complete the power produced by the photovoltaic generator at the same time supplying the total power load. It should be noted that in cases 1 and 2, batteries are not used to produce load power, instead they become a part of the power required. Once the maximum

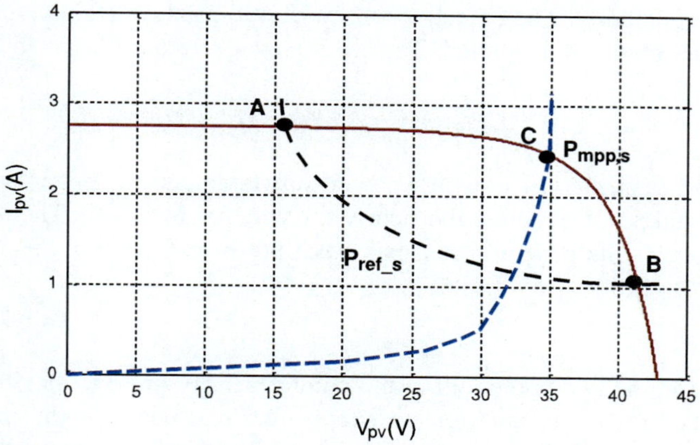

Fig. 7.19 Characteristic $I_{PV}(V_{PV})$ panel with photovoltaic power Reference P_{ref_s} produced [9]

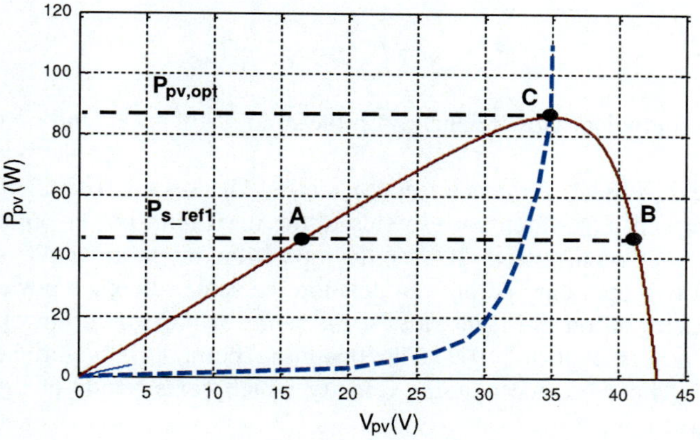

Fig. 7.20 Characteristic $P_{PV}(V_{PV})$ panel with photovoltaic power Reference P_{ref_S} produced (Case 2)

production limit of the hybrid system is reached or exceeded by any power demand, the system switches in the Case 3.

In cases 2 and 3, the PV system produces maximum power at MPPT operation. Different algorithms can be used to extract the maximum power (see Chap. 4). The reference power is given by (Figs. 7.19, 7.20):

$$P_{ref2\_S} = P_{pv}^{opt} = P_{s}^{opt} = V_{pv}^{opt} \cdot I_{pv}^{opt} \tag{7.37}$$

The wind system starts its operation when the PV power is insufficient to supply the total power required. The supervisor controls the wind system by power control or by power operation. The objective in case 2 is to produce the additional power to supply the total power applied. The wind power reference is given by:

$$P_{ref1\_w} = P_{required} - P_s^{opt} = V_{batt} \cdot (I_{load} + I_{batt} - I_s) \tag{7.38}$$

When the contribution of wind power subsystem is no longer sufficient to supply the total power required the supervisor switches in Case 3. The objective of this subsystem is the generation of maximum power extraction.

- Case 3

In this case, the two photovoltaic and wind generator provide maximum power (operating at MPPT). In addition, to supply the load demand, the batteries are charged or discharged. At discharge, Case 3 is maintained as long as the available energy levels of the batteries is sufficient to complete the load demand, after that, the load must be disconnected to charge the batteries. The wind system produces maximum power MPPT, the reference power is given by:

$$P_{ref2\_w} = P_w^{opt} = K_{opt} \cdot \Omega_{opt}^3 \tag{7.39}$$

with K_{opt} a coefficient which depends on the ratio of tip speed and optimal power coefficient.

We note in Fig. 7.21 the intersection of $P_W(\Omega)$ characteristic with reference $P_{ref2\_W}(\Omega)$ (point C′) which corresponds to the maximum power point for a particular value of wind speed. As for the operation of photovoltaic system, we remark that two operating points can develop the same reference power (point A′ and B′). Operation on the right side of the point of maximum power requires a system of power control [191]. The operating point (A′) would be the most appropriate. The reference angular velocity which corresponds to the operating MPPT is given by:

$$\Omega_{ref} = \Omega_{opt} = \sqrt[3]{\frac{P_{ref2-w}}{K_{opt}}} \tag{7.40}$$

Then the supervisor decides the case (1 or 2/3) by comparing the measured mechanical speed with the reference speed.

$$\begin{cases} \text{If} \quad \Omega < \Omega_{opt}, \quad \text{case 1} \quad P_w = P_{ref1\_w} \\ \text{If} \quad \Omega = \Omega_{opt}, \quad \text{case 2/3} \quad P_w = P_{ref2\_w} = P_w^{opt} \end{cases} \tag{7.41}$$

A description of operating cases is shown in Fig. 7.22.

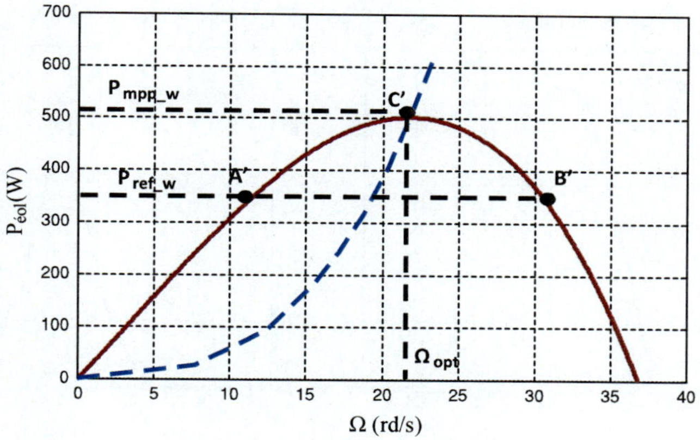

Fig. 7.21 Characteristic $P_W(\Omega)$ with photovoltaic power Reference P_{ref_pv} produced (Case 3)

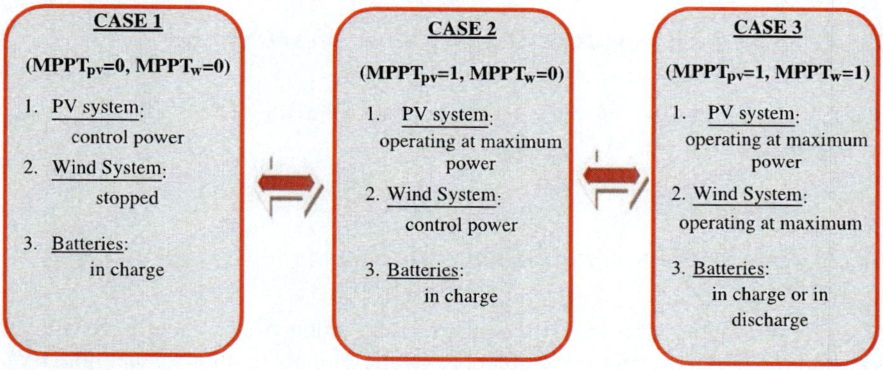

Fig. 7.22 Description of operating cases

7.3.3.4 Application of the Control of Hybrid Photovoltaic/Wind System

We make an application with a hybrid wind/photovoltaic system. It comprises ten photovoltaic panels of 110 W each one (five panels in series in parallel with the five other), a parallel chopper (boost), ten lead acid batteries (12 V, 92 Ah) connected in series and a wind Turbine 600 W with a permanent magnet synchronous generator, a PWM rectifier connected to the battery bank and a load that represents a house with a total load of 3920 Wh/day. The application is made under Matab/Simulink. We note from these results that the controller manages the system functions described above [9] (Figs. 7.23, 7.24, 7.25, 7.26, 7.27, 7.28, 7.29).

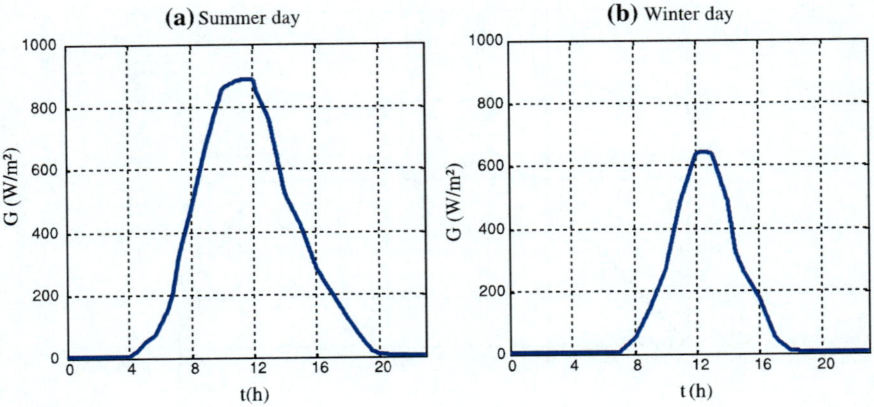

Fig. 7.23 Irradiance (*G*) for a day. **a** Summer day. **b** Winter day [9]

7.3.4 Hybrid Photovoltaic/Wind//Hydro/Diesel System

These systems consist of micro-hydro, solar, wind and Diesel generator and battery as back-up (Fig. 7.30) [192].

7.3.5 Hybrid Photovoltaic-Fuel Cell System

The role of a hybrid (fuel cell-PV) system is the production of electricity without interruption in remote areas. It consists generally of a photovoltaic generator (PV), an alkaline water electrolyser, a storage gas tank, a proton exchange membrane fuel cell (PEMFC) and PCU to manage the system operation of the hybrid system. A PEMFC can be described as two electrodes (anode and cathode) separated by a solid membrane. Energy is produced by a PV generator to supply a user load. Whenever there is enough solar radiation, the user load can be powered totally by the PV energy. During periods of low solar radiation, auxiliary electricity is required. An alkaline high pressure water electrolyser is powered by the excess energy from the PV generator to produce hydrogen and oxygen at a maximum pressure. A PEMFC is used to keep the system reliability at the same level as for the conventional system while decreasing the environmental impact of the whole system. The PEMFC consumes gases which are produced by an electrolyser to supply the user load demand when the PV generator energy is deficient; it works as an auxiliary generator. Power conditioning units dispatch the energy between the components of the system.

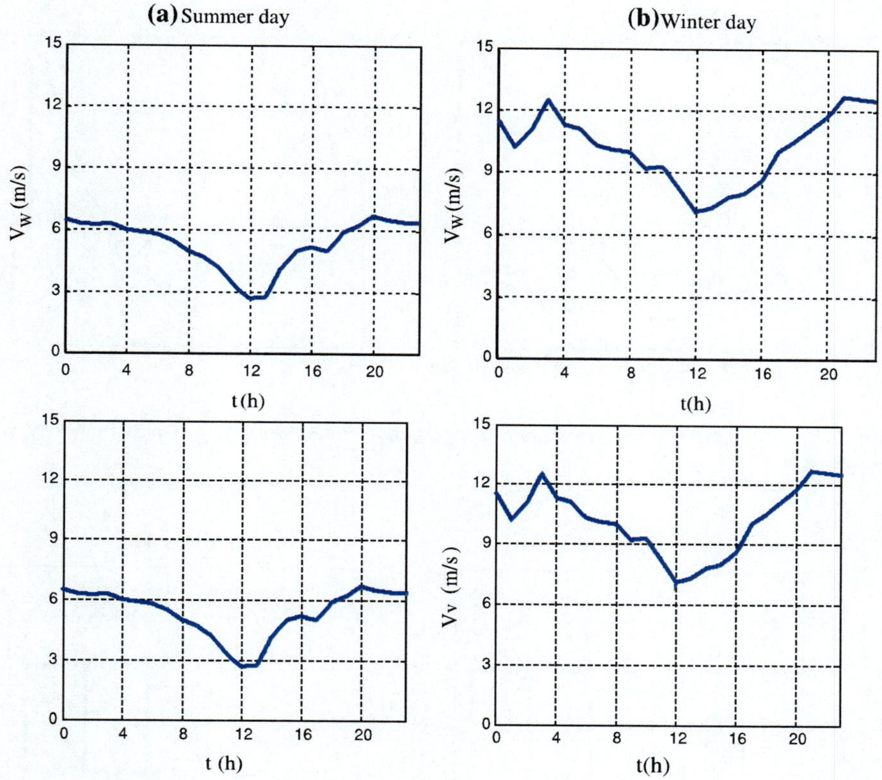

Fig. 7.24 Profile of wind speed (V_w) for a day. **a** Summer day. **b** Winter day

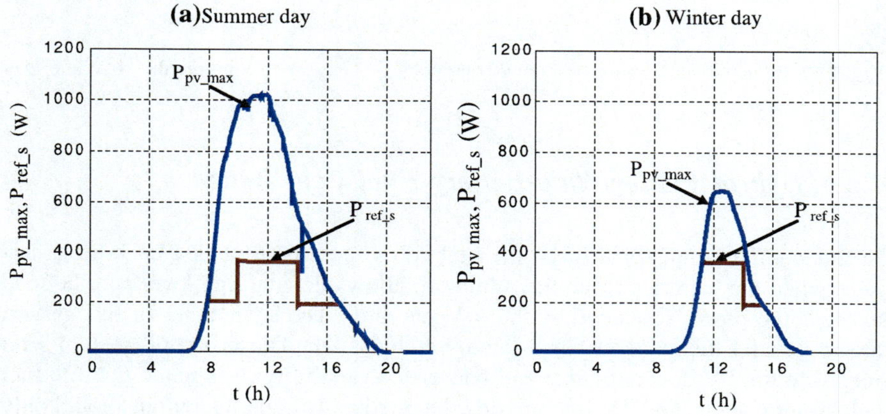

Fig. 7.25 Photovoltaic power output, maximum (P_{pv_max}) and reference power (P_{ref_s}). **a** Summer day. **b** Winter day

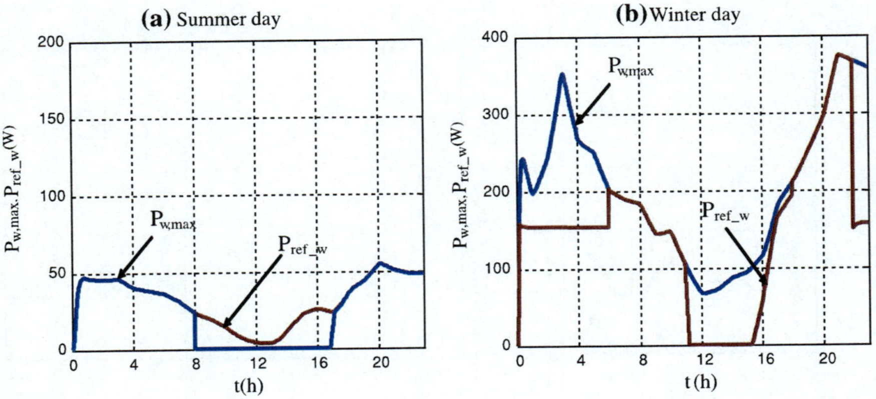

Fig. 7.26 Wind power output, maximum (P_{w_max}) and its reference (P_{ref_w}). **a** Summer day. **b** Winter day

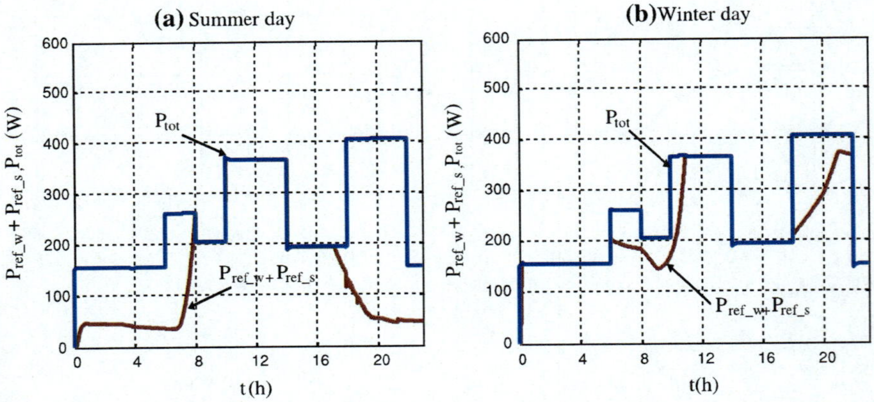

Fig. 7.27 Total power required and power output (P_{ref_s}, P_{ref_w}). **a** Summer day. **b** Winter day

7.3.6 Hybrid Photovoltaic-Battery-Fuel Cell System

In this configuration (Fig. 7.31), the fuel cell system is used as a back-up generator, when the batteries reach the minimum allowable charging level and the load exceeds the power produced by the PV generator. The advantages of this system are in general the same as for a Photovoltaic-Battery-Diesel hybrid system with regard to the PV generator size and batteries, availability. It is noted that the fuel cell system needs more time to provide the rated power and the output should only be increased slowly after startup. The increasing operating temperature which occurs during operation improves the efficiency of a fuel cell significantly [173].

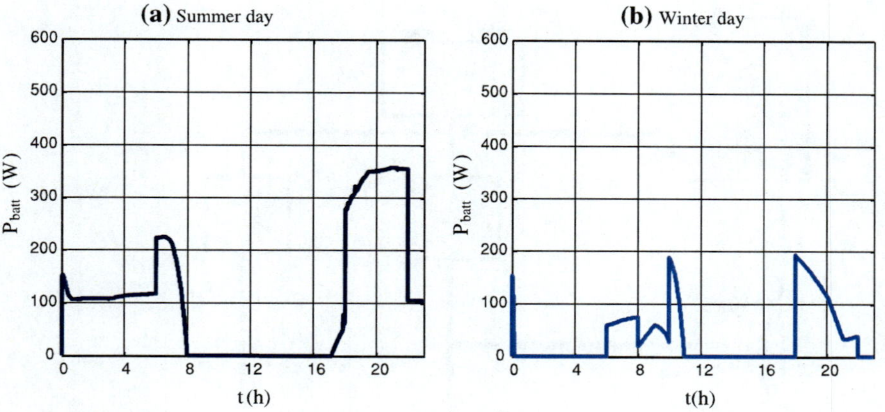

Fig. 7.28 Power supplied by batteries. **a** Summer day. **b** Winter day

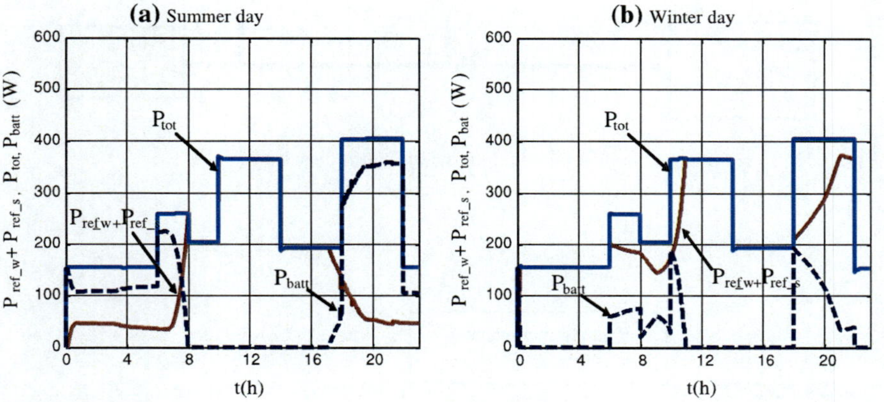

Fig. 7.29 Total power required, power output (Pw + Ps) and power supplied by batteries. **a** Summer day. **b** Winter day

7.3.7 Hybrid Photovoltaic-Electrolyser-Fuel Cell System

In some applications, another source of energy is necessary to realize energy storage. In this system, the excess energy is stored in the form of compressed hydrogen via conversion through the electrolyser. The fuel cell is used to produce power if the load power exceeds that produced from the PV generator. It can also function as an emergency generator, if the PV generator system fails [173] (Fig. 7.32).

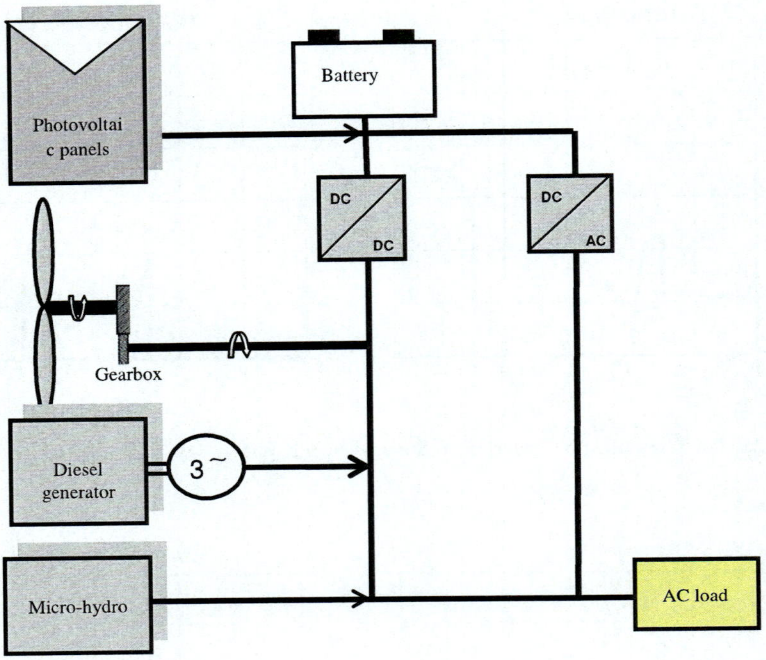

Fig. 7.30 Description of hybrid photovoltaic/wind//hydro/diesel system

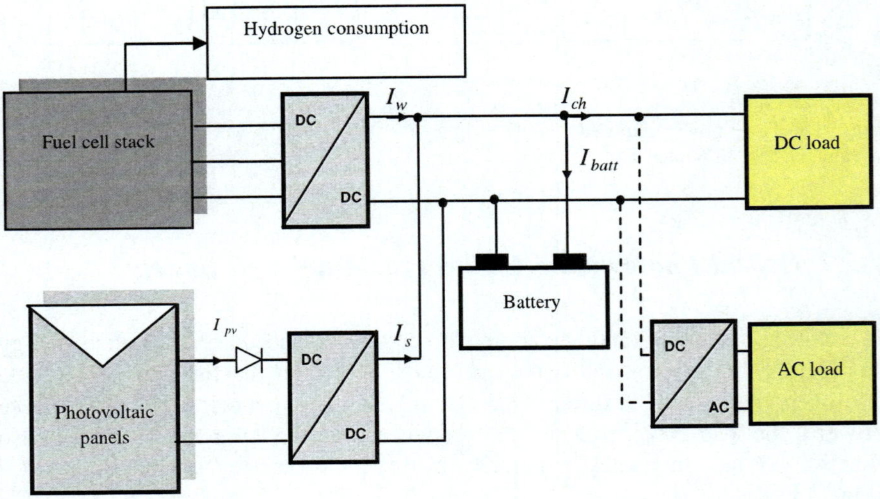

Fig. 7.31 Description of a hybrid photovoltaic-battery-fuel cell

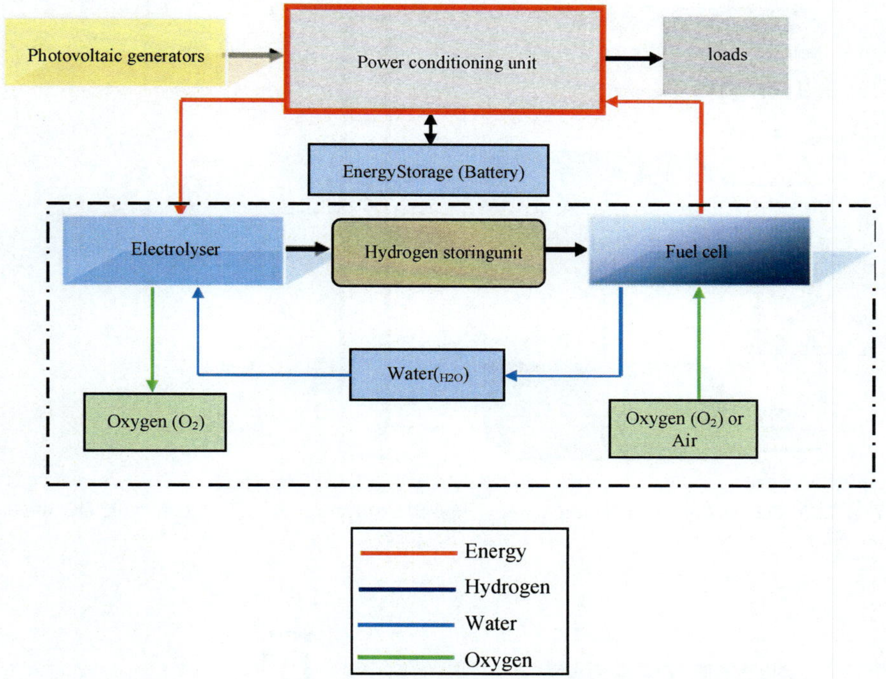

Fig. 7.32 Description of a hybrid photovoltaic-Electrolyzer-fuel cell System [193]

7.3.7.1 Different Topologies of Hybrid Photovoltaic-Electrolyser-Fuel Cell System

Different topologies are competing for an optimal design of the Hybrid Photovoltaic-Electrolyser-Fuel cell System. These topologies are DC and AC coupled systems. The PV generator supply DC voltage to the electrolyser. The AC inverter must also supply DC voltage range of electrochemical components. To obtain correctly the direct coupling of the component, the maximum power point voltage of the PV generator must be equal to the maximum voltage of the fuel cell component and the rated voltage of the electrolyser component [173] (Fig. 7.33).

In this case, the connection between the components and user demand is established through power conditioning unit. It keeps the DC bus voltage almost constant in the event of bus power interruptions (Fig. 7.34).

Components of the hybrid system with AC coupled are connected directly to the AC bus. The inverters can keep the output frequency and voltage stable and allow the energy surplus to flow backwards to be stored into the hydrogen subsystem. This configuration has numerous advantages such as: expandability, utility grid, compatibility, cost reduction, and simple design and installation [173] (Fig. 7.35).

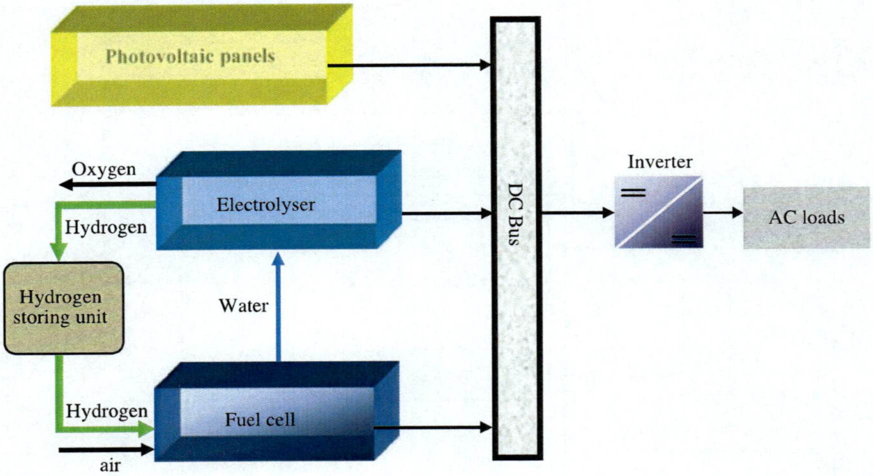

Fig. 7.33 Description of a hybrid photovoltaic-Electrolyzer-fuel cell system with DC direct coupling

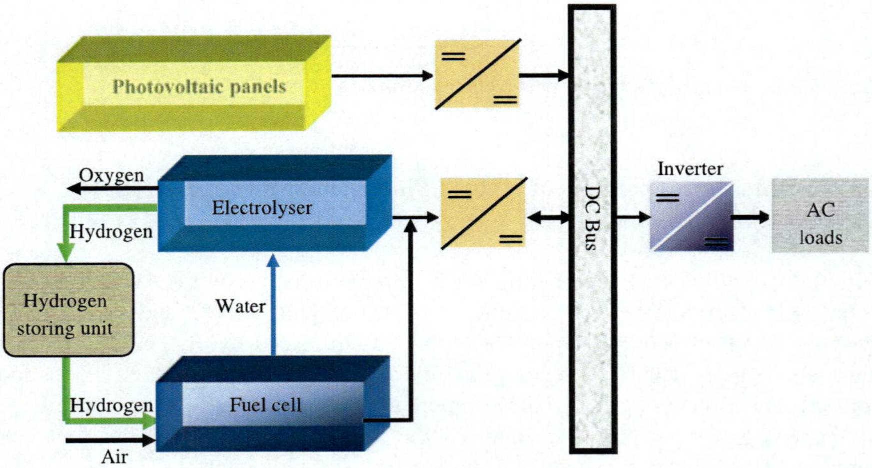

Fig. 7.34 Description of a hybrid photovoltaic-Electrolyzer-fuel cell system with DC indirect coupling

7.3.7.2 Modeling of Hybrid System

Modeling PV

See Chap. 2.

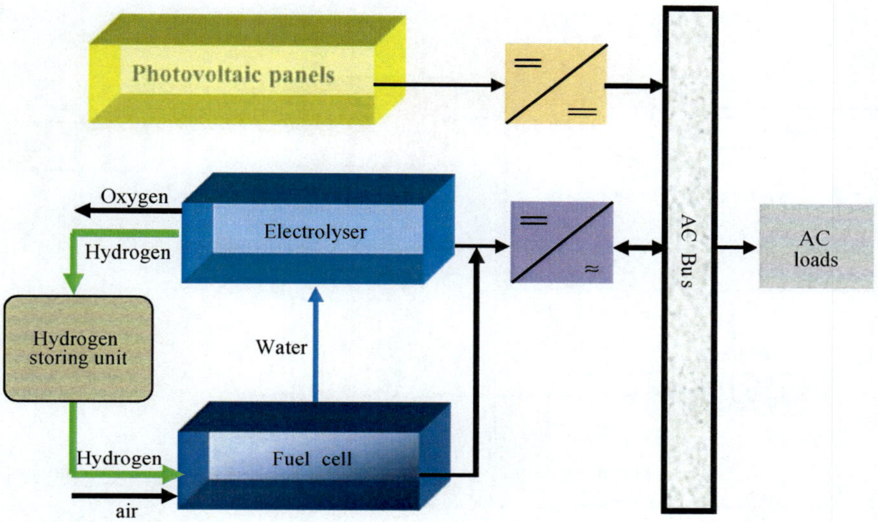

Fig. 7.35 Description of a hybrid photovoltaic-Electrolyzer-fuel cell system with AC coupled [193]

Modeling of Fuel Cell PEMFC

It is necessary to define the different circuits of a fuel cell system to simplify the modeling and control of each circuit. The cell system is composed of the heart cell associated with all necessary ancillaries to the operation of a fuel cell in an embedded application. Figure 7.36 shows all the functions present in a fuel cell system [193].

- Moto-compressor model

The Moto-compressor is composed of an air compressor and an electrical machine. Generally, it is a permanent magnet motor (PMSM).

– Air Compressor model

An air compressor supplies directly each stack, and the flow is regulated through its rotational speed. The compressors used in such applications are volumetric type because they can easily control the outflow. These types of compressors are classified into two categories: reciprocating compressors and rotary compressors. In fuel cell applications, it is the twin-screw rotary compressor types which are used because they do not require lubrication. The inputs of the compressor model are rotating speed ω and discharge pressure P_s (imposed by the pressure control). The outputs are the mass flow F_{mass} and torque compression

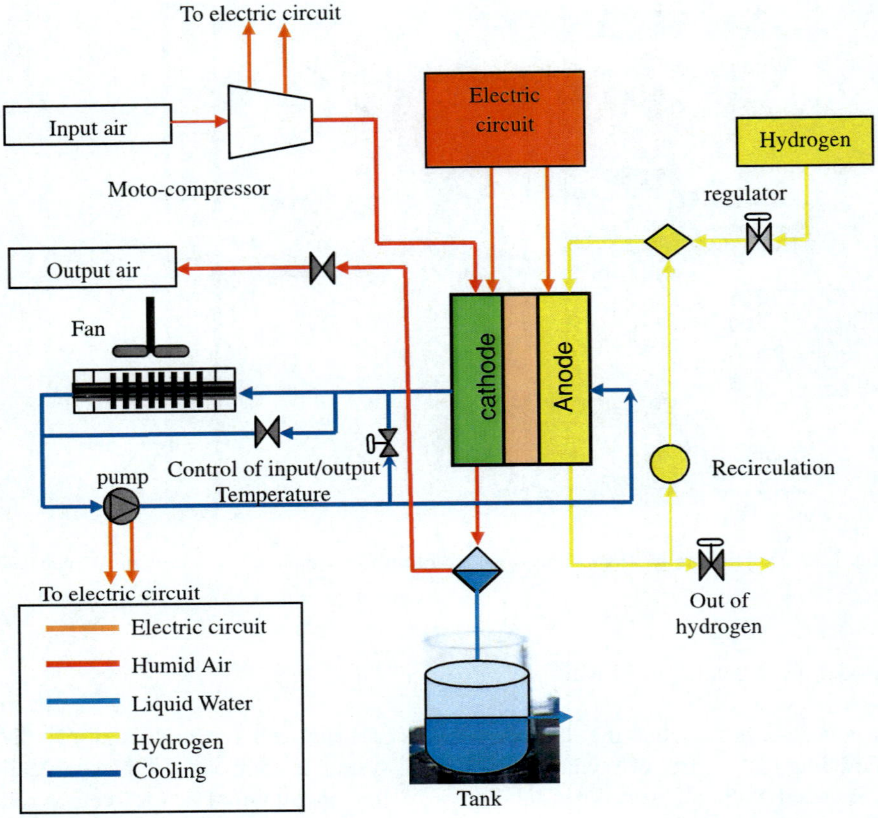

Fig. 7.36 Diagram of a PEMFC

Fig. 7.37 Compressor model

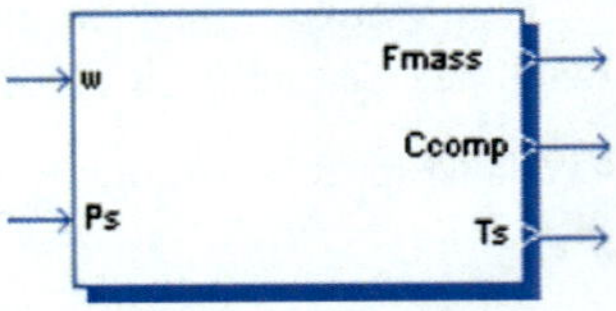

C_{comp}. Another useful parameter for the operation of the stack is the gas temperature at the output of the compressor T_s (Fig. 7.37).

The mass flow F_{mass} is calculated easily with the compressor cylinder C_{yl}, speed ω and density of the gas ρ_e. In the ideal case [193]:

$$F_{\text{mass}} = \rho_e C_{\text{yl}} \frac{\omega}{2\pi} \tag{7.42}$$

with F_{mass} as mass flow compressor (g/s), F_{molar} the Molar flow of the compressor (mol/s), and ρ_e Gas density (kg/m^3).

In the case of a real compressor, we must take into account a leakage rate. This is summarized by involving volumetric efficiency η_v (Π, ω).

$$F_{\text{molar}} = \frac{p_s}{RT_s} C_{\text{yl}} \frac{\omega}{2\pi} \eta_v(\Pi, \omega) \tag{7.43}$$

with T_s a is the output temperature (K °) and Π a the compression ratio.

The cylinder is chosen so that the maximum molar flow rate charged by the cell is reached for the rated rotation speed.

We have [193]

$$W_{\text{is}} = \text{Cyl} p_e^{\frac{1}{\gamma}} \int_s^e \frac{\mathrm{d}p}{p^{1/\gamma}} = p_e \text{Cyl} \frac{\gamma}{1-\gamma} \left(\Pi^{\frac{\gamma-1}{\gamma}} - 1 \right) \tag{7.44}$$

with γ polytropic exponent (for air $\gamma = 1.4$) and p_e the input pressure.

The compression power is deduced as:

$$p_{\text{comp\_is}} = F_{\text{mass}} C_p T_e \left(\Pi^{\frac{\gamma-1}{\gamma}} - 1 \right) \tag{7.45}$$

with C_p as Heat capacity of gases (kJ/K) and T_e: Temperature input (K °).

The various losses are taken into account by introducing an isentropic efficiency of compression. It is defined by:

$$\eta_{\text{is}}(\Pi, \omega) = \frac{p_{\text{comp\_is}}}{p_{\text{comp\_required}}} \tag{7.46}$$

The actual power required for compression is:

$$p_{\text{comp\_required}} = \frac{F_{\text{mass}} C_p T_e}{\eta_{\text{is}}(\Pi, \omega)} \left(\Pi^{\frac{\gamma-1}{\gamma}} - 1 \right) \tag{7.47}$$

and compressor torque is given by

$$C_{\text{comp}} = \frac{p_{\text{comp\_required}}}{\omega} \tag{7.48}$$

with η_{is} isentropic efficiency.

– Modeling of the cathode compartment

Considering the ideal gas law, we can write:

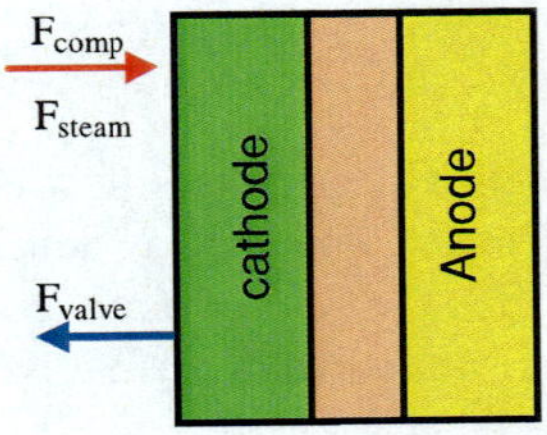

Fig. 7.38 Cathode model

$$P_{\text{cath}} = \frac{RT_{\text{PEMFC}}}{V_{\text{cath}}} n_{\text{cath}} \tag{7.49}$$

The cathode pressure is directly dependent on the moles number of gas contained in the cathode compartment (Fig. 7.38).

We can write [193]

$$\frac{\mathrm{d}n_{\text{cath}}(t)}{\mathrm{d}t} = F_{\text{comp}}(t) + F_{\text{steam}}(t) - F_{\text{O}_2\text{cons}}(t) - F_{\text{valve}}(t) \tag{7.50}$$

with

$$F_{\text{steam}}(t) = \frac{X_{\text{steam}}(P_{\text{steam}}(t), T_{\text{PEMFC}}(t))}{1 - X_{\text{steam}}(P_{\text{cath}}(t), T_{\text{PEMFC}}(t))} F_{\text{comp}}(t)$$

$$X_{\text{steam}}(P_{\text{cath}}(t), T_{\text{PEMFC}}(t)) = \frac{P_{\text{sat}}(T_{\text{PEMFC}}(t))}{P_{\text{cath}}(t)}$$

$$F_{\text{O}_2\text{cons}}(t) = \frac{N_{\text{cell}}}{4F} I_{\text{PEMFC}}(t)$$

$$F_{\text{valve}} = \%\, S_{\text{open}} \sqrt{\frac{2P_{\text{cath}}^2}{MRT_{\text{PEMFC}}} \frac{\gamma}{\gamma - 1} \left[\left(\frac{P_{\text{atm}}}{P_{\text{cath}}} \right)^{\frac{2}{\gamma}} - \left(\frac{P_{\text{atm}}}{P_{\text{cath}}} \right)^{\frac{\gamma+1}{\gamma}} \right]}$$

with S_{open} as the maximum opening of the valve (m^2).

The initial condition of operation is:

$$n_0 = \frac{P_{\text{atm}} V_{\text{cath}}}{RT_{\text{pac}}} \frac{\mathrm{d}P_{\text{cath}}(t)}{\mathrm{d}t} = \frac{RT_{\text{PEMFC}}(t)}{} V_{\text{cath}}$$

$$\left(\left(1 + \frac{X_{\text{steam}}(P_{\text{cat}}(t), T_{\text{PEMFC}}(t))}{1 - X_{\text{Steam}}(P_{\text{cath}}(t), T_{\text{PEMFC}}(t))} \right) F_{\text{comp}}(t) - F_{\text{valve}}(t) - F_{\text{O}_2\text{cons}}(t) \right) \tag{7.51}$$

By simplifying the equation we obtain:

$$\frac{\mathrm{d}P_{\text{cath}}}{\mathrm{d}t} = \left[\frac{RT_{\text{PEMFC}} 10^{-5}}{V_{\text{cath}}} \left(\left(1 + \frac{X_{\text{steam}}}{1 - X_{\text{steam}}} \right) F_{\text{comp}} - F_{\text{valve}} - \frac{N_{\text{cell}}}{4F} I_{\text{PEMFC}} \right) \right] \tag{7.52}$$

with $P_0 = P_{\text{atm}}$ as the initial condition.

We represent the compressor molar flow rate (mol/s) by the following simplified expression in Laplace space:

$$F(p) = \frac{F_{\text{comp}}(p)}{F_{\text{ref}}(p)} = \frac{1}{\left(1 + \frac{1}{\omega_{\text{flow}}}p\right)^2} \tag{7.53}$$

with ω_{flow} as control bandwidth (50 rad/s), n_{cath} moles number in the cathode compartment (mol), P_{cath} pressure in the cathode compartment (P_{a}), T_{PEMFC} cell temperature (K), V_{cath} volume of the cathode compartment (m^3), R molar gas constant: 8.13 J/K/mol, F Faraday constant: 96,485 C/mol, F_{comp} molar flow compressor (mol/s) and M molar mass of the air ($M = 0.029$ g/mol).

Oxygen molar flow is modeled as

$$F_{O_2}(t) = (X_{O_2})_{\text{air}} \cdot F_{\text{comp}}(t) - F_{O_2\text{cons}}(t) - (X_{O_2})_{\text{valve}} F_{\text{valve}}(t) \tag{7.54}$$

with

$$F_{O_2\text{cons}}(t) = \frac{N_{\text{cell}}}{4F_{\text{steam}}} I_{\text{PEMFC}}(t)$$

$$F(p) = \frac{F_{\text{comp}}(p)}{F_{\text{ref}}(t)} = \frac{1}{\left(1 + \frac{1}{\omega_{\text{flow}}}p\right)^2}$$

with $(X_{O2})_{\text{air}}$ as molar fraction of oxygen in the air (21%), $(X_{O2})_{\text{valve}}$ as molar fraction of oxygen in the valve, N_{cell} number of cells in the stack, I_{PEMFC} current in the cells (A), F_{valve} molar flow through the valve (mol/s), F_{steam} corresponds to the steam amount supplied by the humidification system (mol/s), X_{steam} mole fraction of water vapor in the humidification system, F_{ref} molar flow reference in the valve (mol/s)

- Determination of $(X_{O_2})_{\text{valve}} F_{\text{valve}}(p)$

To simplify the model, we can take this constant value

$$(X_{O_2})_{\text{valve}} F_{\text{valve}}(p) = (\text{st}_{O_2} - 1) F_{O_2\text{cons}}(p) = (\text{st}_{O_2} - 1) \frac{N_{\text{cell}}}{4F} I_{\text{PEMFC}}(p) \tag{7.55}$$

with St_{O2} ratio of the stoichiometry (equal to 1.6),

If we replace all in the equation we obtain:

$$F_{O_2}(p) = (X_{O_2})_{\text{air}} F(p) F_{\text{ref}}(p) - \frac{N_{\text{cell}}}{4F} I_{\text{PEMFC}}(p) - (st_{O_2} - 1) \frac{N_{\text{cell}}}{4F} I_{\text{PEMFC}}(p) \tag{7.56}$$

We can write:

$$\frac{F_{O_2}(p)}{F_{ref}(p)} = (X_{O_2})_{air}(F(p) - 1)$$

The number of oxygen moles is given by:

$$n_{O_2}(p) = \frac{1}{N_{cell}p} F_{O2}(p) \tag{7.57}$$

with

$$[n_{O_2}]_{init} = \frac{P_{cath} V_{cath}}{RT_{PEMFC}} [X_{O_2}]_{init}$$

The total number of moles in the cathode compartment is

$$n_{cath} = \frac{P_{cath} V_{cath}}{RT_{PEMFC}} \tag{7.58}$$

The partial oxygen pressure is given by

$$P_{O_2}(p) = \frac{n_{O_2}(p)}{n_{cath}} P_{in}(p) \tag{7.59}$$

The term P_{in} shows the pressure evolution within the cathode compartment.

$$P_{in}(p) = P_{cath} - K_{in} F_{comp}(p)$$

with

$$K_{in}(t) = K_a F_{comp}(t) + K_b$$

K_a and K_b are constants to be determined [193]

- Anodic compartment modeling

We have to determine the partial hydrogen pressure at the membrane and the hydrogen flow within the compartment. The number of steam moles is given by [193]:

$$(n_{H_2O})_{steam} = \frac{P_{H_2Osat}(T_{PEMFC}) V_{anod}}{RT_{PEMFC}} \tag{7.60}$$

with

$$(P_{H_2Osat})(T_{PEMFC}) = e^{23.1961 - \frac{3816.44}{T_{PEMFC} - 46.13}}$$

The number of hydrogen moles is given by:

$$\begin{aligned} n_{H2}(t) &= \frac{V_{anod}}{RT_{PEMFC}} P_{H2}(t) \\ P_{anod}(t) &= \frac{\left(n_{H_2}(t) + (n_{H_2O})_{steam}\right) RT_{PEMFC}}{V_{anod}} \end{aligned} \tag{7.61}$$

with $P_{H_2}(t)$ the partial hydrogen pressure in the compartment,
V_{anod} anode volume of compartment (m^3),
n_{H_2} number of hydrogen moles of in the compartment (mol),
$(n_{H_2O})_{steam}$ number of water steam moles in the compartment (mol)
From these expressions, we can determine the partial hydrogen pressure:

$$P_{H_2}(t) = P_{anod}(t) - P_{H_2Osat}(T_{PEMFC}) \tag{7.62}$$

We have the following equations which can calculate the hydrogen flow $(F_{H2})_{cons}$consumed by the chemical redox reaction [193]:

$$\begin{aligned} F_{H_2}(t) &= (F_{H_2})_{cons} - (F_{H_2})_{purg}(t) \\ (F_{H_2})_{cons} &= \frac{N_{cell} I_{PEMFC}(t)}{2F} \end{aligned} \tag{7.63}$$

with $(F_{H_2})_{purg}$ the purge flow system if the anode compartment is not closed

The purge is a valve and we can model the flow through the following expression:

$$(F_{vanne})_{anod} = \%S_{open} \sqrt{\frac{2P_{anod}^2}{M_{H_2O} RT_{PEMFC}} \frac{\gamma}{\gamma - 1} \left[\left(\frac{P_{atm}}{P_{anod}} \right)^{\frac{2}{\gamma}} - \left(\frac{P_{atm}}{P_{anod}} \right)^{\frac{\gamma+1}{\gamma}} \right]} \tag{7.64}$$

with P_{atm} atmospheric pressure (output pressure) constant at 1.013*times*; 10^5Pa, M_{H2O}molar molecular weight of water (0,018 g/mol)

We can therefore determine $(F_{H2})_{purg}$ by

$$(F_{H_2})_{purg} = (X_{H_2})_{anod} (F_{valve})_{anod} \tag{7.65}$$

with

$$\begin{aligned} (X_{H_2})_{anod} &= 1 - X_{steam} \\ X_{steam} &= \frac{P_{H_2Osat}(T_{pac})}{P_{anod}} \end{aligned}$$

$(X_{H_2})_{anod}$ is the molar fraction of hydrogen in the compartment.

$$P_{H_2in}(t) = P_{anod}(t) - P_{H_2Osat}(T_{PEMFC}) \tag{7.66}$$

– Electrical model of the fuel cell

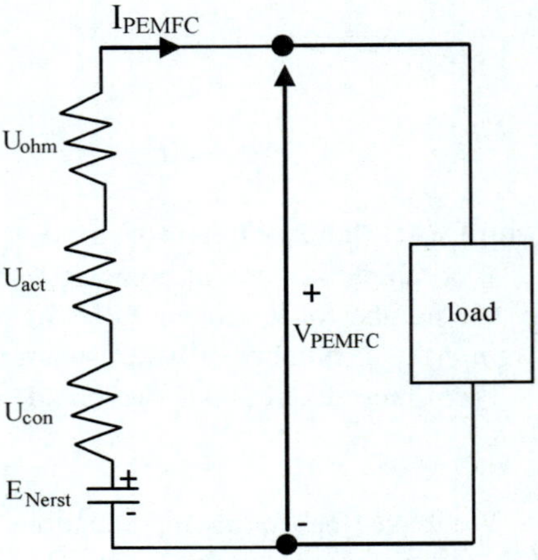

Fig. 7.39 Electrical representation of a PEMFC

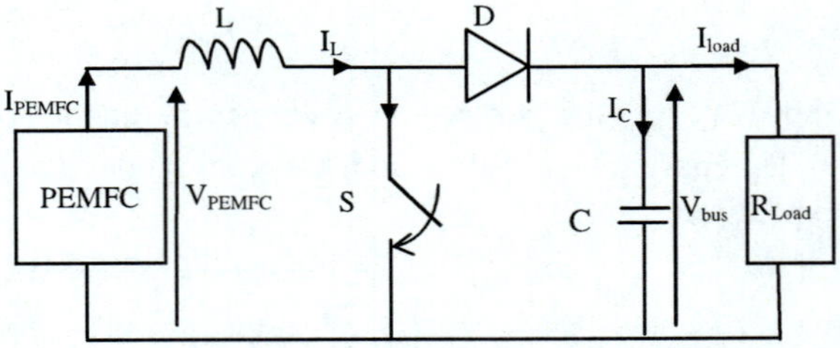

Fig. 7.40 Diagram of the fuel cell associated with converter [193]

The cell voltage V_{PEMFC} is lower than the theoretical voltage E_{Nerst} due to various irreversible loss mechanisms. These losses, which are often called polarization or over-voltage losses, originate primarily from three sources: activation overvoltage U_{act}; concentration or diffusion over-voltage U_{conc} and resistive or ohmic over-voltage U_{ohm} [173] (Fig. 7.39).

$$V_{PEMFC} = E_{Nernst} + U_{act} - U_{ohm} - U_{conc} \tag{7.67}$$

where V_{PEMFC} A unit cell voltage (Volt), E_{Nernst} is the Nernst voltage (Volt), U_{ohm} resistive or ohmic over-voltage (volt)

$$U_{ohm} = \frac{I_{pac}}{S_{cell}} \left(\frac{181.6\left[1+0.03\left(\frac{I_{PEMFC}}{S_{cell}}\right)+0.06\left(\frac{T}{303}\right)^2\left(\frac{I_{PEMFC}}{S_{cell}}\right)^{2.5}\right]}{\left[\lambda-0.634-3\left(\frac{I_{PEMFC}}{S_{cell}}\right)\right]\exp\left[4.18\left(\frac{T-303}{T}\right)\right]} * I_{PEMFC} + S_{cell} * R_c \right) \tag{7.68}$$

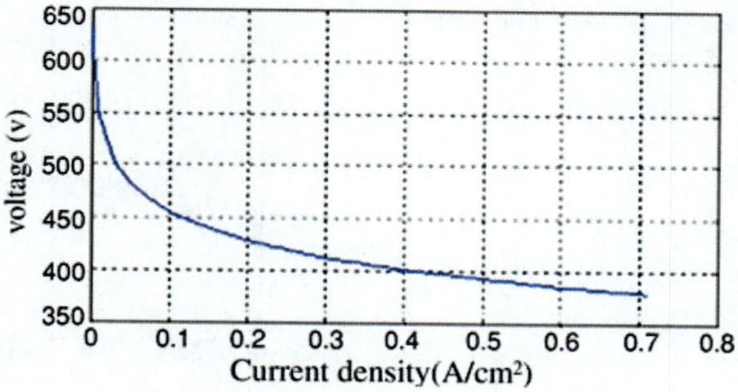

Fig. 7.41 Voltage/current density characteristic of a PEMFC at $T = 25°C$ [193]

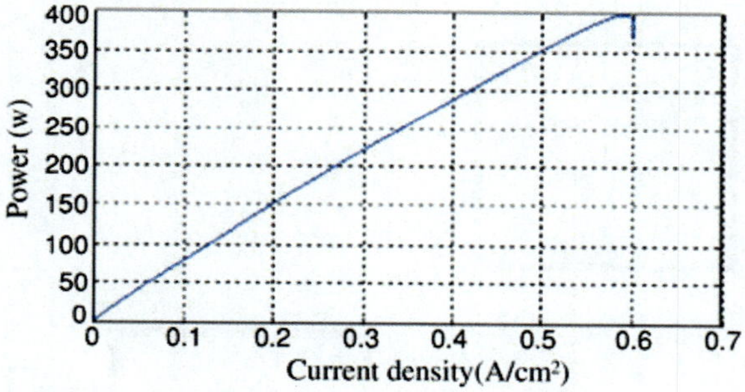

Fig. 7.42 Power density/current density characteristic of a PEMFC at $T = 25°C$ [193]

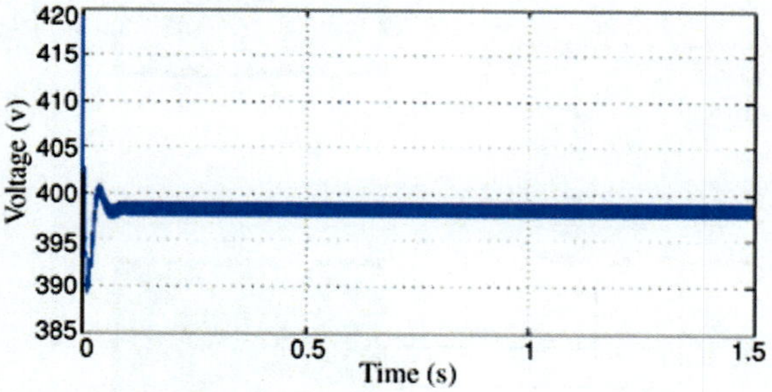

Fig. 7.43 Evolution of the cell output voltage [193]

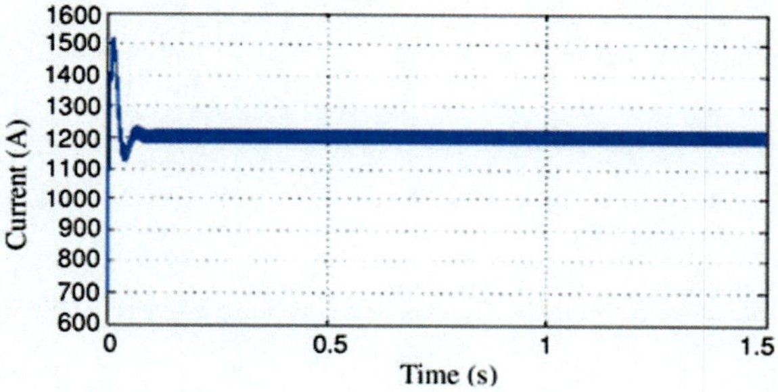

Fig. 7.44 Evolution of the current of the cell [193]

Fig. 7.45 Evolution of the cell power

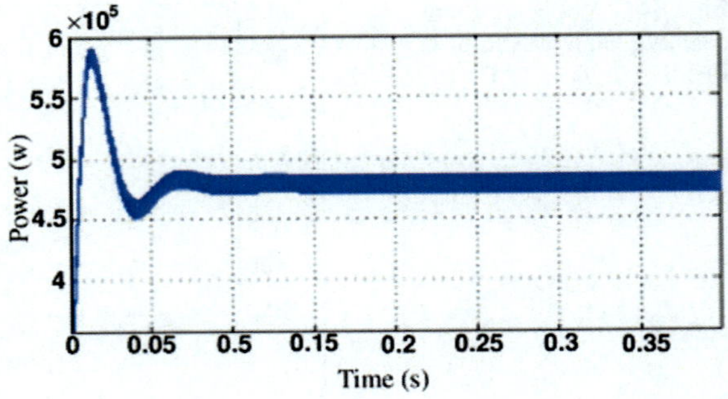

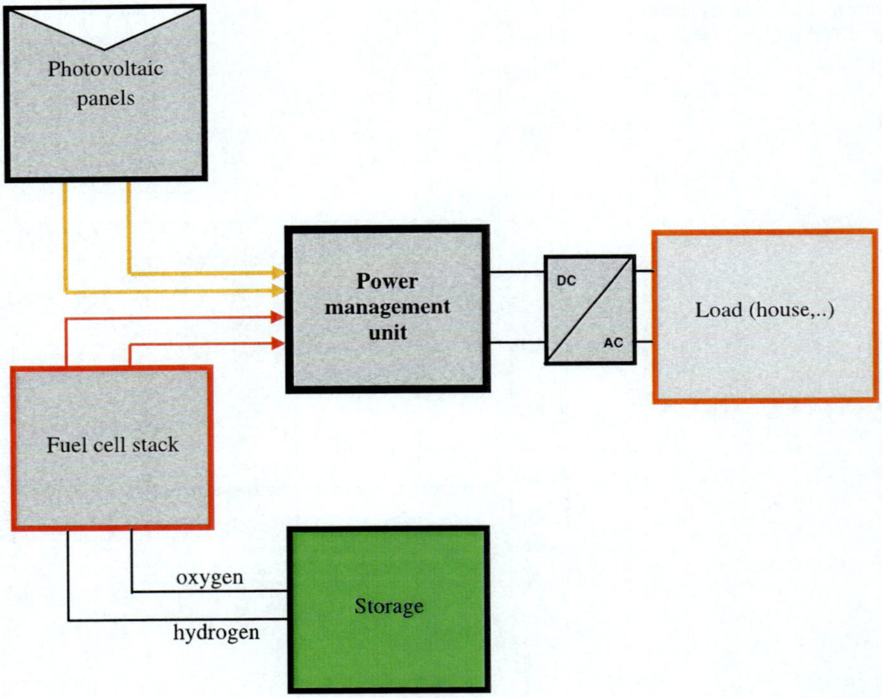

Fig. 7.46 Power management

with U_{conc} concentration or diffusion over-voltage (volt), U_{act} activation over-voltage (volt)

$$U_{\text{act}} = b_1 + b_2 * T_{\text{PEMFC}} + b_3 * T_{\text{PEMFC}} * \ln\left(j * 5 * 10^{-3}\right) + b_4 * T_{\text{PEMFC}} * \ln C^*_{O_2}$$

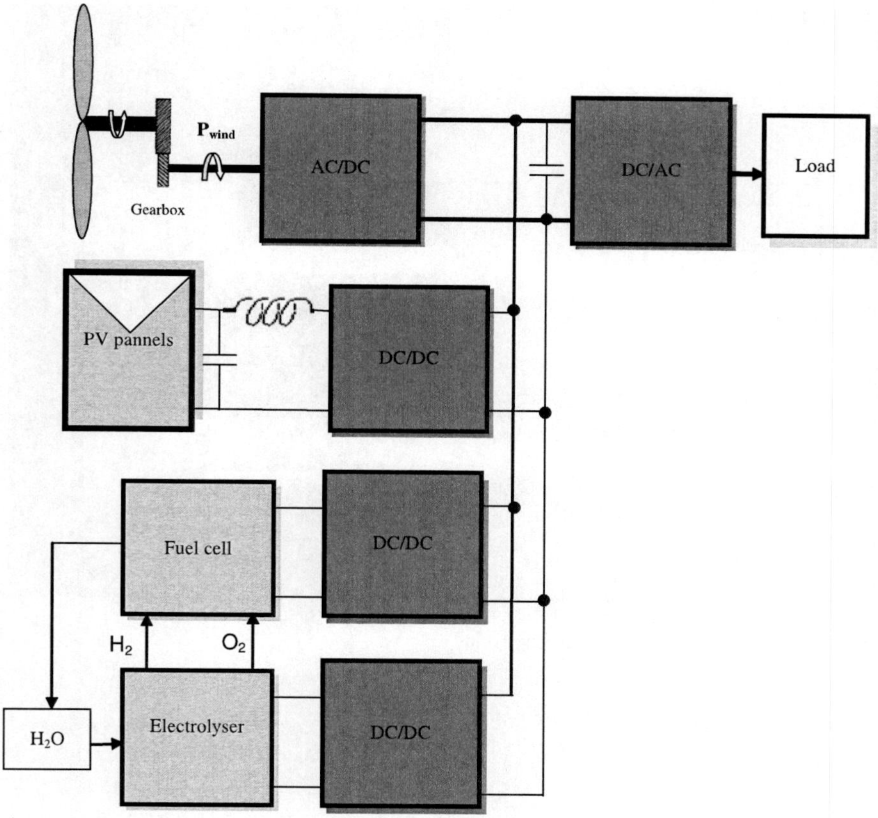

Fig. 7.47 Hybrid wind/Photovoltaic/fuel cell configuration

with $C^*_{O_2}$ as the oxygen concentration in the cathode area (mol/cm^3), b_1, b_2, b_3 and b_4 are constants, S_{cell} cell active area (m^2) and R_c as contact resistance equivalent of the electrodes (Ω)

$$U_{conc} = -B \ln\left(1 - \frac{j}{j_{max}}\right) \tag{7.69}$$

The expression of the Nernst equation according to JC Amphlett is given by:

$$E_{Nernst} = a_1 + a_2 * (T_{PEMFC} - 298.15) + a_3 * T_{PEMFC} * \left(0.5 * \ln P^*_{O_2} + \ln P^*_{H_2}\right) \tag{7.70}$$

with T_{PEMFC} absolute operating temperature of the stack (K), a_1, a_2 and a_3 are constants

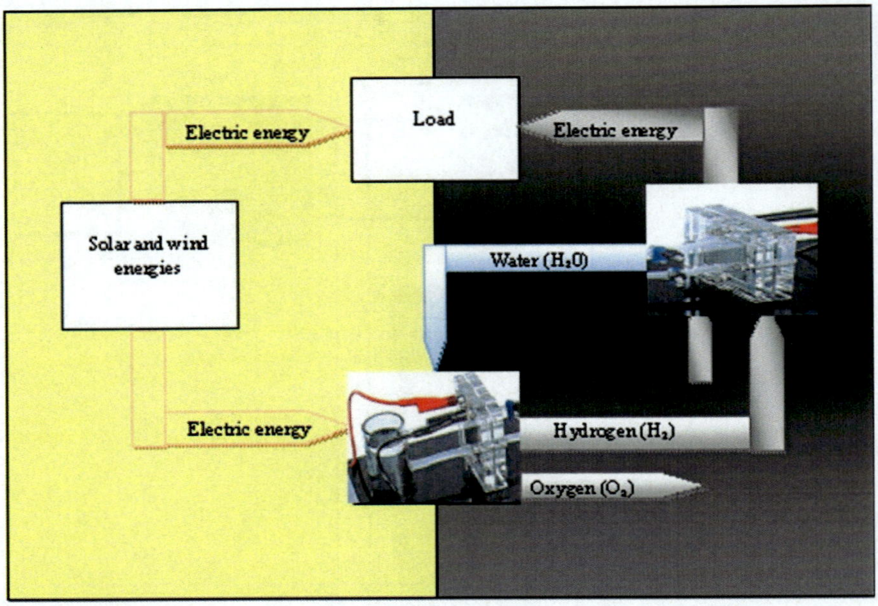

Fig. 7.48 Solar-wind-hydrogen energy cycle

- Association-fuel cell converter
- Application:

We make application under Matlab/simulink (Figs. 7.40, 7.41, 7.42, 7.43, 7.44, 7.45).

Note in this figure that the shape of the power curve follows well the profile of the current.

7.3.7.3 Power Management

Power management unit (PMU) allows the coordination between different energy sources such as PV panels electrolyzes and fuel cells (Fig. 7.46).

Generally, PV subsystem works as a primary source, converting solar irradiation into electricity that is given to a DC bus. The second working subsystem is the electrolyser which produces hydrogen and oxygen from water as a result of an electrochemical process. When there is an excess of solar generation available, the electrolyser is turned on to begin producing Hydrogen which is sent to a storage tank. The produced Hydrogen is used by the third working subsystem (the fuel cell stack) which produces electrical energy to supply the DC bus.

7.3.8 Hybrid Photovoltaic/Wind/Fuel Cell System

The necessary changes in our energy supply system can be accomplished if we use a hybrid system with solar, wind and fuel cell energies. Generally, the overall system comprises a wind subsystem with an AC/DC rectifier to connect the wind generator to the DC bus. It is also consists of a PV subsystem connected to the DC bus via a filter and DC/DC converter. The excess energy is stored as electrolytic hydrogen through an electrolyser and we use a fuel cell to generate electricity during low irradiance and low wind speed (Fig. 7.47).

In fact, when supply and demand do not coincide we need a convenient way to both store and transport renewable energy. This is where hydrogen comes into play as a storage and transport medium. When excess electric energy from wind and solar energy is stored in hydrogen and then converted back to electricity we have a solar-wind hydrogen energy cycle. Electrolysers use wind and solar excess electricity to split water into oxygen and hydrogen. When we need electricity the gases are fed into a fuel cell which converts the chemical energy of the hydrogen (and oxygen) into electricity, water and heat (Fig. 7.48).

References

1. Patel MR (1999) Wind and solar power systems. By CRC Press LLC, Boca Raton
2. http://www.pvresources.com/knowledge/basic-knowledge/photovoltaics.html
3. http://www.eia.doe.gov/cneaf/solar.renewables/page/solarreport/solarpv.html
4. http://en.wikipedia.org/wiki/solar_cell
5. http://www.iapp.de/iapp/agruppen/osol/?research:organic_solar_cells:basics_of_osc
6. Siti NBM (2010) Simulation design of a stand-alone photovoltaic (PV) inverter, a report submitted in partial fulfillment of the requirements for the award of the degree of bachelor of electrical engineering, faculty of electrical engineering, University Technology Malaysia
7. http://www.fsec.ucf.edu/en/consumer/solar_electricity/basics/types_of_pv.htm
8. Photovoltaic applications solar center information. http://www.ncsc.ncsu.edu/include/_upload/.../pvapplication.pdf
9. Lalouni S (2009) Study and control of hybrid electric wind power—photovoltaic, PHD in Science. University of Bejaia, Algeria
10. http://www.altestore.com/howto/how-to-size-a-battery-bank/a94/
11. http://www.photovoltaique.guidenr.fr /cours_photovoltaique_2011/cours_photovoltaique.php
12. Feasibility of photovoltaic systems, renewable energy, fact sheet no 19. http://www.infinitepower.org/pdf/factsheet-24.pdf
13. Matagne E, Photovoltaic Solar Energy (in french:Energie solaire photovoltaïque). http://perso.uclouvain.be/ernest.matagne/SOLAIRE/INDEX.HTM
14. Capderou M (1987) Solar Atlas of Algeria (in french: Atlas Solaire de l'Algérie), 1(1, 2): Theoretical and experimental models, Office of University Publications, Algeria
15. Meeus J (1991) Astronomical algorithms, Willmann-Bell, Richmond, Virginia (in french: Météorologie et energies renouvelable)
16. Gueymard C (1984) Physical modeling of solar radiation received by tritled surfaces with respect to aerosol anisotropic effect, symposium of meteorology and renewable energy, p 303
17. Temps RC, Coulson KL (1977) Solar radiation incident upon slopes of different orientations. Solar Energy 19:179
18. NASA, solar dynamics observatory: The 'Variable Sun' Mission. http://science.nasa.gov/science-news/science-at-nasa/2010/05feb_sdo/ (consulted 08/04/2011)
19. Gueymard C (1995) SMARTS2 a simple model of the atmospheric radiative tranfer of sunshine: algorithms and performance assessment, Florida Solar Energy Center
20. Jacobson MZ (2005) Fundamentals of atmospheric modeling, 2nd edn. Cambridge University Press, Cambridge

D. Rekioua and E. Matagne, *Optimization of Photovoltaic Power Systems*, Green Energy and Technology, DOI: 10.1007/978-1-4471-2403-0,

21. Notton G, Cristofari Ch, Muselli M, Poggi Ph, Heraud N (2006) Hourly solar irradiations estimation: from horizontal measurements to inclined data, First international symposium on environment identities and mediterranean area, ISEIMA'06, Ajaccio, p 234
22. Loutzenhiser PG, Manz H, Felsmann C, Strachan PA, Frank T, Maxwell GM (2007) Empirical validation of models to compute solar irradiance on inclined surfaces for building energy simulation. Solar Energy 81:254
23. Joseph JH, Wiscombe WJ, Weinman JA (1976) The delta-eddington approximation for radiative flux transfer. J Atm Sci 33:2452
24. Jimenez-Aquino JI, Varela JR (2005) Two Stream approximation to radiative transfer equation: an alternative method of solution. Revista Mexicana de Fisica 51(1):82
25. Abramowitz M, Stegun IA (1965) Hanbook of mathematical functions, 3th edn. United States Department of Commerce, Washington
26. Yang J, Qiu J (1992) An easy algorithm for solving radiative transfer equation in clear atmosphere. Adv Atm Sci 9(4):483
27. Matagne E (2011) Enhanced δ-two-stream approximation model of diffuse solar radiation and its application to computation of solar irradiation on tilted surface, analog integrated circuits and signal processing, Online First^{TM}, 24 March
28. Li J, Ramaswamy V (1996) Four-stream spherical harmonic expansion approximation for solar radiative transfer. J Atm Sci 53(8):1174
29. Borowy B, Salameh Z (1996) Methodology for the optimally sizing the combination of a battery bank and PV array in a wind/PV hybrid system. IEEE Trans Energy Convers 11(2):367
30. Jones CP (2002) Underwood a modeling method for building-integrated PV systems. Solar Energy 70(4):349
31. Lalouni S, Rekioua D, Rekioua T, Matagne E (2009) Fuzzy logic control of standalone photovoltaic system with battery storage. J Power Sources 193(2):899
32. Akihiro O (2005) Design and simulation of photovoltaic water pumping system. Master of science in electrical engineering, faculty of California polytechnic state university, San Luis Obispo
33. Laplaze D, Youm I (1985) Solar cells 14:167
34. Wolf M, Rauschenbach H (1963) Adv Energy Convers 3:455
35. Charles JP, Abdelkrim M, Muoy, Mialhe P (1981) Solar cells 4:169
36. Agrawal SK, Muralidharam R, Agrawala AYH, Tewary VK, Jain SC (1981) J Phys 14:1634
37. Cabestany J, Castaner L (1983) Rev Phys Appl 18:565
38. Warashina M, Ushirokawa A (1980) J App Phys 19:179
39. Ouennoughi Z, Chegaar M (1999) Solid-state electron, a simple method for extracting solar cell parameters using the conductance method. Solid-State Electron 43:1985
40. Chegaar M, Ouennoughi Z, Hoffmann A (2001) A new method for evaluating illuminated solar cell parameters. Solid-State Electron 45:293
41. Miahle P, Khoury A, Charles JP (1984) Phys Status Solid (A) 83:403–409
42. Walker GR (2001) Evaluating MPPT converter topologies using a Matlab PV model. J Electr Electron Eng 21(1):49
43. Aberle AG, Wenham SR, Green MA (1993) In: 23th IEEE photovoltaic specialist conference, p 133
44. Gergaud B, Multon B, Ben Ahmed H (2002) Analysis and experimental validation of various photovoltaic system models. In: Proceedings of the 7th international ELECTRIMACS'2002 congress, Montréal
45. Markvard T (2000) Solar electricity, 2nd edn. Willey, New York
46. Rekinger M, Matagne E, El Bachtiri R, Chenni R (2007) A model of solar cell with thermal effect determined on the basis of nominal values. In: Proceedings of days of future electrical—EF'2007. Toulouse
47. Caprasse Q, Matagne E (2008) A model of two-exponential solar cell prepared on the basis of nominal values. In: Proceedings of the conference CEE'08. Batna

48. Kasten F (1999) The linke turbidity factor based on improved values of the integral Rayleigh optical thickness. Solar Energy 56(3):239–244
49. Suri M, Hofierka J (2004) A new GIS-based solar radiation model its application to photovoltaic assessments. Trans GIS 8(2):175–190
50. Kassakian JG, Schlecht MF, Verghese GC (1991) Principles of power electronics. Addison-Wesley, Boston
51. Mohan N, Undeland TM, Robbins WP (1989) Power electronics: converters, applications and design. Wiley, New York
52. Kjaer SB, Pedersen JK, Blaabjerg F (2005) A review of single-phase grid-connected inverters for photovoltaic modules. IEEE Trans Ind Appl 41(5):1292–1306
53. Labouret A, Villoz M (2010) Solar photovoltaic energy. The institution of Engineering and Technology, London, (translated from French, Energie solaire photovoltaïque, 4e éd, Dunod, 2009)
54. Russell R, Norvig P (1995) Artificial intelligence: a modern approach. Prentice-Hall, Englewood Cliffs
55. Maes P (1995) Artificial life meets entertainment: life-like autonomous agents. Commun ACM 38(11):108
56. Foner LN (1997) Entairtaining agents: a sociological case study. In: Proceedings of the 1st international conference Auton agents
57. Wooldridge M (1999) Intelligent agents. In: Weiss G (ed) Multi-agent systems. MIT Press, Cambridge
58. Sycara K (1998) Multiagent systems. AI Mag 19(2):79, August (1987)
59. Deloach SA, Wood MF, Sparkman CH (2001) Multiagent systems engineering. Int J Softw Eng Knowl Eng 11(3):231–258
60. McArthur SDJ, Davidson EM , Catterson VM, Dimeas A, Hatziargyriou N, Ponci F, Funabashi T (2007) Challenges. IEEE Transac Power Syst 22(4):1743
61. Nema S, Nema RK, Agnihotri G (2011) Inverter topologies and control structure in photovoltaic applications: a review. J renew sustain energy 3:012701
62. Ould-Amrouche S, Rekioua D, Hamidat A (2010) Modeling photovoltaic water pumping systems and evaluation of their CO_2 emissions mitigation potential. Appl Energy 87:3451–3459
63. Salameh Z, Taylor D (1990) Set-up maximum power point tracker for photovoltaic arrays. Solar Energy 44(1):57
64. Zhou S, Kang L, Sun J, Guo G, Cheng B, Cao B, Tang Y (2010) A novel maximum power point tracking algorithms for stand-alone photovoltaic system. Int J Control Autom Syst 8(6):1364
65. Yazdani A (2009) A control methodology and characterization of dynamics for a photovoltaic (PV) system interfaced with a distribution network. IEEE Trans Power Deliv 24(3):1538
66. Salas V, Olias E, Barrado A, Lazaro A (2006) Review of the maximum power point tracking algorithms for stand-alone photovoltaic systems. Solar Energy Mat Solar Cells 90(11):1555
67. Hurng-Liahng Jou, Wen-Jung Chiang, Jinn-Chang Wu (2007) A novel maximum power point tracking method for the photovoltaic system, conference international PEDS, p 619
68. Ibrahim HE.-SA, Houssiny FF, El-Din HMZ, El-Shibini MA (1999) Microcomputer controlled buck regulator for maximum power point tracker for DC pumping system operates from photovoltaic system. Fuzzy systems conference FUZZ-IEEE'99, p 406
69. Nishioka K, Sakitani N, Kurobe K, Yamamoto Y, Ishikawa Y, Uraoka Y, Fuyuki T (2003) Analysis of the temperature characteristics in polycrystalline Si solar cells using modified equivalent circuit model. J Appl Phys 42:7175
70. Masoum MAS, Dehbonei H (1999) Design, construction and testing of a voltage-based maximum power point tracker (VMPPT) for small satellite power supply. In: 13th annual AIAA/USU conference, Small Satellite

71. Noguchi T, Togashi S, Nakamoto R (2000) Short-current pulse-based adaptive maximum power point tracking for a photovoltaic power generation system. In: Proceedings of the 2000 IEEE international symposium on ISIE vol 1, p 157
72. Santos LJL, Antunes F, Chehab A, Cruz C (2006) A maximum power point tracker for PV systems using a high performance boost converter. Solar Energy 80(7):772
73. Krauter SCW (2006) Solar electric power generation-photovoltaic energy systems. Springer, New York
74. Al-Atrash H, Batarseh I, Rustom K (2005) Statistical modelling of DSP-based hill-climbing MPPT algorithms in noisy environments, Applied Power Electronics Conf Exposition, APEC'05 IEEE, vol 3, p 1773
75. Jiang J-A, Huang T-L, Hsiao Y-T, Chen C-H (2005) Maximum power tracking for photovoltaic power systems. Tamkang J Sci Eng 8(2):147
76. Won C-Y, Kim DH, Kim SC, Kim WS, Kim H-S (1994) A new maximum power point tracker of photovoltaic arrays using fuzzy controller. In: 25th annual IEEE power electronics specialists conference, vol 396
77. Patcharaprakiti N, Premrudeepreechacharn S, Sriuthaisiriwong Y (2005) Maximum power point tracking using adaptive fuzzy logic control for grid connected photovoltaic system. Renew Energy 30(11):1771
78. Masoum MAS, Sarvi M (2002) Design, simulation and construction of a new fuzzy-based maximum power point tracker for photovoltaic applications, Australian universities power engineering conference, AUPEC, pp 1–6
79. Altas IH, Sharaf AM (2008) A novel maximum power fuzzy logic controller for photovoltaic solar energy systems. Renew Energy 33(3):388–399
80. Gounden NA, Peter SA, Nallandula H, Krithiga S (2009) Fuzzy logic controller with MPPT using line-commutated inverter for three-phase grid connected photovoltaic systems. Renew Energy 34(3):909–915
81. Cheikh MSA, Larbes C, Kebir GFT, Zerguerras A (2007) Maximum power point tracking using a fuzzy logic control scheme. Renew Energies Rev 10(3):387–395
82. Taherbaneh M, Ghafori fard H, Rezaie AH, Karbasian S (2007) Combination of fuzz maximum power point tracker and sun traker for deployable solar panels in photovoltaic systems, Fuzzy systems conference FUZZ-IEEE, pp 23–36
83. Veerachary M, Senjyu T, Uezato K (2003) Neural-network-based maximum-power-point tracking of coupled inductor interleaved-boost-converter-supplied PV system using fuzzy controller. IEEE Trans Ind Electron 50:749–758
84. Simoes MG, Franceschetti NN (1999) Fuzzy optimization based control of a solar array. IEEE Proc Electr Power Appl 146(5):552–558
85. Hua Ch, Lin J, Shen Ch (1998) Implementation of a DSP-controlled PV system with peak power tracking. IEEE Trans Ind Electron 45(1):99
86. Liu F, Duan S, Liu F, Liu B, Kang Y (2008) A variable step size INC MPPT method for PV systems. IEEE Trans Ind Electron 55(7):2622–2628
87. Hohm DP, Ropp ME (2003) Comparative study of maximum power point tracking algorithms. Prog Photovolt: Res Appl 11:47–62
88. Enrique JM, Durán E, Sidrach-de-Cardona M, Andújar JM (2007) Theoretical assessment of the maximum power point tracking efficiency of photovoltaic facilities with different converter topologies. Solar Energy 81:31–38
89. Murray Thomson A (2003) Reverse-osmosis desalination of seawater powered by photovoltaics without batteries, Thesis Doctorat, University of Loughborough, UK
90. Yu GJ, Jung YS, Choi JY, Choy I, Song JH Kim GS (2002) A novel two-mode MPPT control algorithm based on comparative study of existing algorithms. In: Proceedings photovoltaic specialists conference, pp 1531–1534
91. Masoum MAS, Dehbonei H, Fuchs EF (2002) Theoretical and experimental analyses of photovoltaic systems with voltage and current-based maximum power point tracking. IEEE Trans Energy Convers 17(4):514–522

92. Kottas TL, Boutalis YS, Karlis AD (2006) New maximum power point tracker for PV arrays using fuzzy controller in close cooperation with fuzzy cognitive networks. IEEE Trans Energy Convers 21(3):793–803
93. Hiyama T, Kouzuma S, Imakubo T, Ortmeyer TH (1995) Evaluation of neural network based real time maximum power tracking controller for PV system. IEEE Trans Energy Convers 10(3):543–548
94. Gao D, Jin Z, Lu Q (2008) Energy management strategy based on fuzzy logic for a fuel cell hybrid bus. J Power Sources 85(1):311–317
95. Chaabene M, Ammar MB, Elhajjaji A (2007) Fuzzy approach for optimal energy-management of a domestic photovoltaic panel. Appl Energy 84:992–1001
96. Hamdy MA (1994) A new model for the current-voltage output characteristics of photovoltaic modules. J Power Sources 50(1):11–20
97. Arteaga Orozco MI, Vázquez JR, Salmerón P, Pérez A (2009) A sliding maximum power point tracker for a photovoltaic system, 11th Spanish Portuguese conference on electrical engineering (11CHLIE). http://www.aedie.org/11chlie-papers/.../212-arteaga-summary.pdf
98. Ya-Ting Lee, Chian-Song Chiu, Chih-Teng Shen (2010) Adaptive fuzzy terminal sliding mode control of DC–DC buck converters via PSOC, IEEE international conference on control applications part of 2010 IEEE multi-conference on systems and control, Yokohama, Japan, 8–10 Sept
99. Juan José Negroni, Domingo Biel Francesc, Guinjoan Carlos Meza (2010) Energy-balance and sliding mode control strategies of a cascade H-bridge multilevel converter for grid-connected PV systems international conference on industrial technology ICIT2010, 14–17 Mar, Chile
100. Chen-ChiChu, Chieh-LiChen (2009) Robust maximum power point tracking method for photovoltaic cells: a sliding mode control approach. Solar Energy 83:1370–1378
101 Ellouze M, Gamoudi R, Mami A (2010) Sliding mode control applied to a photovoltaic water-pumping system. Int J Phys Sci 5(4):334–344. http://www.academicjournals.org/IJPS
102. Hohm DP, Ropp ME (2003) Comparative study of maximum power point tracking algorithms. Progress in photovoltaic, research and applications 11:47–62
103. Tan CW, Green C (2005) An improved maximum power point tracking algorithm with current-mode control for photovoltaic applications, The sixth international conference on power electronics and drive IEEE PEDS, Malaysia, pp 5–14
104. Chaudhari VA (2005) Automatic peak power tracker for solar bv modules using dspacer software, thesis of the master of technology in energy, Maulana Azad National Institute of Technology (Deemed University)
105. Hassoun MH (1995) Fundamentals of artificial neural networks. The MIT Press, Cambridge
106. Ramaprabha R, Mathur BL (2011) Intelligent controller based maximum power point tracking for solar PV system. Int J Comput Appl 12(10):0975–8887
107. Kamath HR (2008) Modelling of photovoltaic array and maximum power point tracker using ANN. Int J Electr Syst 4(3):1–5
108. Mellit A, Kalogirou SA (2011) ANFIS-based modelling for photovoltaic power supply system: a case study. Renew Energy 36(1):250–258
109. Belhachat F, Larbes C, Barazane L, Kharzi S (2007) Neuro-fuzzy control of a chopper MPPT, 4th international conference on computar integrated manufacturing CIP'2007, 3–4 Nov 2007
110. Chaouachi A., Kamel RM, Nagasaka K (2010) MPPT operation for PV grid-connected system using RBFNN and fuzzy classification, vol 65, World Academy of Science, Engineering and Technology
111. Larbes C, Cheikh SMA, Obeidi T, Zerguerras A (2009) Genetic algorithms optimized fuzzy logic control for the maximum power point tracking in photovoltaic system. Renew Energy 34(10):2093–2100
112. Moubayed N, El-Ali A, Outbib R (2009) A comparison of two MPPT techniques for PV system. WSEAS Trans Environ Dev 5(12):1–10

113. Wu L, Zhao Z, Liu J (2007) A single-stage three-phase grid-connected photovoltaic system with modified MPPT method and reactive power compensation. IEEE Trans Energy Convers 22(4):881–886
114. Comparison of Different Battery Technologies (2006) General electronics battery co., ltd. pp 1–4. www.tradekorea.com/product/file/download.mvc;…TK
115. Doris L, Britton Thomas B (2000) Miller battery fundamentals and operations-batteries for dummies—April
116. Linden D (2002) Handbook of batteries and fuel cells, 3rd edn. Mcgraw-Hill, New York
117. Divya KC, Stergaard J (2009) Battery energy storage technology for power systems—an overview. Electr Power Syst Res 79:511–520
118. Hussien ZF, Cheung LW, Siam MFM, Ismail AB (2007) Modeling of Sodium Sulfur Battery For Power System Applications 9(2):66–72. (http://fke.utm.my/elektrik)
119. London Research International Ltd (2010) Survey of Energy Storage Options in Europe, Report © London Research International Ltd, London, March 2010
120. Vaillant JR (1978) Uses and promises of solar energy, 2nd edn
121. Piemontesi M, Dustmann C, Energy storage systems for ups and energy management at consumer level. (www.battcon.com/papersfinal2010/piemontesipaper2010final_5.pdf
122. Manzoni R, Metzger M, Crugnola G (2008) Zebra electric energy storage system: From R&D to Market, Presented at HTE hi.tech.expo—Milan, 25–28 Nov
123. Rouault H, Blach D (2004) The electricity storage: the batteries, vol 44, Clefs CEA
124. Saisset R (2004) Study contribution systemic including electrochemical energy devices. PHD thesis INP Toulouse specialty electrical engineering. April 2004
125. Sabonnadière JC (2007) New energy technologies and storage technology decarbonizing, Edition Dunod
126. Hendrik Johannes Bergveld (2001) Battery Management Systems Design by Modeling. Thesis University Press Facilities, Eindhoven (doc.utwente.nl/41435/)
127. Lalouni S (2005) Optimizing the quality of electrical energy in the case of a battery charger. Thesis of Master University of Bejaia
128. Hladik J (1977) Storage batteries, first edition: 1st Quarter, Presses Universities France
129. Achaibou N, Haddadi M, Malek A (2008) Lead acid batteries simulation including experimental validation. J Power Sources 185:1484–1491
130. http://www.rollsroyce.com/marine/products/electrical_power_systems/storage/index.jsp
131. Dûrr M (2006) Dynamic model of a lead acid battery for use in a domestic fuel cell system. J Power Sources 161:1400–1411
132. Achaibou N (1999) Introduction to study storage system in a photovoltaic system. J Renew Energy CDER, pp 1–6
133. Zoroofi S (2008) Modeling and simulation of vehicular power systems. Thesis of Master, University of Technology Chalmers
134. Marquet A (2003) Storage of electricity in electric systems, Engineering techniques D4 030
135. Francois CJ (1997/1998) Modeling the art of charge electric vehicle batteries, University of Liege Faculty of Applied Sciences Degree Legal Electricians Civil Engineer
136. Multon B, Ahmed HB, Bernar N, Kerzreho C (2007) The inertial storage electromechanics. 3EI J 48:18–29
137. Salameh ZM, Casacca MA, Lynch WAA (1992) Mathematical model for lead-acid batteries. IEEE Trans Energy Convers 7(1):93–98
138. EA Technology (2004) Review of electrical energy storage technologies and systems and of their potential for the UK, pp 1–34. http://www.wearemichigan.com/JobsAndEnergy/documents/file15185.pdf
139. Kosin L, Usach F (1995) Electric characteristics of lead battery. Russ J Appl Chem 143(3):1–4
140. Labbé J (2006) Electrolytic Hydrogen as a storage of electricity for Photovoltaic Systems Insulated. PhD thesis, School of Mines Paris, Specialty Energetic

141. Francisco M, Longatt G (2006) Circuit based battery models: a review, 2do Congreso Iberoamericano De Estudiantes De Ingeniería Eléctrica (II CIBELEC 2006) pp 1–5
142. Chan HM, Stutanto D (2000) A new battery model for use with battery energy storage systems and electric vehicles power system. IEEE Power Eng Soc Winter Meet 1:470–475
143. Ceraolo M (2000) Dynamical models of lead-acid batteries. IEEE Trans Power Syst 15:1184–1190
144. Gergaud O, Robins G, Multon B, Ben Ahmed H (2003) Energy modeling of a lead-Acid battery within hybrid wind photovoltaic system, EPE'03 Toulouse, France, pp 1–10. http://www.bretagne.ens-cachan.fr/.../EnergiesRenouv/LeadAcidBattery_Gergaud_EPE2003.pdf
145. Ould Mohamed Yahia A, Ould Mhmoud A, Youm I (2007) On modeling integrated storage system in a hybrid system (PV / Wind / Diesel). Renew Energy J 10(2):205–214
146. www.Stielec.Ac-Aix-Marseille.Fr/...2009/Stockage_2009.Pdf
147. Meah K, Fletcher S, Ula S (2008) Solar photovoltaic water pumping for remote locations. Renew Sustain Energy Rev 12(2):472–487
148. Vongmanee V (2002) The vector control inverter for a PV motor drive system implemented by a single chip DSP controller ADMC331. In: Proceedings of IEEE Asia Pacific conference on circuits and systems, p 447
149. Betka A, Moussi A (2004) Performance optimization of a photovoltaic induction motor pumping system. Renew Energy 29(14):2167
150. Odeh I, Yohanis YG, Norton B (2006) Influence of pumping head, insolation and PV array size on PV water pumping system performance. Solar Energy 80(1):51–64
151. Daud AK, Mahmoud MM (2005) Solar powered induction motor-driven water pump operating on a desert well, simulation and field tests. Renew Energy 30(5):701–714
152. Badescu V (2003) Time dependent model of a complex PV water pumping system. Renew Energy 28(4):543–560
153. Ghoneim AA (2006) Design optimization of photovoltaic powered water pumping system. Energy Convers Manag 47(11–12):1449–1463
154. Hadj Arab A, Benghanem M, Chenlo F (2006) Motor—pump system modelization. Renew Energy 31(7):905–913
155. Amer EH, Younes MA (2006) Estimating the monthly discharge of a photovoltaic water pumping system: model verification. Energy Convers Manag 47(15–16):2092–2102
156. Hamidat A, Benyoucef B (2008) Mathematic models of photovoltaic motor—pump systems. Renew Energy 33(5):933–942
157. Rekioua D, Rekioua T, Laporte B, Benmahammed K (2001) Design of a position sensor for torque ripple minimization of VSI fed self synchronous machine. Int J Electron 88(7)
158. Bose BK (1986) Power electronics and variable frequency drives: technology and applications, Wiley-IEEE Press, New York
159. Rekioua T, Rekioua D, Laporte B, Benmahammed K (2001) Optimization and minimization of torque ripple in AC machines supplied by VSI. J Eur Trans Electr Power (ETEP), Germany, Edition Verlag 11(5)
160. Vongmanee V (2002b) the vector control inverter for a pv motor drive system implemented by a single chip DSP controller ADMC331, APCCAS. Denpasar, Bali-Indonesia
161. Abdelli R, Rekioua D, Rekioua T (2011) Performances improvements and torque ripple minimization for VSI fed induction machine with direct control torque 50(2):213–219
162. Chedid R , Chaaban F (2003) Renewable-energy developments in Arab countries: a regional perspective. Appl Energy 74(1–2):211–220
163. Saheb-Koussa D, Haddadi M, Belhamel M (2009) Economic and technical study of a hybrid system (wind-photovoltaic-diesel) for rural electrification in Algeria. Appl Energy 86(7–8):1024–1030
164. Department for Environment, Food and Rural Affairs (2005) Guidelines for company reporting on greenhouse gas emissions, annex 1: fuel conversion factors. http://www.archive.defra.gov.uk/environment/business/envrp/pdf/envrpgasannexes.pdf Accessed Dec 2008

165. Ministère de l'Energie et des Mines. Bilan général des réalisations. <http:// www.mem-algeria.org/fr/enr/energiesolaire/Bilan_generale.htm> Accessed 5 May 2009
166. Boudghene SA (2006) Overview and perspectives of the Algerian renewable energy program. In: Proceedings of National Symposium on Solar Energy CNESOL 06, Bejaia, Algeria, pp 93–99
167. Mcgowan JG, Manwell JF (1999) Hybrid/PV/Diesel system experiences. Rev Renew Energy 16:928–933
168. Belhamel M, Moussa, S Kaabeche A (2002) Production of electricity by means of a hybrid system (Wind-Photovoltaic-Diesel), (traducted from french: Production d'Electricité au Moyen d'un Système Hybride (Eolien- Photovoltaïque -Diesel)). Renew Energy J 49–54
169. El Khadimi A, Bachir L, Zeroual A (2004) Sizing optimization and techno-economic energy system hybrid photovoltaic - wind with storage system. Renew Energy J 7:73–83
170. Koussa DAlem, M, Belhamel M (2002) Hybrid system (Wind, Solar) for the power supply for a load for household. J Renew Energy 1–8
171. Kaldellisa JK, Kavadiasa KA, Koronakis PS (2007) Comparing wind and photovoltaic stand-alone power systems used for the electrification of remote consumers. Renew Sustain Energy Rev 11:57–77
172. Vechiu I (2005) Modelling and analysis of integration of renewable energy In an autonomous network. Ph. D. thesis, University of Havre, France
173. Abou El-Maaty Metwally Metwally Aly Abd El-Aal (2005) Modeling and simulation of a photovoltaic fuel cell hybrid system. A Dissertation in Candidacy For The Degree Of Doctor in Engineering (Dr.-Ing.), Faculty of Electrical Engineering, University of Kassel, Germany
174. Leclerc L (2004) Inertial Storage Contribution From wind turbines in an electrical network in order to ensure system services. Ph. D. thesis, University of Lille
175. Kato N, Kurozumi K, Susuki N, Muroyama S (2001) Hybrid power supply system composed of photovoltaic and fuel cell systems, conférence INTELEC, pp 631–635
176. Mirecki A (2005) Comparative study chain of energy conversion dedicated to a small wind turbine, PhD from the Institute National Polytechnique de Toulouse, France
177. Kaller R, Allenbach JM (1995) Electric Traction 1. Polytechnic and university presses Romande, Lausanne
178. EL Mokadem M, Nichita C, Barkat G, Dakyo B (2002) Control strategy for stand alone wind-diesel hybrid system using a speed model. In: 7th International ELECTRIMACS Congress, Montréal
179. Jurado F, Saenz R (2002) Neuro-Fuzzy control for autonomous wind-diesel systems using biomass. Renew Energy 27:39–56
180. Pouria P (1983) Emergency generator, technical engineering (translated from french: Groupes électrogènes de secours, techniques de l'ingenieur), D 505
181. Muljadi E, Buterfield CP (1999) Pitch controlled variable speed wind turbine generation. IEEE industry applications, society annual meeting, Phoenix, Arizona, 2–7 Oct 1999
182. Saheb–Koussa D, Belhamel M, Benferhat K (2007) Electrify isolated by a hybrid system of renewable energy sources. International conference on renewable energies ICRE'07, pp 1–6
183. Chang L (2002) Systems Conversion of wind energy. IEEE Can Rev, summer, pp 1–5
184. Idjdarene K, Rekioua D, Rekioua T, Tounzi A (2007) Control strategies for an autonomous induction generator taking the saturation effect into account, 12th European conference on power electronics and applications (EPE'2007), Aalborg, Denmark, 2–5 Sept 2007, pp 1–6
185. Taraft S, Rekioua D, Aouzellag D (2010) Control of electric power of a wind generator based on the wound rotor machine connected to the grid, conference ICEEA10, Bejaia (Algeria), 2–3 Nov 2010
186. Gevorgian V, Touryan K, Bezrukikh P, Karghiev V, Bezrukikh P (1999) Wind-Diesel Hybrid Systems For Russia's Northern Territories;Presented at Windpower '99, Burlington, Vermont, 20–23 June 1999

187. Manwell JF, Mcgowan JG, Abdulwahid U (2000) Simplified Performance Model for Hybrid Wind Diesel Systems, Renewable energy: the energy for the 21st century, World renewable energy congress N°6, Brighton , ROYAUME-UNI, pp 1183–1188
188. Lysen EH (2000) hybrid technology. http://climatetechwiki.org/technology/hybrid-technology
189. Prasad AR, Natarajan E (2006) Optimization of integrated photovoltaic-wind power generation systems with battery storage. Energy 31:1943–1954
190. Diaf S, Haddadi M, Belhamel M (2006) Techno Economic Analysis of a hybrid system (Photovoltaic / Wind) Independent Website for The Adrar. Renew Energy J 9(3):127–134
191. Dali M, Belhadj J, Roboam X, Blaquiere JM (2007) Control and energy management of a wind-photovoltaic hybrid system, 12th European conference on power electronics and applications (EPE'2007), Aalborg, Denmark, 02–05 September, pp 1–10
192. Ab Razak J, Sopian K, Ali Y, Alghoul MA, Zaharim A, Ahmad I (2009) Optimization of PV-Wind-Hydro-Diesel Hybrid System By Minimizing Excess Capacity. European J Sci Res 25(4):663–671
193. Bettar N (2008) Study and Modeling of a PEM Fuel Cell. Master's thesis, University of Bejaia, Algeria
194. Rezkallah M Improving quality of energy independent network provided: wind-diesel hybrid, master in electrical engineering, Montréal, 3 May 2010
195. Simulink: dynamic system simulation for Matlab-using Simulink. The MathWorks, Inc. Copyright 1984–2002. Release 13.